Mechatronic and Robotic Systems: Design, Models and Methods

Mechatronic and Robotic Systems: Design, Models and Methods

Edited by
Noel Cole

WILLFORD PRESS

www.willfordpress.com

Published by Willford Press,
118-35 Queens Blvd., Suite 400,
Forest Hills, NY 11375, USA

ISBN: 978-1-64728-534-0

Cataloging-in-Publication Data

Mechatronic and robotic systems : design, models and methods / edited by Noel Cole.
p. cm.
Includes bibliographical references and index.
ISBN 978-1-64728-534-0
1. Mechatronics. 2. Robotics. 3. Microelectronics. 4. Robots. 5. Machine design.
6. Mechanical engineering. I. Cole, Noel.
TJ163.12 .M43 2023
621--dc23

For information on all Willford Press publications
visit our website at www.willfordpress.com

Contents

Preface

Robotic systems are those systems, which interact with their surroundings using actuators, sensors and human interfaces, and provide intelligent services and information. Mechatronics is a superset of robotic technologies and is defined as an interdisciplinary branch of engineering, which combines concepts from various disciplines including electrical and electronic engineering, mechanical engineering, computer science, and robotics. The model of mechatronics system is made up of two interacting submodels, which include a submodel describing the aspects of information flow in the control system and another one describing the aspects of energy flow in the physical system. This book contains some path-breaking studies on mechatronic and robotic systems. It is a collective contribution of a renowned group of international experts. In this book, using studies and examples, constant effort has been made to make the understanding of the difficult concepts of these systems as easy and informative as possible, for the readers.

This book is a result of research of several months to collate the most relevant data in the field.

When I was approached with the idea of this book and the proposal to edit it, I was overwhelmed. It gave me an opportunity to reach out to all those who share a common interest with me in this field. I had 3 main parameters for editing this text:

1. Accuracy – The data and information provided in this book should be up-to-date and valuable to the readers.

2. Structure – The data must be presented in a structured format for easy understanding and better grasping of the readers.

3. Universal Approach – This book not only targets students but also experts and innovators in the field, thus my aim was to present topics which are of use to all.

Thus, it took me a couple of months to finish the editing of this book.

I would like to make a special mention of my publisher who considered me worthy of this opportunity and also supported me throughout the editing process. I would also like to thank the editing team at the back-end who extended their help whenever required.

Editor

Parallel Architectures for Humanoid Robots †

Marco Ceccarelli [1,2,*] ⓘ, **Matteo Russo** [3] ⓘ **and Cuauhtemoc Morales-Cruz** [1,4]

[1]   LARM2: Laboratory of Robot Mechatronics, University of Roma Tor Vergata, Via del Politecnico 1,
      00133 Rome, Italy; cmoralesc0600@alumno.ipn.mx
[2]   Beijing Advanced Innovation Center for Intelligent Robots and Systems, Beijing Institute of Technology,
      Beijing 100081, China
[3]   The Rolls-Royce UTC in Manufacturing and On-Wing Technology, University of Nottingham,
      Nottingham NG8 1BB, UK; matteo.russo@nottingham.ac.uk
[4]   Instituto Politécnico Nacional, GIIM: Group of Research and Innovation in Mechatronics,
      Av. Juan de Dios Bátiz, 07700 Mexico City, Mexico
*   Correspondence: marco.ceccarelli@uniroma2.it
†   This paper is an extended version of our paper published in Marco, C.; Matteo, R. Parallel Mechanism
      Designs for Humanoid Robots. In *Robotics and Mechatronics*; Kuo, C.H., Lin, P.C., Essomba, T., Chen, G.C.,
      Eds.; Springer Nature: Cham, Switzerland, 2020; Volume 78, pp. 255–264.

Abstract: The structure of humanoid robots can be inspired to human anatomy and operation with open challenges in mechanical performance that can be achieved by using parallel kinematic mechanisms. Parallel mechanisms can be identified in human anatomy with operations that can be used for designing parallel mechanisms in the structure of humanoid robots. Design issues are outlined as requirements and performance for parallel mechanisms in humanoid structures. The example of LARMbot humanoid design is presented as from direct authors' experience to show an example of the feasibility and efficiency of using parallel mechanisms in humanoid structures. This work is an extension of a paper presented at ISRM 2019 conference (International Symposium on Robotics and Mechatronics).

Keywords: mechanism design; humanoid robots; biomimetic designs; parallel mechanisms; LARMbot

1. Introduction

The first anthropomorphic humanoid robot, WABOT-1, was built at Waseda University, Tokyo, as part of the WABOT project (1970). WABOT-1 was a full-scale humanoid robot, able to walk, grasp and transport small object with its hands, and equipped with a vision system used for basic navigation tasks. The same research group later built WABOT-2 (1984) and WABIAN (1997), both biped humanoid robots, and is still active in the field [1].

Around 1986, Honda started to develop a biped platform that underwent through several stages, called "E" (1986–1993) and "P" (1993–1997) series, and led to the creation of ASIMO [2]. ASIMO was officially unveiled in 2000 and had a significant impact on the media all around the world. It is a humanoid platform with an advanced vision and navigation system, able to interpret voice or gesture commands and to move autonomously with a semi-dynamic walking mode. In 2008, Aldebaran Robotics launched NAO, a programmable humanoid robot that is now the standard platform for several robotics competitions, such as the RoboCup Standard Platform League [3,4]. NAO has been the most widespread robot in academic and scientific usage, used to teach and develop advanced programming and control in educational and research institutes all around the world. In 2013, Boston Dynamics announced the Atlas robot, a biped robot capable of complex dynamic tasks, such as running, moving on snow, performing a backflip, balancing after being hit by projectiles or jumping on one leg [5].

Several other humanoid structures have been designed both by academy and industry with a variety of solutions as platforms for research and applications for service (as for example in assistance to humans, entertainment, guide operations), exploration, human-robot interaction and learning. Some additional examples include: iCub [6], and for research on learning; WALK-MAN, [7], which is used as a rescue robot for unstructured environments; Pepper, [8], which is used for investigations on human-robot interaction; Ami, [9], which is designed for applications in Domotics; REEM-B, [10], which is designed as service robot for general human assistance; ARMAR, [11,12], which is used as a collaborative robot in Domotics tasks. While there has been a huge development in the control of these robots, they all share a common body architecture, which has not evolved much during time. In fact, most of them are based on a serial kinematic chain with spherical-revolute-universal (SRU) joints for their limbs. This mechanical structure is fairly easy to manufacture and control, while offering a large workspace for its size.

However, serial architectures are outperformed by parallel mechanisms in accuracy, speed, stiffness and payload. Parallel manipulators, also named parallel kinematic machines (PKM), are closed-loop mechanisms that are characterized by high accuracy, rigidity and payload [13]. A parallel manipulator is defined as a mechanical system that allow a rigid body, called from now on an end-effector, to move with respect to a fixed base by means of two or more independent kinematic chains. Very few research groups have proposed robotic limbs with a parallel architecture. As reported in [14], the first robotic legs with parallel architecture were developed in Japan. The first one was the ParaWalker robot, developed at Tokyo Institute of Technology in 1992, while the Waseda Leg (WL) WL-15 was built in 2001 at the Takanishi laboratory of Waseda University, followed by the WL-16 and the WL-16R series from 2002 to 2007 [15]. The last version of the Waseda Leg is the WL-16RV. The WL robots are biped walking chairs, whose legs are based on the architecture of the Gough platform. The WL design inspired other teams in designing novel biped robots built on parallel mechanisms, such as the Reconfigurable Quadruped/Biped Walking Robot of Yanshan University, which is built on four legs with three identical universal-prismatic-universal limbs (3UPU) each. The LARMbot [16–21] was designed as a low-cost, user-oriented leg for a service humanoid robot. Its leg is designed as a 3UPR PKM. Gao used similar lower-mobility parallel legs for a hexapod robot [22–24]. The main drawback for most of these legs, however, is the small dimension of their workspace, which allows for a very small step size when compared to the human one [25–28].

In order to characterize the mechanical performance of robotic limbs that have been developed in the last few decades, the main parameters of representative humanoid robots are listed in Table 1 [27].

Table 1. Characteristic parameters of biped robots.

Name	Type	Weight [kg]	Height [mm]	Step [mm]	Speed [mm·s⁻¹]	DoF [-]	Year
MELTRAN II	Serial	4.7	450	120	200	3	1989
HRP-2	Serial	58.0	1540	-	-	6	2003
WABIAN-2	Serial	64.5	1530	350	360	7	2005
WL-16RV	Parallel	75.0	1440	200	200	6	2008
WABIAN-2RIII	Serial	64.0	1500	500	520	7	2009
HUBO2	Serial	45.0	1300	-	410	6	2009
ASIMO	Serial	54.0	1300	410	750	6	2011
BHR5	Serial	63.0	1620	-	440	6	2012
ATLAS	Serial	150.0	1800	-	-	6	2013
NAO V5	Serial	5.4	574	80	160	6	2016

Each robot is classified according to its kinematic architecture (serial or parallel), system weight and height, maximum step size, maximum gait speed, degrees-of-freedom of each leg and year of production. As reported in Table 1, most of the existing robots have 6-DoFs serial limb architectures. The most famous humanoid robots, such as ASIMO and NAO, are all based on this kinematic scheme. The size and weight of the robots did not change significantly with the years, and even the step size and speed are in a reduced range of values when compared to the size of the robot. Most of the advancements are in the performance of control and gait planning, while the mechanical design has not evolved. Furthermore, the payload capacity of the current structures of humanoid robots is rather small (for example, NAO can lift only 0.15 kg per arm), and they can be often operated with poor dynamics and stiffness. Therefore, challenging design issues can be still identified in improving or designing structures of humanoid robots and parallel mechanisms can be considered a solution or an alternative to achieve a mechanical design with better performance in accuracy, payload and dynamics, not only mimicking human capabilities.

2. Human Anatomy with Parallel Mechanisms

Humanoids are designed with structures and operations replicating human ones. Human nature has a complex design in structure composition, with several kinds of material and architectures that humanoid design can replicate only very partially. The most referenced part of human anatomy for humanoid robot structures is the skeleton system, which inspires solutions mainly with rigid links in serial kinematic chain architectures. However, considering that the functionality of human movable parts is mainly due to a combined/integrated structure of bones and muscles, the reference structure for humanoid robot design can be considered a parallel architecture that combines bones as rigid movable links and muscles as linear actuators. Figure 1 summarizes such an understanding by looking at the bone skeleton structure (Figure 1a) that, together with the muscle complex (Figure 1b), can give a model of functioning mechanisms with parallel mechanisms (Figure 1c). The antagonist functioning of muscles is characterized by the fact that the muscles mainly act with pulling actions when they are contracted, and therefore full mobility requires alternated actions of two muscles in pulling and releasing. For complex motions, such as 3D movements, a bone is actuated by a complex group of muscles that still control the operation through antagonist functionality.

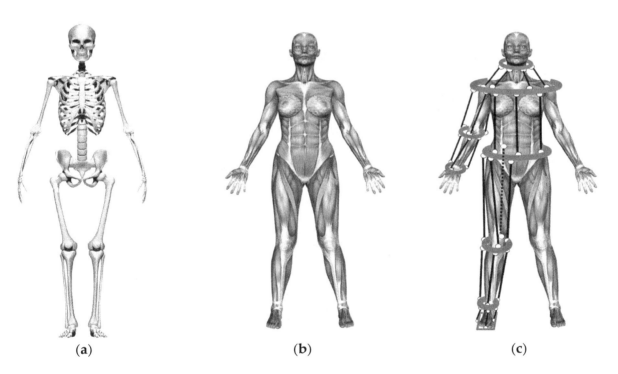

        (a)   (b)   (c)

Figure 1. The anatomy of human body: (**a**) skeleton structure; (**b**) muscle complex; (**c**) a model representing the functionality with parallel mechanisms.

Thus, although a basic principle can be still referred to the example in Figure 1, human anatomy can be of difficult replication in efficient compact designs for humanoid robots. However, the inspiration from human anatomy for designs with parallel mechanisms can be summarized in solutions that are characterized by two platforms with relative motion, which are connected and actuated by a number of pairs of linear actuators working either independently (as rigid variable links) or in antagonism (as cable-driven links). A central rigid link, replicating the bone structure and functionality, can be included in the parallel mechanism design both to keep the size and the load capability.

In particular, Figure 2 gives an example of such an inspiration from human anatomy, with a parallel architecture of bones and muscles for designing a movable arm with a parallel mechanism that is based on the antagonist actuation of a pair of muscles for a planar motion. As per the forearm motion in the sagittal arm plane due to the elbow articulation, Figure 2 shows a solution with a central rigid link L that is connected the platforms with revolute joints whereas the actuation is given by two variable cable links l_1 and l_2. with revolute joints yet. As in Figure 2b, l_2 shortens, simulating the contraction of the muscle, while l_1 is stretched for the release of the corresponding antagonist muscle.

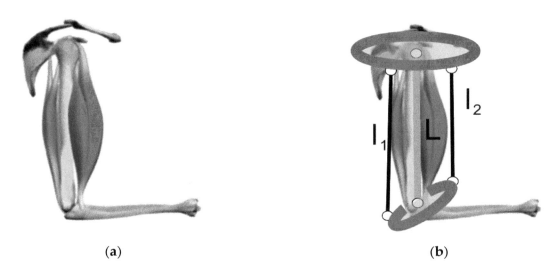

(a)   (b)

Figure 2. An example of parallel architectures with antagonist functioning: (**a**) arm in human anatomy; (**b**) a replicating parallel mechanism.

Figure 3 shows an example of modelling of the complex structure of bones and muscles in a human trunk, in Figure 3a, for designing a humanoid trunk, as in Figure 3b, as based on parallel mechanisms with the above-mentioned design. In particular, the central rigid link L can be designed with a serial chain of links L_i replicating the vertebras, which are connected by spherical joints or 2-DOF revolute joints in a suitable number to ensure the required ranges of flexion and torsion. The complex of the muscles can be replicated with a suitable number of couples of antagonist variable cable links, such as l_1 and l_2, to give a required mobility to the shoulder upper platform with respect to the waist lower platform. The number of those couples of cable links can be limited to only four, and their actuation can be programmed to give some other motion capability when driven in a non-antagonist mode.

The leg structure in Figure 4 can be aimed to the locomotor part of a humanoid robot, with a leg mobility with a large range of motion and a large payload capacity, due to the synergy of the action of the bones and muscle complex. This characteristic can be preserved by using the above-mentioned concept of having the bone load-supporting structure replicated by a central rigid link and the platform relative mobility ensured by the pulling action of the cables working. In addition, the human leg anatomy of the leg-shank structure can be preserved by conceiving two similar parallel mechanisms in series with the mobile platform of the upper-leg part as the fixed platform for the mobile shank platform. The mobility can be designed with a proper number of couples of antagonist cable links as per a required mobility. Thus, the upper-leg parallel mechanism can be designed to give 3 DOFs to the mobile knee platform with a central rigid link L_a and four cable links l_1, l_2, l_3, l_4, which can be activated to give planar motions in sagittal and traversal anatomical planes and even a torsion motion as per a hip joint mobility when the four cables are activated in cooperation (not in antagonism mode). Similarly, the shank motion can be limited to two cables l_5, l_6, for the sagittal motion of the end-effector ankle platform that is connected to the knee platform by a rigid link L_b. Additional cable links l_6, l_7 can be provided to provide the twist motion for the ankle platform.

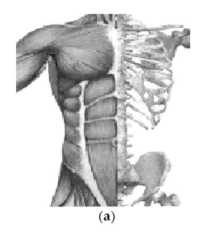
(a)

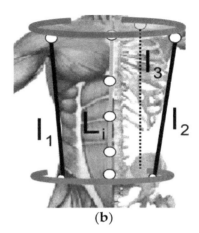
(b)

Figure 3. Trunk structure: (**a**) in human anatomy; (**b**) in a replicating parallel mechanism.

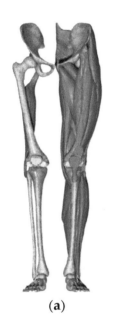
(a)

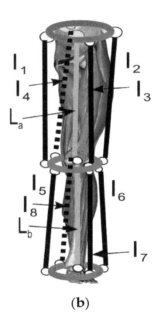
(b)

Figure 4. Leg structure: (**a**) in human anatomy; (**b**) in a replicating parallel mechanism.

The examples in Figures 2–4 give conceptual kinematic designs of the idea of using parallel cable-driven mechanisms for replicating the human anatomy made of bones and muscle complexes in a design of humanoid robots or parts of them. Figure 4 is an example of a combination of parallel mechanisms for the leg structure and how they can be assembled in series.

3. Requirements and Performance for Humanoid Robots

Humanoid robots are aimed at replicating/mimicking human operations mainly in locomotion, manipulation and sensing for human-like tasks. Figure 5 summarizes those main aspects that should be considered for design and operation in mimicking human nature and its functionality, making a humanoid solution efficient, durable and functional, with even better/different performance than those of humans.

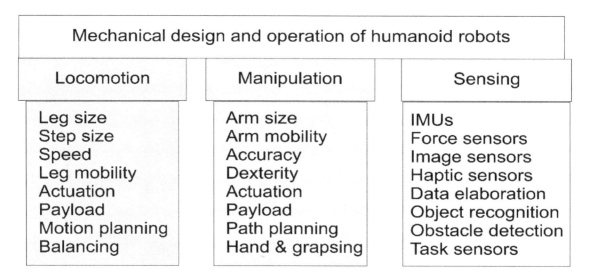

Figure 5. Main requirements for mechanical design and functionality of humanoid robots.

In particular, locomotion requirements can be obtained by analyzing sizing issues and functionality features. Fundamental attention can be addressed to leg size for and as function of the required step size and vice versa. In addition, speed data can be a characteristic of the prescribed task and can affect the previous mentioned aspect. All together they can contribute to design leg workspace, in order to replicate the area of mobility of a human leg, which is usually characterized by suitable values of step length and step height. In addition, the locomotion can be performed in several modes just like humans, such as walking, running and jumping, with characteristic performance in terms of speed and motion smoothness. Among practical requirements, payload capability pays a fundamental role, not only in sustaining the weight of the full humanoid, but also considering the loads and actions that the leg locomotor will have to collaborate with. A locomotion system must be provided with control software and hardware, as well as motion strategies with proper path planning and leg coordination for balancing during bipedal operations.

Similarly, the design and functionality of manipulation system of a humanoid robot can be guided by requirements in terms of sizes and operation performance. Once the size is defined, the constraints for the arm workspace and mobility can be defined to include each point that the arm can reach and all the configurations in which that point can be reached (with different orientations for different manipulation tasks). Accuracy and dexterity are characteristics that can be dictated by the task but also by the flexibility in operation that can be expected by the use of the robot humanoid, The payload of the arm structure should be enough to support a variety of human-like manipulation tasks, which can be linked with a good accuracy and repeatability. Manipulation capability should also be characterized in terms of dexterity, as expressed by multiple reachable arm configurations with suitable motion and dynamics characteristics. Motion planning is also a practical aspect, coming from synergy with the control design and motion programming, with issues that can be determinant in the design and functionality. Finally, the extremity of an arm, such as a hand or grasping systems, needs to be considered as part of the problems for a well-integrated solution in manipulations, including grasp actions.

Sensing in humanoid robots can be considered as being integrated with the capabilities in locomotion and manipulation, as well as additional features, which are nonetheless based on the biomechanics of the structure and their operation, so that sensor equipment is needed with characteristics and composition, as outlined in Figure 5, for the main human-like operation. Those sensors can be useful for the motion and action of a humanoid at proper levels of performance, as well as for monitoring and supervision purposes. Inertial measurement units (IMUs) are useful to have a feedback on the human-like motion, in order to react to external forces or unbalanced configurations with a proper balancing motion. Important sensing is also related to force detection, both in manipulation and locomotion, with or without further control feedbacks. Sensors are significant in grasping tasks that require tactile capability. Image recognition is a sense that makes a humanoid aware of the environment and cameras are needed for autonomous navigation through obstacle detection, and for area inspection through object recognition. Other common sensors in humanoid robots are haptic sensors to perceive the interaction of a humanoid robot with the environment. In addition, a humanoid robot should be equipped with sensors that are sources of information for the task under execution.

Referring to an average human characterization, Table 2 lists an example of numerical evaluation of design requirements for humanoid design, as referred to in the requirements in Figure 5, as linked to solutions with parallel mechanisms.

Table 2. Requirements for humanoid designs as shown in Figure 5, [10–12].

Characteristics	Human Reference Value	Expected Value in Humanoids
Step length (natural)	<94% leg height	50–100% leg height
Step length (fast)	>116% leg height	50–125% leg height
Speed	<105 steps per minute	50–120 steps per minute
Leg mobility	6 D.o.F.	>5 D.o.F.
Leg payload capacity	<200% body weight	>100% body weight
Arm mobility	6 D.o.F.	>6 D.o.F.
Arm payload capacity	<100% body weight	>50% body weight
Torso flexion/extension	30–45°	10–30°
Torso lateral bending	<40°	<30°
Power consumption	<6.00 W/kg	<10.0 W/kg

The expected performance in Table 2 is estimated considering design solutions with parallel architectures, enhancing the whole humanoid design with minimum–maximum ranges that can satisfy task characteristics and/or performance operation in other aspects.

Design solutions with proper dimensions and range of motion can be defined by using computations for the corresponding model and formulation of the kinematics and force transmission of parallel manipulators. In particular, referring to the antagonistic operation mode in the conceptual scheme in Figure 6a, the kinematics of the operation can be formulated with loop-closure equations as

$$\overline{A_0A_i} + l_i + \overline{B_iB_0} = L; \overline{A_0A_j} + l_j + \overline{B_jB_0} = L \tag{1}$$

where the design parameters are the position of the spherical or revolute joints A_i, A_j, B_i and B_j, and the length of the central link L, as well as the motion parameters given by parallel limb lengths l_i and l_j. Given the antagonistic functioning of the system, a single equation for each antagonistic pair of actuators is enough to fully characterize the kinematics of the pair, and the length of the remaining limb can be obtained as a function of its antagonist.

A static or dynamic model can be used to evaluate the actuation forces in the linear actuators or equivalent cable-driven structures, and the equilibrium to translation can be given by

$$F_i + F_j + R + P = 0 \tag{2}$$

where the equilibrium to rotation can be expressed as

$$\overline{B_iB_0} \times F_i + \overline{B_jB_0} \times F_j + M = 0 \tag{3}$$

where F_i and F_j are the actuation forces in the i-th and j-th limb respectively, R is the reaction in the central link and P and M represent external forces and moments applied to the lower platform, as shown in Figure 6b.

The above formulation can be further elaborated for specific cases, as shown in the LARMbot example in Section 4, which is controlled with kinematic and static models that are developed from Equations (1)–(3), as outlined with details in [17].

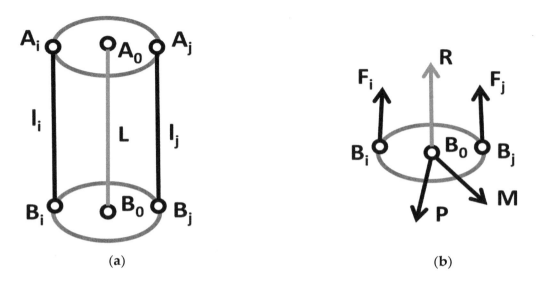

(a)   (b)

Figure 6. A general scheme of a parallel mechanism with antagonistic operation modes: (**a**) kinematic diagram; (**b**) free-body diagram of the lower platform.

4. An Illustrative Example

A direct experience of the authors in using parallel mechanisms in humanoid designs refers to LARMbot design. The LARMbot humanoid robot has been developed in the last decade with contribution of several researchers at LARM laboratory of University of Cassino and South Latium, [29,30]. The LARMbot design, shown in Figure 7, is a service robot for autonomous walking and manipulation tasks, with basic performance in mimicking structure but functionality of humans. As pointed out in [16,29], a first full prototype of LARMbot was assembled in 2015, while a second version, LARMbot II, with a different leg architecture [16,17] is now available for lab testing at LARM2 in University of Rome "Tor Vergata".

The LARMbot design is characterized by two main parallel mechanism systems, namely one for leg-locomotion and one for arm-trunk. The prototype has been built by using commercial components available off-the-shelf and by manufacturing other parts with 3D printing, in order to get a system that is 850 mm tall with a weight less than 3.70 kg. Its payload capability is 0.85 kg for manipulation, and more than 3 kg for the torso/leg operation, whose parallel architectures give a structure that is considerably stronger than traditional humanoids. The payload to weight ratio is 0.23 for manipulation and 0.81 for weightlifting, which is considerably larger than in other existing humanoid robots. For example, the similar-sized NAO humanoid, which is designed with serial kinematic architectures, has a payload to weight ratio of only 0.03 [3,4]. Furthermore, LARMbot is energy-efficient, with a peak 20 W power consumption in LARMbot II prototype, as tested in [17].

The trunk design in Figure 8a is based on the CAUTO solution, [18], as an underactuated cable-driven serial-parallel mechanism whose kinematic scheme is shown in Figure 8b as referring to the conceptual design in Figure 3. The LARMbot trunk is composed of parallel mechanisms with four cables and a central underactuated serial chain whose extremity joints are fixed in the center of the mobile shoulder plate of the parallel mechanism. It is a 4SPS-(3S) parallel mechanism with 4 DOFs, which are actuated by the four motors for varying the length of each cable. The mechanism is inspired by the human torso bone-muscle complex as in the scheme in Figure 3b, with a serial-kinematic compliant spine in the center as a 3S chain, shown in Figure 8b. The cables act as antagonist muscles for motion control in coordination of the cable pairs, according to the kinematic model in Equation (1), with L representing the spine structure. In addition to its main function of load-supporting structure, the LARMbot trunk can be used with its controlled motion to enhance and support walking balance too, as outlined in [17].

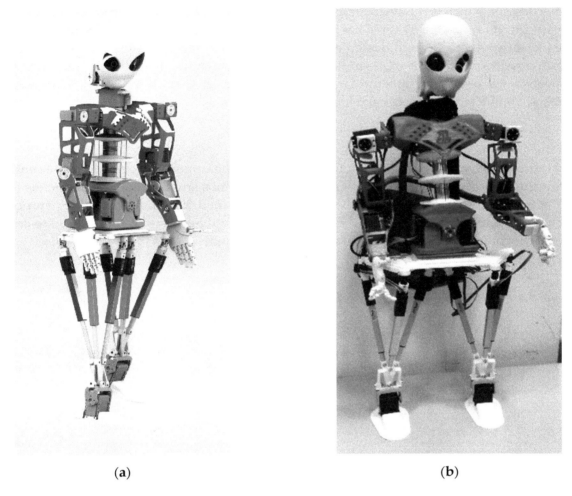

(a)   (b)

Figure 7. LARMbot II humanoid with parallel mechanisms: (**a**) a CAD model; (**b**) a prototype.

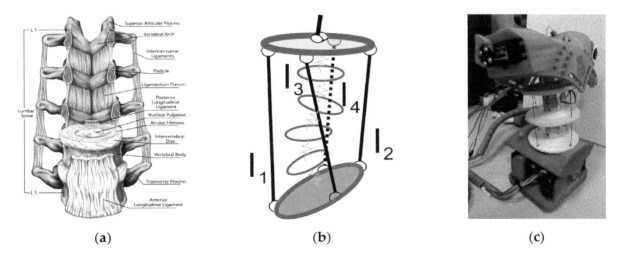

(a)   (b)   (c)

Figure 8. Spinal structure: (**a**) human spine with ligaments; (**b**) a scheme for spine-torso design; (**c**) LARMbot spine-torso prototype.

The leg-locomotor design is shown in Figure 9, with the conceptual design schemes of two legs with human anatomy inspiration, whereas its operational characteristics are presented in [19]. The design is inspired by the human upper-leg structure as in Figure 9a, which refers to a single parallel mechanism in Figure 4, as per gross functionality. Three actuators represent the main muscle groups of the upper leg, namely hamstrings, quadriceps and adductors. Each leg is designed as a 3UPR lower-mobility parallel mechanism, which is shown in Figure 9b as connecting the hip in the waist platform to the ankle mobile platform. It is actuated by three linear actuators in the links, which converge to a single point of the ankle platform. A special joint design ensures the point convergence of the three linear actuators, resulting in a workspace larger than similar parallel manipulators, as well as human-like mobility, which is also characterized by no singular configurations, as discussed in [20,21]. With reference to Figures 6 and 9b, the kinematics of each leg can be expressed as

$$B_0 = \overline{A_0 A_i} + l_i \qquad (4)$$

where A_i represents the position of the spherical joints on the upper platform, A_0 is the center of the upper platform, B_0 is the point of convergence of the three limbs and the motion parameters are given by linear actuator lengths l_i. This formulation can be obtained from Equation (1) by imposing the convergence of the limbs and removing the fixed-length central limb, and it can be used to control the motion of the locomotion system of the robot, as explained with details in [19–21].

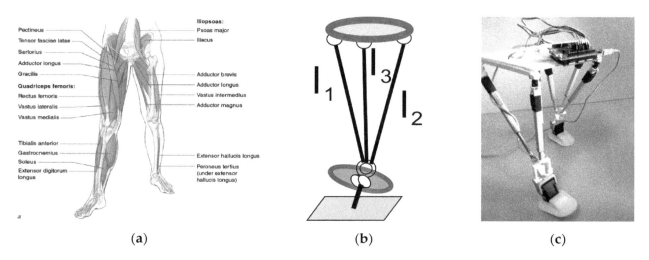

(a)   (b)   (c)

Figure 9. Leg structure: (a) human leg with muscles; (b) a scheme for leg design; (c) LARMbot leg locomotor prototype.

4.1. Experimental Validation

Laboratory tests have been worked out to check the feasibility of the proposed design and to characterize the performance of the LARMbot prototype during its development. In particular, in this paper, experiences of lab testing are reported for motion analysis of the LARMbot leg—locomotor as a parallel biped. Figure 10 shows the structure of the tested LARMbot biped locomotor and a conceptual scheme for the control and test acquisition of biped walking. In Figure 10a, the location of the three sensors that are used for the acquisitions is shown, as well as the orientation of the two IMU (inertial measurement unit) sensors (1) and (2). In Figure 10b, a conceptual scheme is presented for the overall testing frame and connections. The first IMU, denoted as (1), is attached to the left foot, and the second IMU, denoted as (2), is placed on the hip platform. A current sensor (CS) is denoted as (3), and it is fixed on the hip platform where also the rest of the electronic components is installed. The control system is based on an Arduino Mega board to command the eight leg motors, where six of them are linear actuators, and two are rotational motors to drive the ankle joint of each leg. IMU (1) is attached to the left foot to characterize the walking operation cycle's motion in terms of angular velocities and linear acceleration; IMU (2) is used to characterize the hip platform motion; and the current sensor is used to measure the power consumption during walking. An ESP8266-based board is used to acquire and elaborate data from sensors. The information is sent wirelessly to a PC to store data by using Wi-Fi.

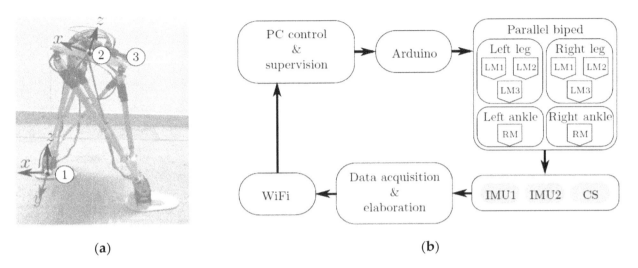

(a)   (b)

Figure 10. A laboratory setup at LARM2 in Rome: (a) Parallel biped and sensors: (1) inertial measurement unit (IMU) at the foot, (2) IMU at the hip, (3) current sensor; (b) a conceptual scheme of the system.

The walking cycle was defined by programming the motion of the biped from point to point. Five different points are stored, which include the corresponding positions of the linear actuators and the rotations of the ankle motors. The first point describes the statically balanced position at the beginning and end of each motion cycle, while the other four points define the main poses of the robot during each step cycle. Thus, the gait is computed by deriving the intermediate positions of the legs through a zero-moment-point approach. Four snapshots of a walking operation test are shown in Figure 11, where the locomotor can be observed while moving from a right-forward double support phase to a right swing, passing through left swing and left-forward double support phase. In this motion, the three linear actuators of each leg contract and extend with a behavior that corresponds to the muscles of the human upper leg, as discussed previously.

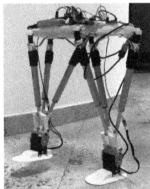

Figure 11. A snapshot of a walking test at LARM2 in Rome.

The acceleration of the hip during this motion has been acquired by IMU (2) and it is reported in Figure 12. The main motion takes place along the x-axis on the horizontal pavement surface and can be observed in the continuous periodic behavior in the graph, which corresponds to the back-and-forth motion of the hip during human-like walking. The y component of the acceleration is instead associated to the lateral balancing motion of the hip, while the acceleration along the z-axis is negligible as referring to vertical displacements during a smooth walking.

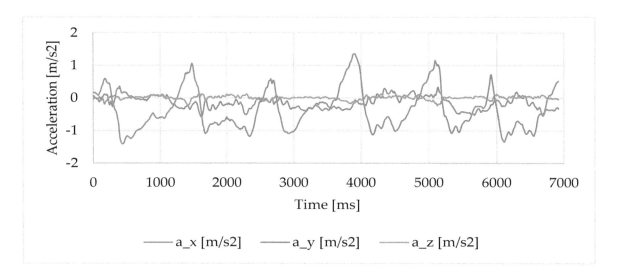

Figure 12. Acquired results of hip motion during a walking test in Figure 11 in terms of acceleration of the hip platform.

The acquisition data of IMU (1) are shown in Figure 13 and illustrate the behavior of the foot during the experimented walking gait. The main component is again in the x direction, with negative peaks corresponding to the foot's dorsiflexion. When the foot is on the ground, the acceleration is negligible, as expected, although some vibration and slipping can be still observed in the data plot. The acquired data for the walking gait test in Figures 12 and 13 show a smooth motion, with acceleration values that are always smaller than 1.5 m/s^2 in absolute value.

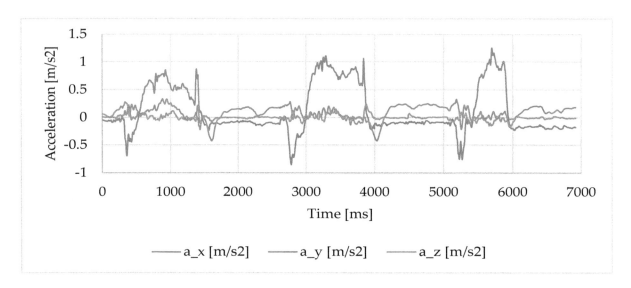

Figure 13. Acquired results of foot motion during a walking test in Figure 11 in terms of acceleration of the foot.

Another significant characteristic of the LARMbot humanoid is its low power consumption. An estimation of the power consumption during the walking gait in Figure 11 can be obtained by the current sensor (3) acquisition together with the power supply voltage (7.4 V), and the results are reported in Figure 14. The static power draw is less than 4.00 W, with a peak of 8.09 W and an RMS (root main square) value is of 5.82 W during the walking operation. Large values of acceleration, up to 4.0 m/s² in absolute value, can be observed instead in the squatting weight-lifting test that was reported in [17], with a 1.00 kg payload and results in Figure 15. The higher value is required by a needed faster balancing action, but the motion is still smooth, and the peak value is well within human-like motion. This acquisition refers to a test of squatting motion with a payload on the arms of the prototype, to show the feasibility and convenience of the parallel mechanisms in LARMbot prototype in a typical high-load task, with the leg design working in coordination with the trunk for balancing. The squatting motion consisted in a vertical up-down displacement of 40 mm at a speed of 10 mm/s, which was obtained by an up-down displacement in the parallel legs that has been properly balanced by the trunk motion with only a pitch adjustment, as shown in Figure 16.

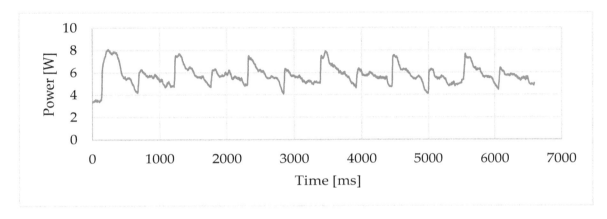

Figure 14. Acquired results of power consumption during a walking test as in Figure 11.

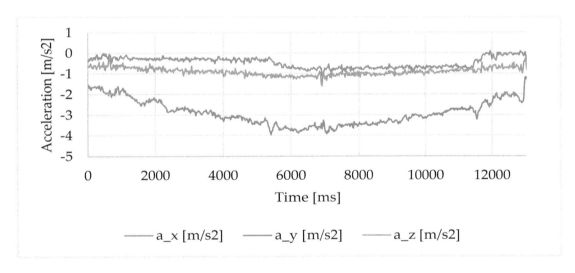

Figure 15. Acquired results of trunk motion during a squat weight-lifting test with a 1.00 kg payload in terms of acceleration at the shoulders [17].

Summarizing the design peculiarities and the laboratory experiences, the parallel mechanisms with human anatomy inspiration in the LARMbot design provide a significant high performance in payload and energy efficiency, as well as the required motion capability for basic humanoid operations.

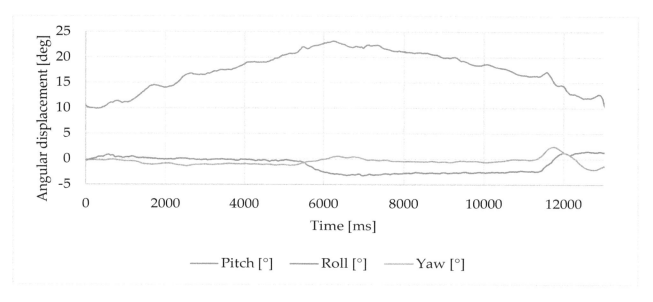

Figure 16. Acquired results of trunk motion during a squat weight-lifting test with a 1.00 kg payload in terms of angular displacement at the shoulders [17].

4.2. Comparison with Serial Architectures

In order to outline the advantages of parallel mechanisms over serial architectures, a comparison between the LARMbot humanoid and the successful NAO design is here reported. The data for LARMbot is obtained either from the experiments presented in Section 4 or from previous works [17,31], whereas the technical specifications of the NAO humanoid can be found on the website of its manufacturer, Aldebaran Robotics [32]. The performance values of both LARMbot and NAO are summarized in Table 3, with reference to the requirements of Table 2.

Table 3. LARMbot performance characteristics [16–19,31,32].

Characteristics	LARMbot	NAO V5
Height	0.85 m	0.57 m
Weight	3.70 kg	5.5 kg
Step length (natural)	100% leg height	50% leg height
Step length (fast)	100% leg height	50% leg height
Speed	100 steps per minute	120 steps per minute
Leg mobility	5 D.o.F.	5 D.o.F.
Leg payload capacity	180% body weight	-
Arm mobility	6 D.o.F.	5 D.o.F.
Arm payload capacity	20% body weight	0.03% body weight
Torso mobility	3 D.o.F.	1 D.o.F.
Torso flexion/extension	0–45°	0
Torso lateral bending	0–45°	0
Power consumption	7.00 W/kg	Battery: 60 min of active use

Even though NAO is able to maintain a faster step rate than the LARMbot, it is outperformed by the latter in payload and efficiency. The parallel architectures of the LARMbot enable a much higher payload and payload to weight ratio, while still maintaining a lightweight design. Usually, serial architectures are preferred to parallel ones for their improved workspace and reach. However, in this case, thanks to its properly designed human-inspired parallel mechanisms, the LARMbot design shows a good performance, not only in payload and energy consumption, but also in workspace, mobility and speed, which can be considered comparable to human ones. The main drawback of parallel design is given by singular configurations and increased control complexity (e.g., kinematics that cannot be solved in closed form or with multiple solutions, force closure for cable-driven mechanisms).

A comparison with NAO only is here reported, since it has a similar size of LARMbot and comparable characteristics. Some other examples may include the iCub [5] and ASIMO [2] platforms that are of a different (larger) size. In general, it can be concluded that parallel mechanisms with antagonist functioning give improved mechanical performance of the humanoid operation in terms of accuracy, stiffness, payload and efficiency, at the not particularly high expense of control complexity.

5. Conclusions

This paper described how parallel mechanisms can be used in humanoid robots in order to improve their structure and operation performance as inspired by human anatomy. Any complex of bones and antagonist muscles in the human body can be modelled as a parallel mechanism whose design can be used for the architecture of humanoid robots. The concept is presented by analyzing the main parts of the human body and extracting a corresponding humanoid robot design with parallel mechanisms, and by discussing requirements and peculiarities for humanoid structures and operations. An illustrative example is reported from the authors' experience with the LARMbot design to show a successful design and implementation of parallel mechanisms in a humanoid robot. The LARMbot design is based on parallel mechanisms to achieve an efficient compact humanoid design with enhanced performance by replicating concepts from human anatomy.

Author Contributions: Conceptualization, M.C.; methodology, M.C.; software, M.R.; experiments, M.R. and C.M.-C.; data curation, M.R.; writing—original draft preparation, M.C. and M.R.; writing—review and editing, M.C.; supervision, M.C. All authors have read and agreed to the published version of the manuscript.

References

1. Lim, H.O.; Takanishi, A. Biped Walking Robots Created at Waseda University: WL and WABIAN Family. *Philos. Trans. R. Soc. Lond. A Math. Phys. Eng. Sci.* **2007**, *365*, 49–64. [CrossRef] [PubMed]
2. Chestnutt, J.; Lau, M.; Cheung, G.; Kuffner, J.; Hodgins, J.; Kanade, T. Footstep planning for the Honda Asimo humanoid. In Proceedings of the 2005 IEEE International Conference on Robotics and Automation, Barcelona, Spain, 18–22 April 2005; pp. 629–634.
3. Kulk, J.; Welsh, J. A low power walk for the NAO robot. In Proceedings of the 2008 Australasian Conference on Robotics & Automation (ACRA-2008), Canberra, Australia, 3–5 December 2008; pp. 1–7.
4. Gouaillier, D.; Collette, C.; Kilner, C. Omni-directional closed-loop walk for NAO. In Proceedings of the 10th IEEE-RAS International Conference on Humanoid Robots (Humanoids), Nashville, TN, USA, 6–8 December 2010; pp. 448–454.
5. Boston Dynamics Atlas. Available online: https://www.bostondynamics.com/atlas (accessed on 18 June 2020).
6. Metta, G.; Natale, L.; Nori, F.; Sandini, G.; Vernon, D.; Fadiga, L.; Bernardino, A. The iCub Humanoid Robot: An Open-Systems Platform for Research in Cognitive Development. *Neural Netw.* **2010**, *23*, 1125–1134. [CrossRef] [PubMed]
7. Tsagarakis, N.G.; Caldwell, D.G.; Negrello, F.; Choi, W.; Baccelliere, L.; Loc, V.G.; Natale, L. WALK-MAN: A High-Performance Humanoid Platform for Realistic Environments. *J. Field Robot.* **2017**, *34*, 1225–1259. [CrossRef]
8. Lafaye, J.; Gouaillier, D.; Wieber, P.B. Linear model predictive control of the locomotion of Pepper, a humanoid robot with omnidirectional wheels. In Proceedings of the 14th IEEE-RAS International Conference on Humanoid Robots (Humanoids), Madrid, Spain, 18–20 November 2014; pp. 336–341.
9. Jung, H.W.; Seo, Y.H.; Ryoo, M.S.; Yang, H.S. Affective communication system with multimodality for a humanoid robot, AMI. In Proceedings of the 2004 4th IEEE/RAS International Conference on Humanoid Robots, Santa Monica, CA, USA, 10–12 November 2004; Volume 2, pp. 690–706.
10. Tellez, R.; Ferro, F.; Garcia, S.; Gomez, E.; Jorge, E.; Mora, D.; Faconti, D. Reem-B: An autonomous lightweight human-size humanoid robot. In Proceedings of the 8th IEEE-RAS International Conference on Humanoid Robots, Daejeon, Korea, 1–3 December 2008; pp. 462–468.

11. Asfour, T.; Regenstein, K.; Azad, P.; Schroder, J.; Bierbaum, A.; Vahrenkamp, N.; Dillmann, R. ARMAR-III: An integrated humanoid platform for sensory-motor control. In Proceedings of the 6th IEEE-RAS International Conference on Humanoid Robots, Genova, Italy, 4–6 December 2006; pp. 169–175.
12. Asfour, T.; Schill, J.; Peters, H.; Klas, C.; Bücker, J.; Sander, C.; Bartenbach, V. Armar-4: A 63 dof torque controlled humanoid robot. In Proceedings of the 13th IEEE-RAS International Conference on Humanoid Robots, Atlanta, GA, USA, 15–17 October 2013; pp. 390–396.
13. Ceccarelli, M. *Fundamentals of Mechanics of Robotic Manipulation*; Springer Science & Business Media: Dordrecht, The Netherlands, 2013.
14. Wang, H.; Sang, L.; Zhang, X.; Kong, X.; Liang, Y.; Zhang, D. Redundant actuation research of the quadruped walking chair with parallel leg mechanism. In Proceedings of the IEEE International Conference on Robotics and Biomimetics (ROBIO), Guangzhou, China, 11–14 December 2012; pp. 223–228.
15. Sugahara, Y.; Carbone, G.; Hashimoto, K.; Ceccarelli, M.; Lim, H.O.; Takanishi, A. Experimental Stiffness Measurement of WL-16RII Biped Walking Vehicle during Walking Operation. *J. Robot. Mechatron.* **2007**, *19*, 272–280. [CrossRef]
16. Ceccarelli, M.; Cafolla, D.; Russo, M.; Carbone, G. LARMBot Humanoid Design Towards A Prototype. *MOJ Appl. Bionics Biomech.* **2017**, *1*, 00008. [CrossRef]
17. Russo, M.; Cafolla, D.; Ceccarelli, M. Design and Experiments of A Novel Humanoid Robot with Parallel Architectures. *MDPI Robot.* **2018**, *7*, 79. [CrossRef]
18. Cafolla, D.; Ceccarelli, M. Design and Simulation of A Cable-Driven Vertebra-Based Humanoid Torso. *Int. J. Hum. Robot.* **2016**, *13*, 1650015. [CrossRef]
19. Russo, M.; Ceccarelli, M.; Takeda, Y. Force Transmission and Constraint Analysis of A 3-SPR Parallel Manipulator. *Proc. Inst. Mech. Eng. Part C J. Mech. Eng. Sci.* **2017**. [CrossRef]
20. Russo, M.; Ceccarelli, M. Kinematic design of a tripod parallel mechanism for robotic legs. In *Mechanisms, Transmissions and Applications, Mechanism and Machine Science*; Springer: Cham, Switzerland, 2018; pp. 121–130.
21. Russo, M.; Herrero, S.; Altuzarra, O.; Ceccarelli, M. Kinematic Analysis and Multi-Objective Optimization of A 3-UPR Parallel Mechanism for a Robotic Leg. *Mech. Mach. Theory* **2018**, *120*, 192–202. [CrossRef]
22. Yang, P.; Gao, F. Kinematical Model and Topology Patterns of a New 6-Parallel-Legged Walking Robot. In Proceedings of the ASME 2012 International Design Engineering Technical Conferences and Computers and Information in Engineering Conference, Chicago, IL, USA, 12–15 August 2012; pp. 1197–1205.
23. Yang, P.; Gao, F. Leg Kinematic Analysis and Prototype Experiments of Walking-Operating Multifunctional Hexapod Robot. *Proc. Inst. Mech. Eng. Part C J. Mech. Eng. Sci.* **2014**, *228*, 2217–2232. [CrossRef]
24. Pan, Y.; Gao, F. A New Six-Parallel-Legged Walking Robot for Drilling Holes on the Fuselage. *Proc. Inst. Mech. Eng. Part C J. Mech. Eng. Sci.* **2014**, *228*, 753–764. [CrossRef]
25. Knudson, D. *Fundamentals of Biomechanics*; Springer Science & Business Media: Dordrecht, The Netherlands, 2007.
26. Farris, D.J.; Sawicki, G.S. The Mechanics and Energetics of Human Walking and Running: A Joint Level Perspective. *J. R. Soc. Interface* **2011**, rsif20110182. [CrossRef] [PubMed]
27. Winter, D.A. Kinematic and Kinetic Patterns of Human Gait: Variability and Compensating Effects. *Hum. Mov. Sci.* **1984**, *3*, 51–76. [CrossRef]
28. Russo, M. Design and Validation of a Novel Parallel Mechanism for Robotic Limbs. Ph.D. Thesis, University of Cassino and Southern Latium, Cassino, Italy, 2019.
29. Ceccarelli, M. An Illustrated History of LARM in Cassino. In Proceedings of the RAAD 2012 International Workshop on Robotics in Ale-Adria-Danube Region, Naples, Italy, 10–13 September 2012; pp. 35–42.
30. Ceccarelli, M. LARM PKM Solutions for Torso Design in Humanoid Robots. *Front. Mech. Eng.* **2014**, *9*, 308–316. [CrossRef]
31. Cafolla, D.; Ceccarelli, M. An Experimental Validation of A Novel Humanoid Torso. *Robot. Auton. Syst.* **2017**, *91*, 299–313. [CrossRef]
32. Aldebaran Robotics' NAO. Available online: http://doc.aldebaran.com/index.html (accessed on 31 August 2020).

2

Design, Simulation, and Preliminary Validation of a Four-Legged Robot

Stefano Rodinò, Elio Matteo Curcio⑩, Antonio di Bella, Mattia Persampieri, Michele Funaro and Giuseppe Carbone *

Department of Mechanical, Energy and Management Engineering (DIMEG), University of Calabria, 87036 Rende, Italy
* Correspondence: giuseppe.carbone@unical.it

Abstract: This paper outlines the design process for achieving a novel four-legged robot for exploration and rescue tasks. This application is also intended as an educational mean for masters' students aiming at gaining skills in designing and operating a complex mechatronic system. The design process starts with an analysis of the desired locomotion and definition of the main requirements and constraints. Then, the paper focuses on the key design challenges, including analytical/numerical modeling and simulations of kinematic and dynamic performances. Specific attention is addressed to the manufacturing of a proof-of-concept prototype, including mechanical and control hardware, as well as the development of the needed software for an autonomous operation. Preliminary tests were carried out, to validate the main features required by the final prototype, to prove its feasibility and user-friendliness, as well as the effectiveness of this complex mechatronic design task for successfully engaging students towards learning complex theoretical, numerical, and practical skills.

Keywords: design procedure; kinematics; dynamics; legged robots; quadrupeds

1. Introduction

The constant increase in natural catastrophic events such as earthquakes, fires, and hurricanes has pushed research towards the development of robotic devices capable of performing patrolling and rescue tasks, as discussed, for example, in References [1,2]. BIGDOG [3], Momaro [4], DOGGO [5], and Mini-Mini Cheetah [6] are valuable design examples within the rich literature on the topic. Some designs have also evolved into commercial products such as SPOT (Boston Dynamics), ANYMAL (ANYbotics), and AlienGo (Unitree Robotics), as reported, for example, in Reference [7]. The intended operation tasks for the abovementioned robots consist mainly on identification of the environment, inspection of key features, detection of dangerous materials, and locomotion on unstructured environments with the presence of obstacles of different sizes and shapes. In this context, legged locomotion has been found to be the most effective strategy, since it allows us to overcome obstacles, whose size is comparable with a leg size. Nevertheless, legged locomotion is quite challenging from design and operation viewpoints, as it is requiring high-level multidisciplinary knowledge in mechanics, control, programming, and sensing, as also outlined in Reference [8].

Multiple-legged locomotion has attracted specific attention from the research team led by Professor Carbone in the last 15 years, also aiming at developing proper design procedures and innovative designs for cost-oriented and user-friendly solutions in specific locomotion tasks, as well as proposing locomotion architectures ranging from biped to hexapod, as reported, for example, in References [8–14].

This paper proposes the design of a four-legged robot for exploration and rescue tasks as a challenging mechatronic concept to be developed as a master students teamwork within the classes of Mechatronics delivered at University of Calabria, Italy. Accordingly, this paper proposes a general design procedure to address the specific technical challenges of quadruped locomotion. Moreover, it also aims at demonstrating the effectiveness of a "learning-by-doing" teaching approach where master students are stimulated to gain in-depth knowledge on a challenging multidisciplinary topic, learning skills and expertise which is often difficult to master through other traditional teaching approaches.

With the above premise, this work describes the key design and operation challenges on the development of a full quadruped robotic prototype including all hardware, control, sensors and software components. Namely, in this paper, Section 2 describes the attached problem and the design steps towards a conceptual design. Then, Section 3 provides kinematics modeling and analyses for a proper type and size synthesis of a leg. Section 4 addresses the dynamic modeling and analyses also aiming at a proper selection and sizing of the required actuators. The last section describes the manufacturing of a proof-of-concept prototype including all hardware and sensory components. This is followed by a description of the proposed control hardware and software followed by results of preliminary tests, which have been carried out for validating the feasibility and effectiveness of the proposed design choices.

2. The Attached Problem and the Proposed Conceptual Design

Inspection and rescue tasks are characterized by several design requirements and constraints, which are briefly outlined, for example, in References [1,2]. Accordingly, it is fundamental to undergo a systematic design approach for being able to fulfil the expectations and needs of specific application tasks.

A feasible design procedure has been defined through key design steps as outlined in the flow-chart of Figure 1. In particular, the very first step consists of an in-depth analysis of the available literature for identifying quantitative data regarding the prescribed design requirements and constraints. These quantitative data can be collected into a product design specification (PDS) table, as reported, for example, in Table 1. This first step can be followed by a team brainstorming session for identifying the main challenges and potential design concepts/solutions, which can be technically allow the prescribed product design specification. Quantitative and qualitative comparisons among multiple design concepts or choices may lead on reaching a consensus on a specific design concept. The next design step consists of quantitative calculations (i.e., kinematics and dynamics modeling) and software aided simulations aiming at identifying the most appropriate materials, hardware components, sensors as well as for performing mechanism type and size syntheses.

The specific proposed attached problem is providing a valuable engineering design learning benchmark, as it requires us to address several key challenging theoretical and practical aspects, including theoretical kinematics and dynamics aspects. Furthermore, the proposed design procedure and approach can be understood as a valuable general-purpose learning-by-doing educational means, which can be used and replicated as an effective learning approach for mechanical engineering masters' students, who are engaged and stimulated towards gaining complex theoretical, simulation, operation, and experimental testing skills with the development of a complex mechatronic system. Furthermore, the proposed specific design solution has the merit and novelty to propose a topology with just 2-dofs kinematic architecture for each leg, in combination with a crank and rope driving mechanism and a balancing spring. This proposed solution is demonstrated to be a very effective solution for achieving dynamic walking and trotting gaits while keeping a reasonable cost-oriented and user-friendly design.

This phase might require multiple iterations until a proper solution is confirmed to fulfil design performance expectations based on the product-design-specifications checklist. This phase can be conveniently followed by a rapid prototyping phase in which one can achieve preliminary experimental proof of the proposed design concept. This phase might require reiterating the whole design process in case any key performance results are below the expectations. Moreover, there might be multiple design refinement/optimization phases until the proposed design reaches its final stage.

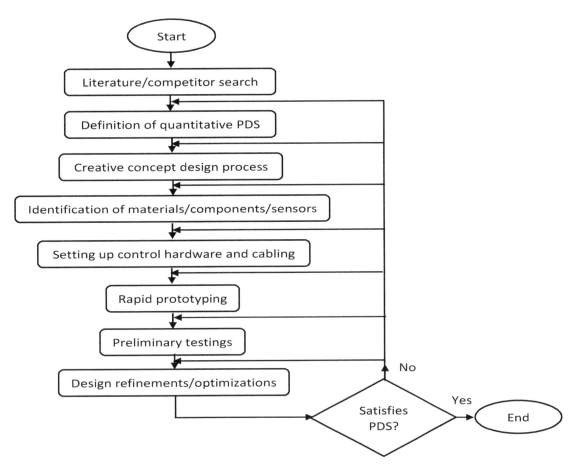

Figure 1. A flowchart of the proposed design procedure.

Table 1. The proposed product design specifications (PDS).

	FEATURES
FUNCTION	• Exploration on uneven terrain • First aid • Identifying victims • Detection of harmful substances in the air

PERFORMANCE	• Size:
	  – Height: 20 ÷ 40 cm
	  – Width: 30 ÷ 60 cm
	  – Length: 20 ÷ 80 cm
	• Weight: 5 ÷ 60 kg
	• Maximum speed: 1.5 m/s
	• Payload: 0.5 ÷ 1 kg
	• Power supply: Onboard batteries (DC)

MANUFACTURING	• Cost: 4000€ ÷ 40,000€ • Amount: 1 Prototype
ENVIRONMENT	• Presence of water or moisture • Possible gases or corrosive chemicals • Medium–high heat sources • Dust presence
MATERIALS	• Carbon fibre • Steel • Aluminum alloys (Nickel and Magnesium)

3. Kinematic Analysis

As shown in Figure 2, a four-bar mechanism is used to actuate the robot's leg, operated by using a crank and a rope. The mechanism is obtained through the assembly of four links with revolute joints. The degrees of freedom of this kinematic mechanism are two, one relating to the rotation of the crank connected to the frame with an additional revolute joint connecting to the rope that activates the movement of the four-bar mechanism, as reported in Figure 2a. In this section, the direct kinematics model is developed and solved by following the design variable that are reported in Figure 2 and Table 2. This model provides the position of the paw in terms of X and Y coordinates once the values of the joint variables q_1 and q_2 are assigned.

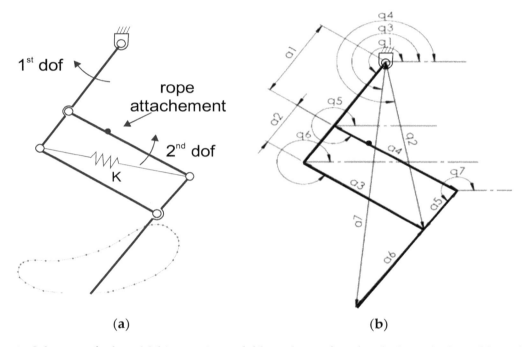

Figure 2. Schemes of a leg: (**a**) kinematic model based on a four-bar linkage (a dotted line shows a feasible motion path of the foot) and (**b**) main design parameters.

Table 2. Link dimension for the proposed mechanism.

a_1	a_2	a_3	a_4	a_5	a_6
0.085 m	0.035 m	0.130 m	0.130 m	0.035 m	0.115 m

The direct kinematics model was developed by using the closing polygon method or close-loop method [14], identifying three of the possible polygons in the kinematic scheme of Figure 3.

$$\begin{cases} \overleftarrow{a_1} + \overleftarrow{a_2} + \overleftarrow{a_3} - \overleftarrow{q_2} = 0 & (1) \\ \overleftarrow{a_2} + \overleftarrow{a_3} - \overleftarrow{a_4} - \overleftarrow{a_5} = 0 & (2) \\ \overleftarrow{q_2} + \overleftarrow{a_6} - \overleftarrow{a_7} = 0 & (3) \end{cases}$$

where $a_1, a_2, a_3, a_4, a_5,$ and a_6 are the dimensions of the various links that make up the kinematic scheme; a_7 is the virtual segment that traces the position of the point of contact with the ground; and q_2 indicates the length of the rope that activates the four bar mechanism. A Cartesian reference frame was chosen, as reported in Figure 4. Accordingly, one can consider the projections along X and Y axes of the vector Equations (1)–(3). Using the abovementioned projections, it is possible to obtain a system of six equations in six unknowns, namely q_3, q_4, q_5, q_6, q_7, and a_7, as reported in Equation (4).

$$\begin{cases} a_1 \cos q_1 + a_2 \cos q_1 + a_3 \cos q_6 - q_2 \cos q_4 = 0 \\ a_1 \sin q_1 + a_2 \sin q_1 + a_3 \sin q_6 - q_2 \sin q_4 = 0 \\ a_2 \cos q_1 + a_3 \cos q_6 - a_4 \cos q_5 - a_5 \cos q_7 = 0 \\ a_2 \sin q_1 + a_3 \sin q_6 - a_4 \sin q_5 - a_5 \sin q_7 = 0 \\ q_2 \cos q_4 + a_6 \cos q_7 - a_7 \cos q_3 = 0 \\ q_2 \sin q_4 + a_6 \sin q_7 - a_7 \sin q_3 = 0 \end{cases} \tag{4}$$

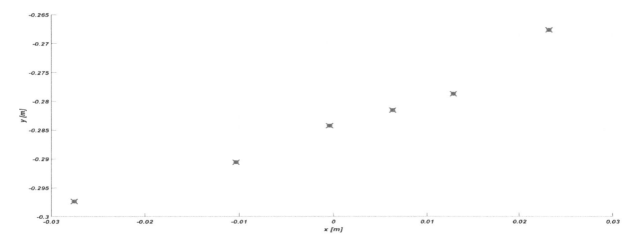

Figure 3. Comparison of results between direct kinematic model and SimMechanics model.

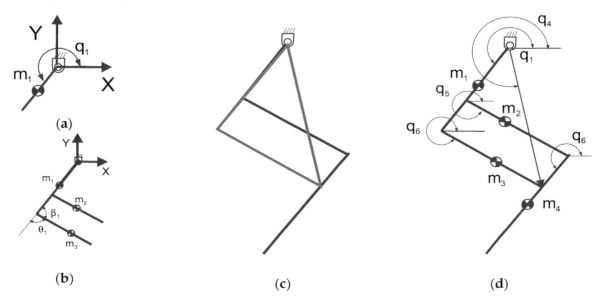

Figure 4. Schemes of the proposed leg design: (**a**) first link mass center, (**b**) mass centers of links two and three, (**c**) closing polygon scheme, and (**d**) mass center of link four.

The system of Equations (4) was solved by means of the Matlab symbolic calculation software, in order to obtain the end-effector positions as function of the input angle, q_1, and the length of the rope, q_2.

$$P_x = a_7 \cos q_3 \tag{5}$$

$$P_y = a_7 \operatorname{sen} q_3 \tag{6}$$

By solving the system of Equation (4), it is possible to obtain the positions P_x and P_y, as indicated in Equations (5) and (6). These equations represent the X and Y positions plotted by the virtual link a_7 indicated in Figure 3. The validation of the kinematic model developed in this section was carried out by comparing the respective values obtained through a multi-body model created on SimMechanics. Figure 3 shows the overlap of the points obtained from the solved kinematic model Equation (4) and the points extracted from the multi-body model created on the SimMechanics simulation software. Thus, this validates the correctness of the proposed analytical model.

Figure 4 shows the geometric relationships used to evaluate the position of the centers of mass of the links that compose the whole mechanism. The position of the mass m1 coincides with the center of mass of the link of length $l_1 = a_1+a_2$, and the position of the center of mass P_{m_1} can be obtained as follows:

$$P_{m_1} = \begin{bmatrix} \frac{l_1}{2}c_1 \\ \frac{l_1}{2}s_1 \\ 0 \end{bmatrix} \tag{7}$$

where c_1 and s_1 stand for $\cos(q_1)$ and $\sin(q_1)$, $l_1 = a_1+a_2$ is the length of the first link, and P_{m_1} indicates the position of the center of mass m_1 along the components X, Y and Z. The position of the masses m_1 and m_2 indicated in Figure 4b is obtained starting from the geometric considerations referring to the triangle shown in Figure 4c. By applying Carnot's theorem, it is possible to obtain the angle θ_1 as a function of the length of the segment q_2 as follows:

$$\theta_1 = Acos(\frac{q_2^2 - l_1^2 - a_3^2}{-2l_1 a_3}) \tag{8}$$

where q_2 is the length of the segment in Figure 5c, a_3 is the length of link three in Figure 4c, and θ_1 is the angle formed by links one and three visible in Figure 5c. Once the angle θ_1 is known, it is possible to obtain the angle β_1 shown in Figure 4b and obtain the position of the centers of mass P_{m_2}, P_{m_3}, and P_{m_4} as follows:

$$P_{m_2} = \begin{bmatrix} a_1 c_1 + \frac{a_4}{2}\cos(q_1 + \beta_1) \\ a_1 s_1 + \frac{a_4}{2}\sin(q_1 + \beta_1) \\ 0 \end{bmatrix} \tag{9}$$

$$P_{m_3} = \begin{bmatrix} l_1 c_1 + \frac{a_3}{2}\cos(q_1 + \beta_1) \\ l_1 s_1 + \frac{a_3}{2}\sin(q_1 + \beta_1) \\ 0 \end{bmatrix} \tag{10}$$

$$P_{m_4} = \begin{bmatrix} a_1 c_1 + \frac{a_4}{2}\cos(q_1 + \beta_1) + \frac{(a_5+a_6)}{2} c_1 \\ a_1 s_1 + \frac{a_4}{2}\sin(q_1 + \beta_1) + \frac{(a_5+a_6)}{2} s_1 \\ 0 \end{bmatrix} \tag{11}$$

where a_3, a_4, a_5 and a_6 are the lengths of the segments visible in Figure 5c. The position of the center of mass m4 in Figure 5d is determined by considering that geometric constraint is given by the parallelism of link 1 and link 4. Accordingly, the angles q_1 and q_7 are always kept at the same value.

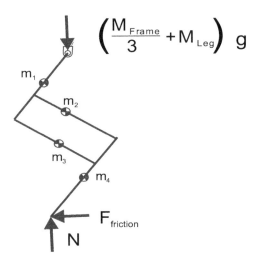

Figure 5. A scheme of the external forces acting on a leg.

4. Dynamic Analysis

A dynamic analysis was developed for the kinematic model that operates the robot paw for making a suitable choice of servomotors. The approach used is the energy approach as based on the Euler–Lagrange formulation. Starting from Equations (7) and (9)–(11), the kinetic and potential energy of the kinematic scheme in Figure 2 was evaluated, taking into account the presence of a spring installed at the diagonal of the four bar (a_2, a_3, a_4, and a_5), between the joints 7 and 6 of Figure 5. This spring has the function of a gravity balancing element, useful to bring the paw back into contact with the ground when the rope corresponding to the segment has stretched q_2, as shown in Figure 5.

The kinetic and potential energy values shown in Figure 6a,b were obtained by assigning to the joint variables q_1 and q_2 of the third order polynomial paths, so as to trace a closed circular path in the paw–ground contact area. The assigned trajectories are shown in Figures 7 and 8, where one can note a maximum lift of about 0.22 m and a maximum step size of about 0.14 m for a maximum leg size of about 0.5 m at nominal standing configuration.

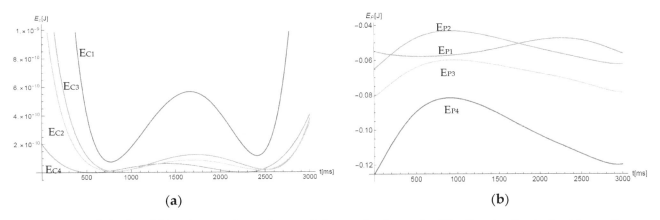

Figure 6. Calculated kinetic energies and potential energies for each of the four bodies in the kinematic model of Figure 4: (**a**) kinetic energies for bodies one to four, and (**b**) potential energies for bodies one to four.

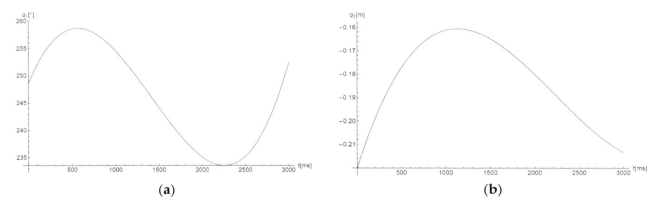

Figure 7. A plot of a set of input q_1 and q_2 joints' trajectory used to evaluate the dynamical model: (**a**) q_1 versus time; (**b**) q_2 versus time.

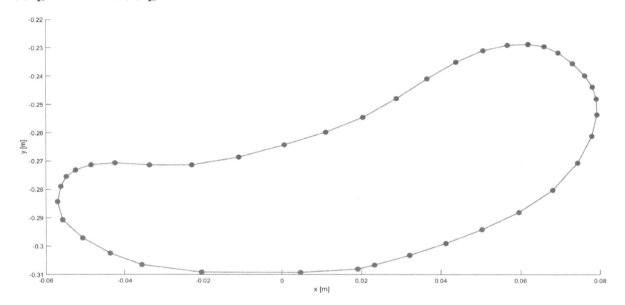

Figure 8. A path described by a point attached to the foot end point, as obtained by the model in Figure 5, when using the input variables given in Figure 7.

The forces N and $F_{Friction}$ are present in the model of one leg, as shown in Figure 5, and were evaluated according to Equations (12) and (13), as follows:

$$N = \left(\frac{M_{Frame}}{3} + M_{Leg}\right) g \tag{12}$$

$$F_{Friction} = K_d \, N \tag{13}$$

where N is the constraint reaction deriving from contact with the ground, M_{Frame} is the mass of the frame of the entire robot, M_{Leg} is the weight of the single leg, g is the gravitational acceleration constant (9.81 m/s^2), $F_{Friction}$ is the force of friction, and K_d is the dynamic friction coefficient considering a contact of the rubber-cement type and with a value equal to 0.5. One should note that Equation (12) refers to the case of three legs in contact with the ground while only a single leg is lifted up. Accordingly, the mass of the frame, M_{Frame} , is distributed equally among three legs, and the mass of the single lifted leg, M_{Leg} , sums up to compute the total normal force, N, acting on each leg in contact with the ground.

The weight force of the frame weighs on the movement of the paw only when it appears to be in contact with the ground; therefore, in reference to the trajectory traced in Figure 8, a time of contact with the ground equal to 0.4 s was estimated. From these considerations, we can build a dynamic model based on Lagrange's equations. After calculating the potential and kinetic energy of the system, as shown in Figure 6a,b the Lagrangian function is calculated according to Equations (12) and (13).

$$L = E_{c_{Total}} - E_{P_{Total}} \tag{14}$$

where $E_{c_{Total}}$ and $E_{P_{Total}}$, respectively, represent the total and potential kinetic energy of the entire system in Figure 6. Lagrange's equations were obtained starting from Equation (14), as follows:

$$\frac{d}{dt}\left(\frac{\partial L}{\partial \dot{q}_1}\right) - \left(\frac{\partial L}{\partial q_1}\right) = \tau + \tau_N + \tau_a \tag{15}$$

$$F_{Friction} = K_d\, N \tag{16}$$

where τ_N is the resisting torque due to the constraint reaction, N, and τ_a is the resisting torque generated by friction during the time of contact. Equations (15) and (16) were solved through the Wolfram Mathematica software, providing the torque and actuation force results needed to follow the trajectory in Figure 8.

As can be seen from Figure 9b, the torque to be applied is maximum at the time interval during which the paw touches the ground. Subsequently, there is a drastic decay of the torque values, due to a short transient during which the paw detaches from the ground. This is justified by the fact that the torque required to rotate the paw in the air must essentially contrast only its own weight. The transient duration was defined from practical viewpoint, with a very small value, equal to 0.1 s. Furthermore, a linear trend of the forces was assumed during the transient phase. As far as the tension of the rope is concerned, it can be noted (in Figure 9a) that the constraint reaction contributes positively to the lifting of the leg during contact with the ground. Thus, this is decreasing the force necessary to lift it.

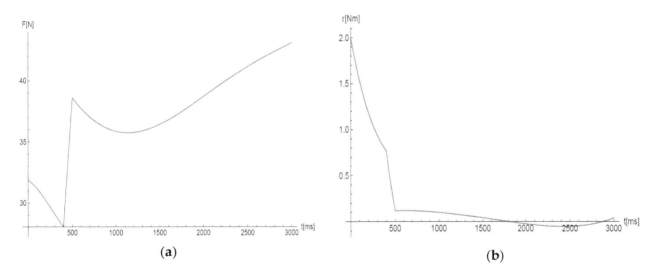

(a)   (b)

Figure 9. Plots of simulation results: (**a**) actuation force for q_2 joint variable and (**b**) actuation torque for q_1 joint variable.

5. Hardware Selection and Prototyping

Eight motors are needed to actuate the robot, i.e., two on each leg. One actuates hip movement, and the other one actuates knee movement, as detailed in Figure 10.

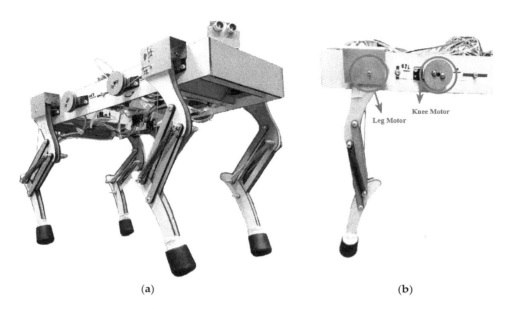

(a)   (b)

Figure 10. Pictures of the built prototype: (**a**) full robot and (**b**) a detail of the built full leg actuation mechanism.

In this case, it is necessary to precisely control the position of the motors. This is achieved by using commercial servo motors (Figure 10b). They allow us to control the angular position in a range between 0° and 180°, in a close-loop control, including an absolute encoder. Moreover, they are easily programmable with an Arduino board. The choice of motors was made starting from the results obtained in Section 4, with the results of the inverse dynamics in Figure 9a,b. The maximum actuation torque values of 2 Nm were detected for the leg motor, while the maximum force required to move the rope is about 45 N. Since the radius of the pulley is 2 cm, the maximum torque that the motors must exert to rotate the pulley will be approximately 0.9 Nm. Eight identical motors were selected, to simplify the control architecture. A JX Servo PDI-HV5932MG [15] was selected with a 6.0 V power supply. In this case, the servomotor is delivering a torque of 2.5 Nm, which is higher than the maximum required peak torque, as reported in Figure 9b.

The robot is equipped with several sensors allowing the acquisition of data relating to both the robot external and internal environment. There is an IMU (inertial measurement unit) sensor providing accelerometers and gyroscopes, which allow monitoring of the dynamics of a moving robot body. The chosen IMU sensor is the GY-521 MPU-6050 (Figure 11a). The InvenSense GY-521 MPU-6050 sensor [16] contains, in a single integration, a three-axis MEMS accelerometer and a three-axis MEMS gyroscope. Its operation is very accurate, as based on a 16-bit analog/digital converter for each channel. As a result, it manages to simultaneously capture the values of the X, Y, and Z axes.

The measurement of temperature and humidity of the external environment is carried out via a DHT22 sensor (Figure 11b). This sensor allows us to detect the relative humidity and temperature of an environment and to transmit it digitally through a single wire (in addition to the two necessaries for power supply) to a microcontroller [17]. The humidity sensor is of the capacitive type. It uses a maximum of 2.5 mA in work and allows us to acquire a maximum of one sample each 2 s. The humidity measurement varies from 0 to 100%, with a typical accuracy of ±2%. As for the temperature, the DHT22 has a measuring range between −40 and + 80 °C, with an accuracy of 0.5 °C, by using a DS18B20 sensor integrated inside the robot body. To detect the presence of objects in the immediate vicinity, an ultrasound SRF05 sensor is used (Figure 11c). This sensor is equipped with a microcontroller that provides all the processing functions. It is enough to send an impulse and read the return echo, to establish the distance of the obstacle or object in front. It can identify even small obstacles and objects from a minimum distance of 3 cm up to 3 m [18]. Like all ultrasonic sensors, it is insensitive to light, so it can also be used outdoors. It is also ideal for applications requiring greater precision or range. The sensitive workspace of this sensor is quite wide, and it can be optimized by adjusting its sensitivity threshold.

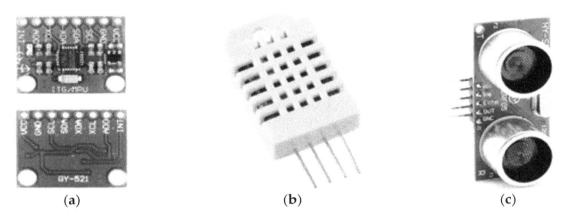

(a)   (b)   (c)

Figure 11. (a) Inertial measurement unit (IMU) sensor, (b) temperature and humidity sensor, and (c) proximity sensor.

With reference to Figure 12, it can be observed that the proof-of-concept prototype was made by using low-cost off-the-shelf components. The structure is mostly made of wood; the components were laser-cut and assembled through bolts and gluing. Steel parts were also made to support the load on the leg motors, while the rope that moves the knee is a nylon fishing line.

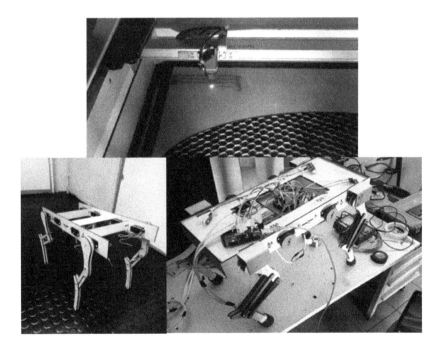

Figure 12. Prototype of manufacturing phases.

After generating the complete CAD model, the 2D files were exported for cutting, and the pieces were assembled, considering the moving parts and the games necessary for moving the links. The springs were chosen based on the torque delivered by the servo motors and based on the weight of the robot pitch; they aim at returning the robot to an upright configuration.

Figure 13 outlines the electrical wiring of the prototype with a 12 V input power supply. Voltage regulators are used to provide 6 V as input power for the servomotors. Voltage regulators were doubled, to guarantee a maximum 15 A output current. This value was calculated by considering that each servomotor must be powered at 6V −5A (30 W), to provide the required torque (2.5 Nm). The sensors, instead, require a 5 V power supply, which is always supplied by using a stabilized voltage regulator. An Arduino MEGA 2560 R3 controller is also powered through the same regulator. As shown in Figure 13, two Arduino controllers were used: an Arduino MEGA 2560 R3 in combination with an Arduino NANO. Communication between the two Arduinos was achieved via I2C communication. The Arduino Nano has the master role and communicates with a PC, serially, sending the sensor data and receiving input commands by a user, from a specifically developed graphic user-interface. The Arduino NANO transmits the commands to the Arduino MEGA, which begins the calculation for the gaits. The master-and-slave Arduinos are connected via two SDA (serial data)–SCL (serial clock) standard communication cables. Only the pins with PWM (pulse-width modulation) outputs are connected to the Arduino MEGA for controlling the servomotors, while the Arduino NANO is connected via I2C not only to the Arduino MEGA but also with the 3-axis accelerometer (IMU), the radar sensor, and the temperature and humidity sensor.

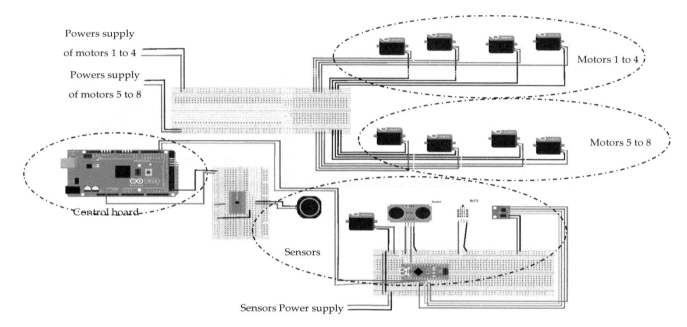

Figure 13. A logical scheme for the control hardware, motors, and sensors.

6. Main Programming Architecture

The control and sensors' feedback architecture is entirely managed by the Arduino NANO controller, as shown in Figure 14. The program compiled and loaded on the sensor controller acts as the master for the whole system. In fact, it simulates multi-thread processes that allow it to be able to quickly acquire information from the sensors without ever stopping (as in the case of a single process). The reading of the outputs of all the peripherals is carried out simultaneously, and all the data are subsequently sent to the various viewers. The *delay()* function can be avoided through a technique based on the system clock, which slows down the data transmission speed. This technique therefore simulates a parallel execution of all reading processes. The process begins with the creation of a class that is used for the creation of objects in the programming. It includes attributes and methods that will be shared by all objects created (instances), starting with the class. The objects that can be instantiated by the class are events (I2C reading from the sensors, reading, and writing on the serial port). The attributes related to the objects of the class define the time that elapses between the execution of an action and its subsequent repetition. This time is evaluated by counting the milliseconds, starting from the system activation time, using the *millis()* function. The logic of the object ensures that the system does not stop until a predetermined time interval is exceeded; once exceeded, the process linked to the object itself is reactivated. In this case, the I2C communication requires the master to take the action command from the serial port that the user wants the robot to do and send it back to the motor controller, in the form of Byte.

As shown in Figure 14, the servomotors are controlled through the Arduino MEGA 2560 R3 controller. The #*servo* library was used to control the movement of the servomotors, which automatically manages PWM communication. To facilitate programming and control logic, a special class was created that manages each leg and all legs of the robot.

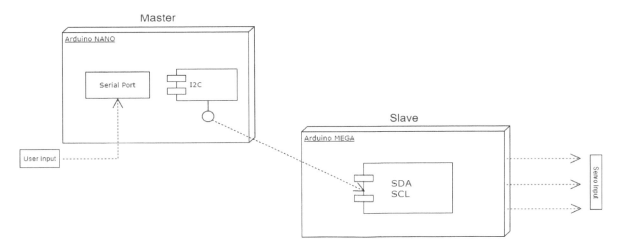

Figure 14. Master–slave communication between the Arduino MEGA and Arduino NANO controllers.

7. Preliminary Testing Results

A dedicated user interface was developed in Java programming language, as shown in Figure 15, with the aim of making the output signal from the sensors as user-friendly as possible. This allowed us to conduct a series of preliminary tests and a preliminary comparison of the robot's performance. For each test, the robot covered a total distance of about three meters, demonstrating the synchronization speed, this led to defining an optimal walking speed that allowed us to obtain maximum walking stability.

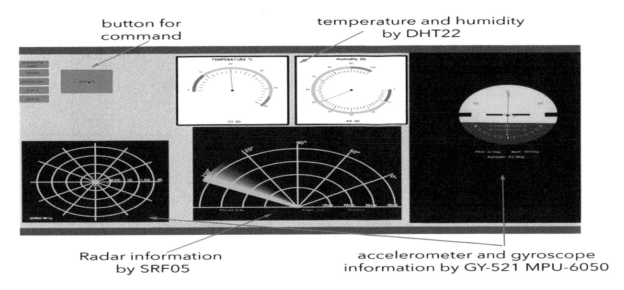

Figure 15. GUI (Graphical User Interface) for controlling robot and acquiring data.

A specific motion strategy was implemented proposed in the scheme of Figure 16 with a sequence of diagonally opposite legs can rest or move alternatively. The gait sequence starts from the right rear and front left and then vice versa, as also discussed in References [14,19]. Figure 17 reports some snapshot frames taken from videos of the experiments. Through the frames, one can clearly see that the synchronization movements of the legs are allowing the robot to walk, for instance, at very low speed. In such a case, the center of gravity must always be within the stability triangle area described by connecting the three support points. When increasing the speed, we achieve a dynamic motion by using a trot gait strategy. In this case, the support points shift from three to two, and a dynamic equilibrium needs to be achieved by considering the inertia of the robot body. Figure 17 confirms that the desired dynamic gait was successfully achieved.

Figure 16. A footfall scheme of trotting gait, [8].

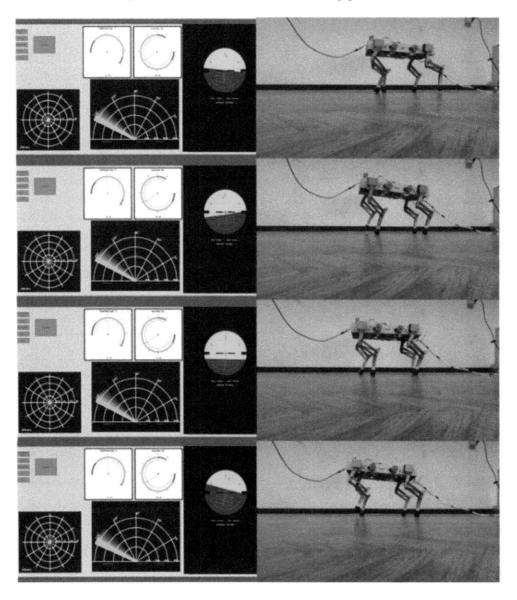

Figure 17. Experimental data during a trotting experiment.

The reported experiments have preliminarily demonstrated the proper operation of the prototype in achieving complex dynamic gaits. They also demonstrate the effectiveness of the attached very challenging topic for successfully engaging students towards learning very complex theoretical, numerical, and practical skills. Further studies will be carried out in future, to explore the potential use of the built prototype for multiple applications including exploration, patrolling, surveillance of unstructured environments, as proposed, for example, in References [20–22].

8. Conclusions

This paper outlines the conceptualization, design, and proof-of-concept manufacturing of a low-cost and user-friendly four-legged robot that is fulfilled within a proposed general design procedure. Each design phase is discussed for identifying the main features and key points as part of a multidisciplinary approach. Namely, the proposed procedure includes (but it is not limited to) mechanism design, kinematic, and dynamic modeling/simulation, to inform a proper design procedure and to achieve a proper selection and adaptation of off-the-shelf components. The proposed design procedure results in developing a specific leg design topology with 2-dofs kinematic architecture for each leg, in combination with a crank and rope driving mechanism and a balancing spring. A prototype was built based on the proposed procedure, and it was integrated with all the mechatronic components, including control and sensory hardware. A specific user-interface and software was also developed based on a dedicated master–slave Arduino architecture. An experimental validation was successfully reported to demonstrate the fulfilment of the desired key performances in terms of static and dynamic gaits, as well as in terms of real-time collection of environment data. These successful results were obtained by a team of masters' students within a teamwork project. This also demonstrates the effectiveness of the proposed challenging multidisciplinary mechatronic design task as an effective means for non-conventional learning of multiple complex topics. In particular, it demonstrates that the proposed interesting and very challenging topic can successfully engage students towards learning very complex theoretical, numerical, and practical skills.

Author Contributions: All authors contributed equally to conceptualization, methodology, software, validation, and writing; G.C. additionally provided overall supervision. All authors have read and agreed to the published version of the manuscript.

References

1. Liu, J. Current Research: Key Performances and Future Development of Search and Rescue Robot. *J. Mech. Eng.* **2006**, *42*, 1–12. [CrossRef]
2. Technical Report of Urban Search and Rescue Robotics. Available online: https://warwick.ac.uk/fac/sci/eng/meng/wmr/projects/rescue/reports/1516reports-copy/submission_tech_report.pdf (accessed on 15 October 2020).
3. Raibert, M.; Blankespoor, K.; Nelson, G.; Playter, R. BigDog, the Rough-Terrain Quaduped Robot. In Proceedings of the 17th IFAC World Congress, Seoul, Korea, 6–11 July 2008; pp. 10822–10825.
4. Schwarz, M.; Schreiber, M.; Behnke, S. Hybrid Driving-Stepping Locomotion with the Wheeled-legged Robot Momaro. In Proceedings of the IEEE International Conference on Robotics and Automation (ICRA), Stockholm, Sweden, 16–21 May 2016.
5. Kau, N.; Schultz, A.; Ferrante, N.; Slade, P. Stanford Doggo: An Open-Source, Quasi-Direct-Drive Quadruped. In Proceedings of the International Conference on Robotics and Automation (ICRA), Montreal, QC, Canada, 20–24 May 2019; pp. 6309–6315.
6. Katz, B.; Carlo, J.D.; Kim, S. Mini Cheetah: A Platform for Pushing the Limits of Dynamic Quadruped Control. In Proceedings of the International Conference on Robotics and Automation (ICRA), Montreal, QC, Canada, 20–24 May 2019; pp. 6295–6301.

7.   The Robot Report Homepage. Robotic Business Review. Available online: https://www.therobotreport.com/ (accessed on 21 November 2020).

8.   Ceccarelli, M.; Carbone, G. *Design and Operation of Human Locomotion Systems*; Elsevier Academic Press: Amsterdam, The Netherlands, 2019.

9.   Orner, A.M.M.; Ogura, Y.; Kondo, H.; Morishima, A.; Carbone, G.; Ceccarelli, M.; Lim, H.-O.; Takanishi, A. Development of a humanoid robot having 2-DQF waist and 2-DOF trunk. In Proceedings of the 5th IEEE-RAS International Conference on Humanoid Robots, Tsukuba, Japan, 5–7 December 2005; pp. 333–338.

10.   Orozco-Madaleno, E.C.; Gomez-Bravo, F.; Castillo, E.; Carbone, G. Evaluation of Locomotion Performances for a Mecanum-wheeled Hybrid Hexapod Robot. *IEEE/ASME Trans. Mechatron.* **2020.** [CrossRef]

11.   Tedeschi, F.; Carbone, G. Design of a novel leg-wheel hexapod walking robot. *Robotics* **2017**, *6*, 40. [CrossRef]

12.   Carbone, G.; Ceccarelli, M. A low-cost easy-operation hexapod walking machine. *Int. J. Adv. Robot. Syst.* **2008**, *5*, 161–166. [CrossRef]

13.   Carbone, G.; Shrot, A.; Ceccarelli, M. Operation strategy for a low-cost easy-operation Cassino Hexapod. *Appl. Bionics Biomech.* **2007**, *4*, 149–156. [CrossRef]

14.   Tedeschi, F.; Carbone, G. Design issues for hexapod walking robots. *Robotics* **2014**, *3*, 181–206. [CrossRef]

15.   Data Sheet of JX Servo PDI-HV5932MG. Available online: https://www.banggood.com/JX-Servo-PDI-HV5932MG-30KG-Large-Torque-360-High-Voltage-Digital-Servo-p-1074871.html?cur_warehouse=CN (accessed on 19 September 2019).

16.   GY-521 MPU-6050 Module. Available online: http://win.adrirobot.it/sensori/MPU-6050/sensore_MPU-6050.htm (accessed on 25 September 2019).

17.   DHT22 Module. Available online: https://www.sparkfun.com/datasheets/Sensors/Temperature/DHT-22.pdf (accessed on 22 October 2020).

18.   HY-SRF05 Datasheet. Available online: http://www.datasheetcafe.com/hy-srf05-datasheet-sensor (accessed on 22 October 2020).

19.   Geva, Y.; Shapiro, A. A combined potential function and graph search approach for free gait generation of quadruped robots. In Proceedings of the IEEE International Conference on Robotics and Automation, St. Paul, MN, USA, 14–19 May 2012; pp. 5371–5376.

20.   De Almeida, J.P.L.S.; Nakashima, R.T.; Neves, F., Jr.; de Arruda, L.V.R. Bio-inspired on-line path planner for cooperative exploration of unknown environment by a Multi-Robot System. *Robot. Auton. Syst.* **2019**, *112*, 32–48. [CrossRef]

21.   Corah, M.; O'Meadhra, C.; Goel, K.; Michael, N. Communication-efficient planning and mapping for multi-robot exploration in large environments. *IEEE Robot. Autom. Lett.* **2019**, *4*, 1715–1721. [CrossRef]

22.   Romano, D.; Donati, E.; Benelli, G.; Stefanini, C. A review on animal–robot interaction: From bio-hybrid organisms to mixed societies. *Biol. Cybern.* **2019**, *113*, 201–225. [CrossRef] [PubMed]

3

Local CPGs Self Growing Network Model with Multiple Physical Properties

Ming Liu [1], Mantian Li [1], Fusheng Zha [1,2,*], Pengfei Wang [1], Wei Guo [1] and Lining Sun [1,*]

[1] State Key Laboratory of Robotics and System, Harbin Institute of Technology, Harbin 150001, China; 14B308017@hit.edu.cn (M.L.); limt@hit.edu.cn (M.L.); wangpengfei1007@163.com (P.W.); wguo01@hit.edu.cn (W.G.)

[2] Shenzhen Academy of Aerospace Technology, Shenzhen 518057, China

* Correspondence: zhafusheng@hit.edu.cn (F.Z.); lnsun@hit.edu.cn (L.S.)

Abstract: Compared with traditional control methods, the advantage of CPG (Central Pattern Generator) network control is that it can significantly reduce the size of the control variable without losing the complexity of its motion mode output. Therefore, it has been widely used in the motion control of robots. To date, the research into CPG network has been polarized: one direction has focused on the function of CPG control rather than biological rationality, which leads to the poor functional adaptability of the control network and means that the control network can only be used under fixed conditions and cannot adapt to new control requirements. This is because, when there are new control requirements, it is difficult to develop a control network with poor biological rationality into a new, qualified network based on previous research; instead, it must be explored again from the basic link. The other direction has focused on the rationality of biology instead of the function of CPG control, which means that the form of the control network is only similar to a real neural network, without practical use. In this paper, we propose some physical characteristics (including axon resistance, capacitance, length and diameter, etc.) that can determine the corresponding parameters of the control model to combine the growth process and the function of the CPG control network. Universal gravitation is used to achieve the targeted guidance of axon growth, Brownian random motion is used to simulate the random turning of axon self-growth, and the signal of a single neuron is established by the Rall Cable Model that simplifies the axon membrane potential distribution. The transfer model, which makes the key parameters of the CPG control network—the delay time constant and the connection weight between the synapses—correspond to the axon length and axon diameter in the growth model and the growth and development of the neuron processes and control functions are combined. By coordinating the growth and development process and control function of neurons, we aim to realize the control function of the CPG network as much as possible under the conditions of biological reality. In this way, the complexity of the control model we develop will be close to that of a biological neural network, and the control network will have more control functions. Finally, the effectiveness of the established CPG self-growth control network is verified through the experiments of the simulation prototype and experimental prototype.

Keywords: CPG; self-growing network; quadruped robot; trot gait

1. Introduction

Compared with traditional control methods, Central Pattern Generator (CPG) network control has been widely used for the motion control of bionic robots because it can significantly reduce the dimension of control variables without losing the complexity of its output motion mode; the method also has some other advantages [1–3]. However, the study of CPG control network to date has been polarized: one direction has focused on the control function [4–6] and ignored the biology rationality.

In this direction, a control network is designed only for certain fixed situations, which leads to the poor functional adaptability of the control network. Once there are new control requirements, it is difficult for the control network to be developed into a new, qualified one according to the previous research; instead, it has to be researched again from the basic link. Especially when applied to the motion control of a bionic robot, the above-mentioned CPG control network will have the following disadvantages: with excessive prior knowledge and excessive factors given by humans in each link of the control system, which includes the design of the CPG network, the setting method of network parameters and the establishment of the neuron model in the control network, the control function of the CPG network has lower complexity, worse adaptability and worse inheritance capabilities in future research.

Another direction has focused on biological rationality, but not the function of CPG control, which means that the control network is only similar to a real neural network in form but of no practical use [7–9]. Dehmamy et al. [10] modeled axon growth with driven diffusion. Their simulation showed that with background potential derived from the concentration guiders, even coupled random walkers can generate realistic paths for axons. Marinov et al. [11] proposed an ad-hoc growth model built upon the diffusion principle to reproduce the shape of Micro-Tissue Engineered Neural Networks (Micro-TENNs). The proposed model imposed various rules for individual neuronal growth to a 3D diffusion equation. Fard et al. [12] proposed a generative growth model to investigate growth rules for axonal branching patterns in cat area 17. The model achieves better statistical accuracy both locally and globally than the commonly used Galton–Watson model. Gafarov et al. [13] proposed a closed-loop growth model in which neural activity affects neurite outgrowth in an interactive way. Experiments showed that the model can be potentially used to form large-scale neural networks by self-organization. However, it is very difficult to study how the scattered neurons in different parts of the growth region form the whole network through interactive behaviors with biological methods, because the growth environment of cells is complex, and so it is difficult to precisely control their growth conditions to verify the specific growth mechanism. Therefore, the study of the growth process of a neural network by numerical simulation is very effective. In a simulation environment, researchers can set strict growth conditions and analyze the growth result of the network in a variety of ways and then compare it with a real biological neural network to verify the validity of the designed model [9,14–16].

However, the numerical simulation of the growth of a biological neural network has often been unable to simulate the control function of the network, and it can only compare the simulation model with the real biological network in terms of morphology and statistics. At the same time, in the control based on a neural network, it is precisely the complexity of the formation process of the biological neural network and the simplifications and assumptions made for the simulation model that make the network control model unrealistic and not universally applicable [17]. Therefore, it is necessary to combine the growth process with the control function in the control of bionic robots based on the CPG network to allow the two to develop in a coordinated way in the long term.

Therefore, we propose some physical characteristics (including axon resistance, capacitance, length and diameter, etc.) that can determine the corresponding parameters of the control model to combine the growth process and the function of the CPG control network. By coordinating the two, we aim to realize the control function of the CPG network under the condition of biological reality as far as possible. In this way, the complexity of the developed control model will be closer to that of a biological neural network, and the control network will have more control functions.

2. Establishment of Self-Growing Model of Neurons

In the formation process of a neural network, the position of each neuron is fixed, axons and dendrites grow from the neuron, and the axon of the former neuron and the dendrite of the latter neuron form a synaptic contact and finally realize the connection of the whole network. Studies have shown that, with the continuous elongation and swerving of the axon, the growth of the tail end of the neuron will not change the shape of the other part [9]. Therefore, the process of growing and eventually forming synaptic connections with another neuron in the neuron model can be simplified into the following model:

$$\vec{F} = \sum_{j=1}^{k} \vec{F}_j = \sum_{j=1}^{k} gm \frac{M_j}{d_j^2} \cdot \vec{e}_j \tag{1}$$

where k is the number of neurons in the gravitational field, d_j is the distance from the growth cone to the jth neuron and $\vec{e}_j$ is the direction vector of gravitation on the growth cone generated by the jth neuron.

In Equation (1), the growth cone can be regarded as a particle with a mass of m. Regard the neuron as a sphere with a mass M concentrated in the center; the distribution range of dendrites is understood as a sphere with radius r because the dendrites are numerous and dense. The guiding effect of the chemical molecules on the growth cone can be equivalent to the gravitational effect on particle m in the gravitational field generated by mass M, and the motion trajectory of the particle in the gravitational field can be regarded as the axon. The position of growth cone particle in the gravitational field is $P_i(x_i, y_i, z_i)$ at a certain time, with a growth rate of $V_i(v_{xi}, v_{yi}, v_{zi})$, and the gravitation on the growth cone can be determined from Newton's law of gravitation. Thus, the acceleration is

$$\vec{a} = \sum_{j=1}^{k} \vec{a}_j = \sum_{j=1}^{k} g \frac{M_j}{d_j^2} \cdot \vec{e}_j = (a_{xi}, a_{yi}, a_{zi}) \tag{2}$$

After a short period of time Δt (Δt is defined as 0.01 s in this paper), the displacement of the growth cone is

$$\begin{cases} \Delta x_i = v_{xi} \cdot \Delta t + \frac{1}{2} a_{xi} \cdot \Delta t^2 \\ \Delta y_i = v_{yi} \cdot \Delta t + \frac{1}{2} a_{yi} \cdot \Delta t^2 \\ \Delta z_i = v_{zi} \cdot \Delta t + \frac{1}{2} a_{zi} \cdot \Delta t^2 \end{cases} \tag{3}$$

Thus, the new position $P_{i+1}(x_{i+1}, y_{i+1}, z_{i+1})$ of the growth cone is

$$\begin{cases} x_{i+1} = x_i + \Delta x_i \\ y_{i+1} = y_i + \Delta y_i \\ z_{i+1} = z_i + \Delta z_i \end{cases} \tag{4}$$

and the new growth rate V_{i+1} $(v_{xi+1}, v_{yi+1}, v_{zi+1})$ is

$$\begin{cases} v_{xi+1} = v_{xi} + a_{xi} \cdot \Delta t \\ v_{yi+1} = v_{yi} + a_{yi} \cdot \Delta t \\ v_{zi+1} = v_{zi} + a_{zi} \cdot \Delta t \end{cases} \tag{5}$$

$P_{i+1}(x_{i+1}, y_{i+1}, z_{i+1})$ and V_{i+1} $(v_{xi+1}, v_{yi+1}, v_{zi+1})$ can be used for the next iteration. In this way, the continuous growth process of axons is discretized into a programmable iterative process.

Although it is reasonable to use the gravitational field model to describe the chemotaxis of growth cones, the growth of the model's axon is quite different from that in the actual situation. Experiments show that the overall deflection of the axon does not become more obvious as the concentration gradient increases gradually, because the axon grows faster on the side of the inverse gradient than the side along the gradient (growth rate regulation), which counteracts the chemotaxis. The results show that when the concentration gradient is large, the chemotactic deflection plays a leading role [8]. In addition, the guiding effect of chemical factors on the growth cone is either attraction or repulsion, but we only consider the case of attraction when modeling, because the mathematical model of the repulsion case is the same as that of the attraction case except with the opposite direction of force.

In fact, the growth direction of the growth cone depends on many intracellular and extracellular signals, which may lead to large fluctuations in the growth direction [9]. For example, the noise gradient in the chemoattractant orientation perception stage mentioned in [18,19] is one of the factors making the deflection direction of the growth cone uncertain when proceeding. In order to describe the uncertain component of the deflection direction of the growth cone, many studies on the mathematical modeling of the axon growing process introduce a random motion component based on the deterministic motion generated by gravity [20,21], as shown in Figure 1.

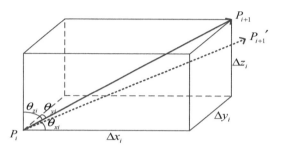

Figure 1. Diagram of a single growth iteration of an axon.

In Figure 1, P_i is the position of the growth cone after the i-1th growing process. P_{i+1} is the position of the growth cone after the ith growing process without the effect of a random motion component. Δx_i, Δy_i and Δz_i are the projected length of the ith growth length of the growth cone on the x, y, and z axes, and P'_{i+1} is the real position of the growth cone after the ith growing process under the effect of random motion component.

As can be seen from Figure 1, we can calculate the angle between the ith growth segment and the coordinate axis, $(\theta_{xi}, \theta_{yi}, \theta_{zi})$ using the following formula:

$$\begin{cases} \theta_{xi} = arctan\dfrac{\sqrt{\Delta y_i^2 + \Delta z_i^2}}{\Delta x_i} \\ \theta_{yi} = arc\tan\dfrac{\sqrt{\Delta x_i^2 + \Delta z_i^2}}{\Delta y_i} \\ \theta_{zi} = arc\tan\dfrac{\sqrt{\Delta y_i^2 + \Delta x_i^2}}{\Delta z_i} \end{cases} \tag{6}$$

Furthermore, we add a random motion component to this growing process:

$$\begin{cases} \theta_{xi}' = \theta_{xi} + \Delta\theta_x \\ \theta_{yi}' = \theta_{yi} + \Delta\theta_y \\ \theta_{zi}' = \theta_{zi} + \Delta\theta_z \end{cases} \tag{7}$$

where $\Delta\theta_x$, $\Delta\theta_y$ and $\Delta\theta_z$ meet uniform distribution in the interval $(-c\delta, c\delta)$. $\delta > 0$, which ensures that the random motions lead to the largest deflection angle, $c \in (0, 1)$ implies the intensity of random motion, and $(\theta_{xi}', \theta_{yi}', \theta_{zi}')$ is the real deflection angle of this growing process considering the random motion. Then, we can calculate the real displacement of the growth cone in this process, $(\Delta x_i', \Delta y_i', \Delta z_i')$, using the following formula:

$$\begin{cases} l_i = \sqrt{\Delta x_i^2 + \Delta y_i^2 + \Delta z_i^2} \\ \Delta x_i' = l_i \cdot \cos\theta_{xi}' \\ \Delta y_i' = l_i \cdot \cos\theta_{yi}' \\ \Delta z_i' = l_i \cdot \cos\theta_{zi}' \end{cases} \tag{8}$$

where l_i is the length of the axon segment of the ith growing process; the random motion only changes the deflection of each growing process, but not the step size. Finally, we calculate the real position of the growth cone after the ith growing process, $P'_{i+1}(x'_{i+1}, y'_{i+1}, z'_{i+1})$, using the following formula:

$$\begin{cases} x_{i+1}' = x_i + \Delta x_i' \\ y_{i+1}' = y_i + \Delta y_i' \\ z_{i+1}' = z_i + \Delta z_i' \end{cases} \tag{9}$$

Then, we use $P_{i+1}'(x_{i+1}', y_{i+1}', z_{i+1}')$ and $V_{i+1}(v_{xi+1}, v_{yi+1}, v_{zi+1})$ calculated from Equations (2)–(5) for the next iteration.

Three intensities of random motion values are selected as $c = 0.05$, $c = 0.15$ and $c = 0.56$, and 20 of the corresponding 200 axons are selected to draw the growing diagram of the growth cone under the effect of random motion of different intensities, as shown in Figure 2.

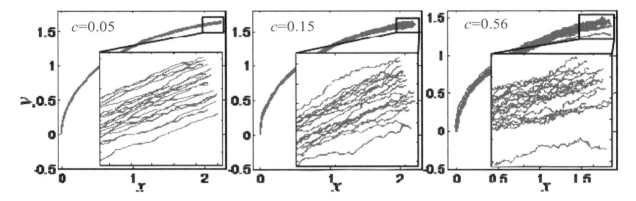

Figure 2. Growth of growth cone under different intensities of random motion.

According to the growth details that can be seen in Figure 2, the growth of the growth cone is relatively smooth around $c = 0.05$ with small random fluctuations. The random fluctuations increases gradually around $c = 0.15$, but the deflection angle of the growth cone does not change extremely, and the growth of the growth cone is still in order. However, when $c = 0.56$, the trajectory of the growth cone changes abruptly and the partial shape of the growth cone is obviously out-of-order. Therefore, the randomness of the growing process of the growth cone increases gradually with the increase of the effect intensity of random motion.

3. Establishment of Neuronal Axonal Signal Transmission Model

In order to describe the signal transmission process of the neurons listed above, a simplified model of neuron signal transmission, as shown in Figure 3, is established in this paper.

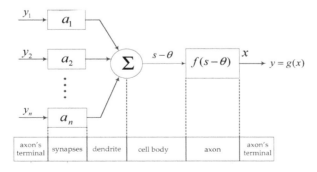

Figure 3. The signal transmission model of the neuron.

If there are n neurons on a neuron's dendrite to form synaptic connections with it, the output signals of these n neurons are $y_1, y_2, \ldots, y_n$, the connection weights of synapses are $a_1, a_2, \ldots, a_n$, the change of membrane potential caused by the input is s and the excitation threshold is θ; the membrane potential is x when reaching the axon's terminal after being transmitted through the whole axon, and the neuron output signal is y. Then, the signal transmission process of the neuron is

$$\begin{cases} s = \sum_{i=1}^{n} a_i y_i \\ x = f(s - \theta) \\ y = g(x) \end{cases} \tag{10}$$

where $f(.)$ represents the function of an axon on membrane potential and $g(.)$ represents the nonlinear output function of neurons, which is different for different biological neurons. In this paper, the commonly used threshold function is taken; i.e., $g(x) = \max(0, x)$.

In 1957, W. Rall proposed the equivalent cable model of an axon during signal transmission [22], which regarded the cable as composed of many identical resistance capacitance circuits, considered the current flow in extracellular fluid and the distribution of extracellular potential and finally obtained the distribution of membrane potential in time space $\lambda^2(\partial^2 V/\partial x^2) = \tau(\partial V/\partial t)+V$, where V is the potential of the membrane and λ and τ are the time constant and length constant of the thin film of the cable joint. This model is a second-order partial differential equation, and the solution is relatively complex. In this paper, we only consider the input and output characteristics of the neuron as a signal processing unit instead of the distribution of membrane potential on the whole axon, so we assume that the extracellular potential along the whole axon is always 0—equivalent to grounding—as can be seen in Figure 4. Thus, U is equal to the membrane potential. Given the output change of an input signal after the membrane potential is transmitted through the whole axon, we can calculate $f(.)$ as shown in Formula (10).

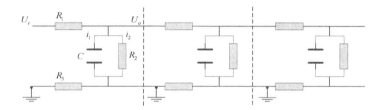

Figure 4. The cable model without considering the extracellular potential.

First, we analyze the basic resistance capacitance circuit in Figure 4 separately and calculate the transfer function between the output U_0 and the input U_i of each section. In Figure 4, i is the current, R is the resistance and C is the capacitance [18]. The following equation can be obtained from Kirchhoff's law and Ohm's law.

$$i_1 = C\frac{dU_o}{dt} \tag{11}$$

$$U_o = i_2 R_2 \tag{12}$$

$$U_i = (i_1 + i_2)(R_1 + R_3) + U_o \tag{13}$$

Equations (11) and (12) are substituted into Equation (13):

$$U_i = (C\frac{dU_o}{dt} + \frac{U_o}{R_2})(R_1 + R_3) + U_o \tag{14}$$

The Laplace transform of Equation (14) is

$$U_i(s) = (sCU_0(s) + \frac{U_o(s)}{R_2})(R_1 + R_3) + U_o(s) \tag{15}$$

Finally, we obtian the transfer function between U_0 and U_i:

$$G(s) = \frac{U_o(s)}{U_i(s)} = \frac{1}{C(R_1 + R_3)s + \frac{R_1+R_2+R_3}{R_2}} \tag{16}$$

Assume τ and b as Equation (17):

$$\begin{cases} \tau = C(R_1 + R_3) \\ b = \frac{R_1+R_2+R_3}{R_2} \end{cases} \tag{17}$$

Then, Equation (16) is

$$G(s) = \frac{1}{\tau s + b} \tag{18}$$

As can be seen from Figure 4, the total transfer function of the axon to the membrane potential is $F(S) = G^N(S)$. N is the number of basic links on the whole axon. Thus far, the signal transmission model of the neuron has been determined. Finally, the amplitude–frequency and phase–frequency characteristics of $G(S)$ are analyzed to determine the physical characteristics of the neuron self-growing model:

$$G(j\omega) = \frac{1}{\tau j\omega + b} \tag{19}$$

where ω is the angular frequency.

The amplitude–frequency characteristic of Equation (19) is

$$|G(j\omega)| = \frac{1}{\sqrt{\tau^2\omega^2 + b^2}} \tag{20}$$

The phase–frequency characteristic of Equation (19) is

$$\angle G(j\omega) = -\arctan(\frac{\tau\omega}{b}) \tag{21}$$

(1) Influence of axon length on amplitude and phase lag

Suppose an axon cable is composed of N basic links as shown in Figure 4; then, the amplitude of the input signal is attenuated to $(1/\sqrt{\tau^2\omega^2 + b^2})^N$ after being transmitted through the axon, and the phase lag is $-N\arctan(\tau\omega/b)$. The longer the axon is, the more basic links in the corresponding cable model there should be; in other words, the larger N is. We can see that $b > 1$ from Equation (17), and so the longer the axon is, the larger N is and the larger the amplitude attenuation and phase lag are. The length of the axon can be obtained by the iteration in Equation (8). If the length of the axon corresponding to the basic link of a resistance capacitance loop is l_1 and the total length of the axon is L [23], then the amplitude of the signal will decay to $(1/\sqrt{\tau^2\omega^2 + b^2})^{L/l_1}$ and the phase lag will be $-N\arctan(\tau\omega/b)$ after passing through the whole axon.

(2) Diameter of the axon

As can be seen from Equations (20) and (21), the amplitude–frequency and phase–frequency characteristics of the neuron signal transmission are determined by the two parameters τ and b. In Equations (16) and (17), R_2 is the electrical leakage resistance between the inside and outside of the cell within a basic link, whose resistance value is much higher than R_1 and R_3, which is equivalent to the myelin sheath of the outer layer of the axon of biological neurons and plays an insulating role. C is the leakage capacitance between the inside and outside of the cell within a basic link. R_3 is the conductive resistance of the extracellular fluid within a basic link. R_2, R_3 and C are the same for every neuron and do not change with the gradual growth of axons. R_1 represents the resistance of the axon core within a basic link, which can be obtained according to the resistance law:

$$R_1 = \rho \frac{l_1}{s} \tag{22}$$

where ρ is the resistivity of the inner core of the axon, s is the cross-sectional area of the axon—$s = \pi^2 d/4$ since the cable model is cylindrical—and d is the diameter of the axon. The axon diameters of different neurons determine the amplitude–frequency and phase–frequency characteristics of the basic link in the cable model.

(3) Connection weight to the postsynaptic neuron

Since $|G(j\omega)| < 1$, the cable model shown in Figure 4 must cause the attenuation of signal amplitude during signal transmission. However, we occasionally want a single neuron to transmit the signal without changing its amplitude when controlling, only changing its phase. For example, in the CPG control network of the joint of a legged robot, each joint is controlled by a motor neuron with joint angular displacement output, and the output signals of the motor neuron in the symmetrical position of the body need to have the same amplitude, because the joint angle of these joints has the same range. If all the intermediate neurons in the network do not change the amplitude of the signal during signal transmission, the output of each joint with the same amplitude can be obtained from the input with the same amplitude. In this paper, a proportional amplification link is artificially added after each basic link to prevent the change of signal transmission amplitude, as shown in Figure 5.

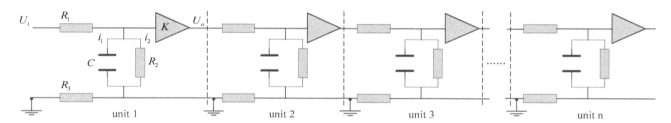

Figure 5. Cable model with a proportional amplification link artificially added.

According to Equation (20), we can obtain K in Figure 5:

$$K = \frac{1}{|G(j\omega)|} = \sqrt{\tau^2 \omega^2 + b^2} \tag{23}$$

As can be seen from Equation (23), the amplitude of the output signal from the original single basic link remains constant after passing the proportional amplification link. The output signal of the whole axon is equivalent to multiplying the original output by K^N. According to the definition of connection weight, this is equivalent to the connection weight between the neuron and the postsynaptic neuron:

$$a = K^N = \left(\sqrt{\tau^2 \omega^2 + b^2}\right)^N \tag{24}$$

If the input signal of the neuron is an ideal impulse signal or step signal, then we can say that $\omega = 0$. According to Equation (17), Equation (24) is

$$a = b^N = \left(\frac{R_1 + R_2 + R_3}{R_2}\right)^N \tag{25}$$

Because R_2 is much larger than R_1 and R_3, we know that b tends to 1. According to $N = L/l_1$, we can select l_1 reasonably so that the order of magnitude of N will not be too large, and this will mean that the connection weights between all the neurons in the network meet $a \in (1, 2)$.

In Equation (18), the transition time of the first-order system's response to the impulse signal and the step signal—called the delay—is proportional to the time constant τ, which is assumed to be $k\tau$, where k is a constant coefficient related to the allowable error value of the response. If there are N basic links on the whole axon, as shown in Figure 5, the total delay of the output signal relative to the input signal will be $Nk\tau = Lk\tau/l_1$, which is equivalent to the delay of the output of a basic link with a time constant of $L\tau/l_1$ to the input:

$$Tr = L\tau/l_1 \tag{26}$$

Therefore, when the input signal is an impulse signal or a step signal, the self-growing model of a neuron with multiple physical properties can be simplified as follows: the whole neuron is regarded as a basic link, and the length of the axon obtained from the self-growing is used as the total time constant T_r of the neuron to the input signal. A random number from (1, 2) is selected as the connection weight value a between the neuron and the postsynaptic neuron. In addition, the ends of the axons of biological neurons diverge and form synaptic connections with dendrites of many other neurons; thus, multiple axons of the neurons are made to grow directly from the cell body without considering the choice of branching points to simplify the model.

4. Structure of Local CPG Network

(1) Study of neuron model

In 1987, Matsuoka proposed a mathematical neuron monomer model and studied the "oscillator" network output with different numbers of neurons and different connection structures based on it [24]. The mathematical model of the neuron monomer is

$$\begin{cases} T_r\frac{dx}{dt} + x = s - bf \\ y = g(x - \theta), \quad (g(x) = \max\{0, x\}) \\ T_a\frac{df}{dt} + f = y \end{cases} \tag{27}$$

where x represents the membrane potential of the neuron; s stands for neuron input; y is the output of the neuron; f is the quantity reflecting the fatigue effect of neurons; b is a constant coefficient representing the size of fatigue effect; θ presents the membrane potential threshold that activates the neuron, and without losing generality, we can take $\theta = 0$; $g(x)$ is the nonlinear output function; and T_r represents the time constant of the rise of the membrane potential caused by the input signal. T_a represents the time constant of the neuronal fatigue effect. Since the physical characteristics of neurons of the model were not taken into account, there is no concept of the axon, and thus the membrane potential will not change after axon transmission. We continue to regard x as the membrane potential at the end of the axon, and so this model differs from the model established in Equation (10) by only one fatigue characteristic characterized by f and T_a. Set s as the step input signal with n amplitude of 1, $T_r = 1$, $b = 10$, $T_a = 10$, and calculate the change of the output signal of the neuron monomer model over time; the results are shown in Figure 6.

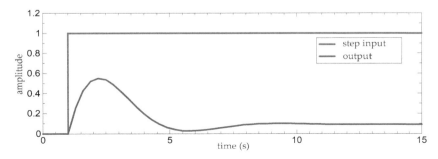

Figure 6. Step response of the neuron model.

As can be seen from Figure 6, first, the neuron is activated and the membrane potential rises; then, the membrane potential declines slowly due to the fatigue effect and remains at a low level. Although the step signal continues, the neuron will not be activated again. We can see that, under appropriate parameters, the mathematical model of the neuron monomer conforms to the signal transmission characteristics of biological neurons.

(2) Double neural oscillator model

Connect two neuron monomer models and make the two inhibit each other as shown in Figure 7. If N1 and N2 are the neurons, the hollow endpoint connection presents the excitatory connection and the solid endpoint connection presents the inhibitory connection; S1 and S2 are the step input signals with an amplitude of 1 with slightly different step times. If S1 and S2 have the same step time, the two neurons will have the same output and will not produce an oscillation because of the symmetrical structure of the network.

Figure 7. Network of two neurons that inhibit each other.

The mathematical model of each neuron monomer is

$$\begin{cases} T_{ri}\frac{dx_i}{dt} + x_i = s_i - b_i f_i - \sum_{j=1}^{n} a_{ji} y_j \\ y_i = \max\{0, x_i\} \\ T_{ai}\frac{df_i}{dt} + f_i = y_i \end{cases} \tag{28}$$

where $\sum a_{ji} y_j$ are the inputs from other neurons. If the input is excitatory, the operator preceding it has a plus sign, with a minus sign for the inhibitory input.

Assume $T_{r1} = T_{r2} = 1$, $b_1 = b_2 = 2.5$, $a_{21} = a_{12} = 1.5$ and $T_{a1} = T_{a2} = 12$ and calculate the variation of the output signal of the oscillator model with time. The results are shown in Figure 8.

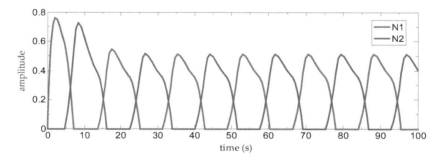

Figure 8. Reciprocal output of two neurons.

In the spinal CPG network of mammals, the interactivity of neurons responsible for the coordination of the left and right sides of the body and the coordination of the extensor and flexor muscles of each joint inhibits the output [25]. It can be seen in Figure 5 that the reciprocal inhibition outputs of the two neurons have the same form as the reciprocal inhibition outputs of the neurons in the biological CPG network, indicating that the neuron monomer model in Equation (26) can reflect the characteristics of the reciprocal inhibition output caused by the reciprocal inhibition structure in the biological CPG network.

(3) Connection structure of local CPG network

As can be seen from the analysis above, as long as the neuron model represented by Equation (27) is composed of inhibitory connections among the neurons and appropriate parameters are taken, the interactive inhibitory output between the neurons will be obtained. Since each neuron monomer in the neuron model represented in Equation (27) has two time constants, Tr and Ta, and each neuron in the simplified model of self-growing neurons established in Section 3 has only one time constant, Tr, we cannot make the growth algorithm of single neuron grow into a local CPG network which has the function of controlling a robot joint. In order to solve this contradiction, we modify the original neuron monomer model, as shown in Equation (29), according to the characteristics of the neuron connections in a biological CPG network.

$$\begin{cases} T_{ri}\frac{dx_i}{dt} + x_i = s_i - \sum_{j=1}^{n} a_{ji}y_j \\ y_i = \max\{0, x_i\} \end{cases} \tag{29}$$

The original neuron monomer model is described with the structure shown in Figure 6. We can see that the excitatory connection from the input neurons to the output neurons in Figure 9 is equivalent to the input signal S of the original model. The inhibition from the Renshaw Cell to the output neurons is equivalent to the fatigue characteristic in the original model; thus, the input and output characteristics of signal transmission of the structure shown in Figure 9 are essentially the same as those in the Matsuoka neuron monomer model.

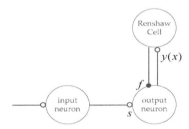

Figure 9. Equivalent model of the original neuron monomer model.

According to Equation (29), the modified neuron monomer model only contains one time constant, Tr, and each axon contains a connection weight a to the postsynaptic neuron, which is consistent with the simplified model of self-growing neurons established in Section 3. Moreover, the connection structure shown in Figure 6 has the anatomical basis of the biological CPG network. Known as an inhibitory intermediate neuron directly connected to motor neurons, the Renshaw Cell is involved in the coordination of most extensor and flexor muscles as part of the ipsilateral inhibitory network [14,25]. Moreover, the structure in Figure 6 can be seen in many models of spinal motor nerve circuits [26,27].

A local CPG network for the control of a quadruped robot's hip joint can be formed with four identical structures, as shown in Figure 10. In Figure 10, N1, N2, N3 and N4 are the output neurons, respectively corresponding to the four hips of the quadruped robot. N5, N6, N7 and N8 are input neurons, and S1, S2, S3 and S4 are step input signals of the same amplitude. N9, N10, N11 and N12 are four output neurons connected with Renshaw Cells, Tri is the time constant of each neuron and $a_{i,j}$ is the connection weight of the synapse between the axon of neuron i and the neuron j; the connection weights between peripheral neurons are not shown in Figure 10. The network structure is the growth target of the local CPG network self-growing algorithm.

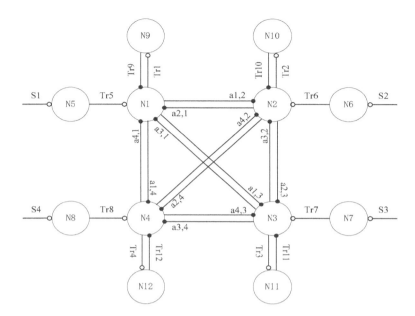

Figure 10. Local Central Pattern Generator (CPG) network connection structure for the hip joint control of a quadruped robot.

(4) Output of local CPG network

One of the basic features of a biological CPG network is that it can obtain a periodic output signal with a certain rhythm from a simple non-rhythmic input signal. In order to verify that the network growth model proposed in this paper can obtain growth results with biological CPG network characteristics, a CPG network growth experiment was carried out and is presented in this section. Assume that the radius r of a single neuron in the growth model is 0.1 units of length and the growth space is a cube with a side length of 5; 12 neurons were randomly distributed in the growth space as shown in Figure 11.

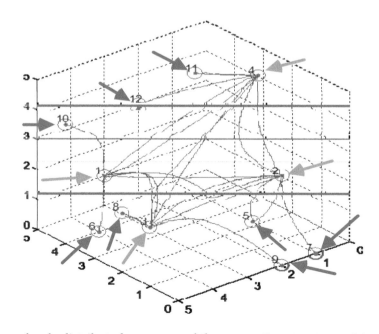

Figure 11. Twelve randomly distributed neurons and the connection structure arising from the network.

In Figure 11, a circle represents a neuron, and the number (1, 2, ... , 12) next to it shows the number of neurons. The neurons numbered 1, 2, 3 and 4 are the output neurons, represented by N1, N2, N3 and N4 (as shown by the green arrow in Figure 11); the neurons numbered 5–12 are peripheral neurons, represented by N5–N12. Numbers N5, N6, N7 and N8 are input neurons (as shown by the red arrow in Figure 11). The solid blue lines are the growing connections, as shown in Figure 11.

For simplicity, take $l_1 = r = 0.1$, $\tau = 10$. Putting the length of the blue solid line in Figure 11 into Equation (26), the value of Tr can be obtained as shown in Table 1.

Table 1. The time constant of the axon in the network.

Time Constant	Length of Corresponding Axon	Value
T_{r1}	Length of axon between output neuron N1 and Renshaw Cell N6	2.1801
T_{r2}	Length of axon between output neuron N2 and Renshaw Cell N5	2.5868
T_{r3}	Length of axon between output neuron N3 and Renshaw Cell N8	1.1984
T_{r4}	Length of axon between output neuron N4 and Renshaw Cell N11	1.9691
T_{r5}	Length of axon of Renshaw Cell N5	2.8071
T_{r6}	Length of axon of Renshaw Cell N6	2.2821
T_{r7}	Length of axon of input neuron N7	3.7401
T_{r8}	Length of axon of Renshaw Cell N8	1.2096
T_{r9}	Length of axon of input neuron N9	5.4620
T_{r10}	Length of axon of input neuron N10	4.8037
T_{r11}	Length of axon of Renshaw Cell N11	1.8388
T_{r12}	Length of axon of input neuron N12	4.6752
$T_{r13}-T_{r24}$	Length of interactional axon between N1, N2, N3 and N4	0

Step input signals with amplitude of 1 were input to the four input neurons. Observe the output signals of the network; a representative growth result was taken for discussion. The connection structure of the network growth is shown in Figure 11. In Figure 11, each neuron axon contains two parameters: the time constant Tr and the connection weight a. Since there are 24 axons in the network, there are 48 parameters with 24 time constants and 24 connection weights in the network. Such a large number of parameters in the network is not conducive to the study of the rule between the output and growth results of the network. In order to facilitate the analysis, the axons in the network are divided into two categories: in the first category are the axons that inhibit the connection between the output neurons, with a total of 12; the second category are the axons connected with the peripheral neuron (the input neuron and the Renshaw Cell) between the output neuron, with a total of 12. For the first type of axons, only the connection weights in the growth results are used, and the time constants are all set to 0. For the second type of axon, only the time constant in the growth result is used, and the connection weights are all fixed values. This halves the network's parameter variables. The time constants of the axons of each neuron after a certain growing process are shown in Table 1.

According to Equation (25), the connection weight $a \in (1, 2)$ can be obtained. In order to simplify the calculation, a random number between (1, 2) is taken as the value of a. The connection weights obtained are shown in Table 2, and the connection weights between the output neuron and the peripheral neuron are set to 1 and 2, which meet $a \in [1, 2]$. The output result of the entire network is calculated according to Equation (29), as shown in Figure 12.

Table 2. The connection weight of neuronal axons in the network.

$a_{10,1}$	$a_{7,2}$	$a_{9,3}$	$a_{12,4}$	$a_{1,6}$	$a_{2,5}$	$a_{3,8}$	$a_{4,11}$	$a_{6,1}$	$a_{5,2}$	$a_{8,3}$	$a_{11,4}$
1	1	1	1	1	1	1	1	2	2	2	2
$a_{1,2}$	$a_{1,3}$	$a_{1,4}$	$a_{2,1}$	$a_{2,3}$	$a_{2,4}$	$a_{3,1}$	$a_{3,2}$	$a_{3,4}$	$a_{4,1}$	$a_{4,2}$	$a_{4,3}$
1.24	1.49	1.64	1.90	1.23	1.72	1.78	1.15	1.92	1.77	1.16	1.31

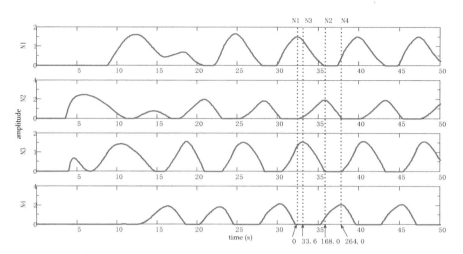

Figure 12. Output signal of the local CPG network.

As can be seen from Figure 12, the output signals of the four output neurons have the same period and phase difference and a similar range of amplitude after a period of stability. Furthermore, it is verified that the self-growing network model established in this paper can obtain a local CPG network which can satisfy the biological CPG network input and output characteristics, and the output of the network can be used for the control of the hip joint of quadruped robot. If the output signal of neuron N1 is used as the basis to calculate the phase difference, it can be determined that the output phase difference between N2 and N1 is 168.0°, the output phase difference between N3 and N1 is 33.6° and the output phase difference between N4 and N1 is 264.0°. The ideal phase differences of the corresponding hip joints of the quadruped robot at the walk gait are 180°, 90° and 270°. We can see that the phase difference between N3 and N1 is large, but the other two phase differences are very close, which shows that the local CPG network obtained by self-growing can output the corresponding control signals for the hip joint to the quadruped robot at a walking gait.

5. Experiment and Analysis

(1) Quadruped robot prototype

The experimental platform of the quadruped robot is shown in Figure 13. There are three joints on each leg: the hip joint α, knee joint β and ankle joint θ. The size of the platform is 1.2 m × 0.5 m × 1.4 m, and the weight is 150 kg. Figure 13a is the simulation prototype of the quadruped robot, and Figure 13b is the experiment of the quadruped robot.

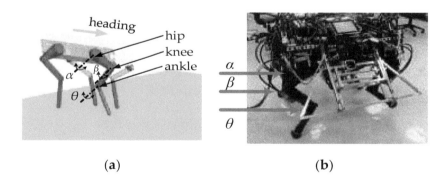

(a)   (b)

Figure 13. Quadruped robot. (**a**) Simulation prototype of quadruped robot. (**b**) Experimental prototype of quadruped robot.

(2) Foot trajectory and joint trajectory planning

Since the ideal angular displacement of the hip joint can be approximated as a cosine function [14], we assume that the control signal of the hip joint is $\alpha = 3\cos(2\pi t/T)$. The trajectory of the foot is a cycloid, as shown in Figure 14b. The height of the robot's hip joint is $h = 100$ cm from the ground, and the step length of each stride is 30 cm. The former leg is taken as an example to calculate the joint angular displacement and that of the knee and ankle joints, as shown in Figure 14a; thus, we obtain the following formula:

$$\left. \begin{array}{l} l_1 \cos\alpha - l_2\cos(\beta-\alpha) + l_3\cos[\theta-(\beta-\alpha)] = x \\ l_1 \sin\alpha + l_2\sin(\beta-\alpha) + l_3\sin[\theta-(\beta-\alpha)] = h - y \end{array} \right\} \tag{30}$$

where $l_1 = 45$ cm, $l_2 = 60$ cm and $l_3 = 50$ are the lengths of the robot's thigh, middle leg and calf; x and y are the analytic coordinates of the foot trajectory, as shown in Figure 14c, where α and h are known. Thus, Equation (30) is a system of two equations with two unknowns, β and θ, whose solution is unique. The calculated relationship between β, θ and α is shown in Figure 15. α ranges from 30° to 50°, which can be seen from the solid blue line in Figure 15, in order to make sure that the quadruped robot has a large stability margin when walking. The joint trajectory planning process of the hind leg of the robot is the same as that of the foreleg except for the different range of joint angles and the different shape of the curve.

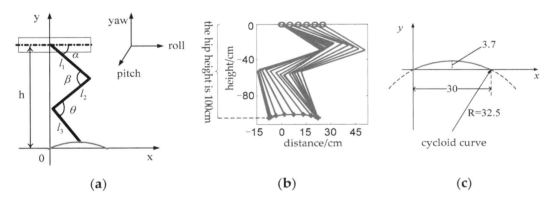

Figure 14. Foot trajectory of robot. (**a**) Structure diagram of one leg. (**b**) Experimental prototype of quadruped robot. (**c**) The analytic coordinates of foot trajectory.

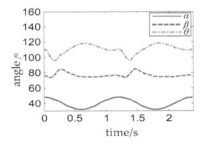

Figure 15. Trajectory planning of the knee and ankle joints.

Thus, the experimental prototype of the quadruped robot used for local CPG network gait control has been completed. It is only necessary to input periodic control signals corresponding to the angular displacements of the four hip joints into the prototype, and then the control signals of the knee and ankle joints of each leg will be obtained according to the corresponding relations in Figure 15; thus, we can drive the robot prototype to walk.

(3) Experiment on simulation prototype and experimental prototype

The random selected parameters $a \in [1, 2]$ of the self-growing network are shown in Table 3. The time constant Tr is shown in Table 1. After calculating the output of the network and cosine to fit the output signal according to Equation (29), the result is shown in Figure 16. We can see that the output of the network is definitely the interaction oscillations between N1, N3, N2, and N4. N1 and N3 are input to the left front and the right hind hip joint of the simulation prototype, while N2 and N4 are input to the right front and the left hind hip joint of simulation prototype; then, the trot gait of the robot is implemented, and the simulation decompositions are shown in Figure 17a. As can be seen from Figure 17a, from $t = 0.2$ s to $t = 1.7$ s, the left front leg and the right hind leg are in the swing phase and swing forward; from $t = 1.9$ s to $t = 3.4$ s, the right front leg and the left hind leg are in the swing phase and swing forward. Each swing time is about 1.6 s.

Table 3. The connection weight of neuronal axons in the network.

$a_{9,1}$	$a_{8,2}$	$a_{12,3}$	$a_{11,4}$	$a_{1,10}$	$a_{2,6}$	$a_{3,7}$	$a_{4,5}$	$a_{10,1}$	$a_{6,2}$	$a_{7,3}$	$a_{5,4}$
1	1	1	1	1	1	1	1	2	2	2	2
$a_{2,1}$	$a_{3,1}$	$a_{4,1}$	$a_{1,2}$	$a_{3,2}$	$a_{4,2}$	$a_{1,3}$	$a_{2,3}$	$a_{4,3}$	$a_{1,4}$	$a_{2,4}$	$a_{3,4}$
1.19	0	1.22	1.25	1.56	0	0	1.41	1.88	1.84	0	1.64

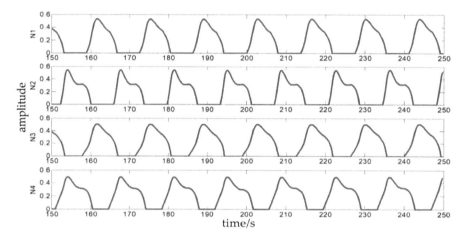

Figure 16. Output of self-growing network in trot gait.

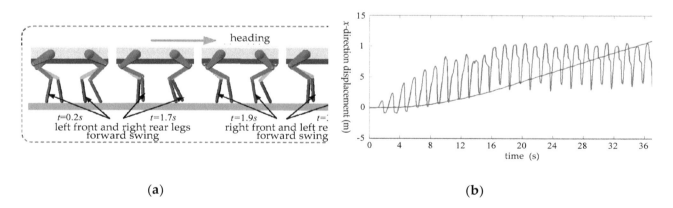

(a)   (b)

Figure 17. Simulation decomposition diagram of the trot gait of the simulation prototype. (**a**) Simulation diagram of trot gait. (**b**) The y-direction velocity and x-direction displacement curve of a quadruped robot.

The y-direction velocity and x-direction displacement of the simulation prototype are shown in Figure 17b, where the red solid line is the y-direction velocity curve and the blue solid line is the x-direction displacement curve. It can be seen from Figure 17b that the simulation prototype enters a stable walking state gradually, and the y-direction velocity and x-direction displacement reach a stable change when $t = 16$ s.

In Figure 16, N1 and N3 in the output of the self-growing network are input to the left front and the right hind hip joint of the experimental prototype, respectively, while N2 and N4 are input to the right front and the left hind hip joint of the experimental prototype respectively. The trot gait decompositions of the experimental prototype obtained are shown in Figure 18. Combining Figure 17a and Figure 18, we can see that the experimental prototype and the simulation prototype have the same gait cycle (about $T = 1.7$ s) under the control of the same self-growing network, and the self-growing network can achieve stable control of the trot gait of the simulation prototype and the experimental prototype.

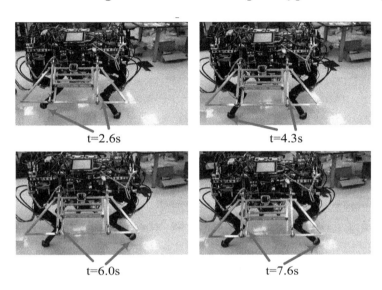

Figure 18. Trot gait diagram of experimental prototype.

The experiment above shows that the quadruped robot control system based on a CPG neural network self-growing algorithm can accomplish the growth of the network, the signal output of the network and the control of a robot's joint, and then realize the rhythmic movement of a quadruped robot. This shows that the CPG neural network self-growing algorithm can be well applied to the rhythm control of a robot. The experiment preliminarily simulated the process from growth to mastery of a certain rhythmic movement of the CPG neural network and verifies the feasibility of the idea of realizing robot rhythmic movement control by combining the microscopic mechanism of the growth and the macro characteristics of the output of the neural network.

6. Conclusions

In this paper, the biological neuron axon could continuously elongate and turn, and the growth of the back part did not change the shape of the previously used grown part. Combined with the law of gravitation, random motion, the Rall cable model and the Matsuoka oscillator model, a local CPG self-growing network with multiple physical characteristics was established. The fusion of a self-growing network and CPG has been realized, the control ability of self-growing network was presented, and the motion control of a quadruped robot was completed. Our conclusions are as follows.

(1) An axon self-growing algorithm based on universal gravitation and random motion was established, the Rall Cable Model for studying the potential distribution of an axon membrane was analyzed and simplified, and multiple physical properties of the synapse (including resistance, capacitance, axon length and diameter, etc.) were combined with the Rall Cable Model to establish a simplified model of single neuron signal transmission, as shown in Equation (19). The multi-physical properties obtained by the neuron self-growth model (such as the synapse length calculated by Equation (9) and the resistance R1 calculated by Equation (22)) were added to Equation (17) to calculate parameters τ and b of Equation (19). On this basis, the entire synapse system of the neuron's self-growth is taken as a whole, as well as the key parameters in the control model: the delay time constant Tr and the synapse, as shown in Equations (25) and (26). The connection weight corresponds to the axon

length and axon diameter in the growth model, which allows the combination of the growth and development process of the neuron and the control function.

(2) By analyzing the Matsuoka oscillator model (shown in Equation (27)) and comparing the difference between the mathematical model of its neuron monomer and the model built in this article, a local CPG network topology is proposed as shown in Figure 10. This topology is regarded as the goal of network self-growth. A neural network growth model based on gravitational field control is established, meaning that the network can grow the target topology from any initial position of the neuron, as shown in Figure 11, to obtain the rhythm output signal corresponding to the angular displacement of the hip joint of the quadruped robot, as shown in Figure 12.

(3) The quadruped robot motion control simulation prototype and experimental prototype are used as shown in Figure 13: the output signal of the local CPG network (as shown in Figure 16) is used as the hip joint control signal of the simulation prototype and the experimental prototype to control the quadruped robot simulation. The prototype and the experimental prototype realized walking with a trot gait, as shown in Figures 17 and 18, which verified the feasibility of the local CPG network self-growth model with multi-physical characteristics proposed in this paper for the gait control of a quadruped robot.

Since the gait control of the quadruped robot by the CPG network in this article is still an open-loop control system, it was not possible to change the gait of the robot through environmental feedback information or to adapt to the complex road conditions and change the rhythm and mode of the network output accordingly. Therefore, in order to give the self-growth-based network model higher practical value when applied to the motion control of a footed robot, the research work that needs to be done in the next stage is as follows:

(1) Continue to improve the self-growth model of neurons, considering more physical characteristics that can be linked to the control model parameters, so that the built model can be constantly close to the complexity of the biological neural network in terms of its microstructure and control function. For example, in this paper, the diameter of the neuron axon is constant. The fundamental reason for this is that only a one-dimensional size of the axon can be obtained by using the trajectory of the particle in the gravitational field as the axon model. The biological mechanism that determines the diameter must be introduced into the self-growth model to obtain a biologically reasonable diameter, so that the control model does not lose its general applicability.

(2) Study how many parameters obtained by network self-growth affect the network output. In the simulation experiment, we found that the time constant only affects the shape, amplitude and period of the output signal and does not affect the mutual inhibition mode, which is mainly affected by the connection weight. Moreover, not every time the network grows can the interactive inhibition output between output neurons be obtained. On occasion, individual or all output neurons will output a constant value signal after reaching stability; that is, the acquisition of the interactive inhibition output is important for the network growth parameters. This is conditional, but we still do not know the necessary condition.

(3) Study the micro-mechanism of interaction with the environment and the adjustment of the upper central system during the development of the CPG network and construct feedforward and feedback pathways for network development, so that the growth and development of the network can change the connection weights between synapses in biological neural networks by learning. Based on the research of (2), weight change is carried out in the direction in which the desired network output can be obtained to realize closed-loop control.

Author Contributions: Conceptualization, M.L. (Ming Liu) and F.Z.; methodology, M.L. (Ming Liu) and F.Z.; software, M.L. (Ming Liu); validation, M.L. (Ming Liu), M.L. (Mantian Li) and F.Z.; formal analysis, M.L. (Ming Liu); investigation, M.L. (Ming Liu); resources, P.W.; data curation, W.G.; writing—original draft preparation, M.L. (Ming Liu), M.L. (Mantian Li) and F.Z.; writing—review and editing, L.S.; visualization, M.L. (Ming Liu); supervision, M.L. (Mantian Li) and L.S.; project administration, M.L. (Mantian Li) and F.Z.; funding acquisition, F.Z. All authors have read and agreed to the published version of the manuscript.

References

1. Endo, G.; Morimoto, J.; Matsubara, T.; Nakanishi, J.; Cheng, G. Learning CPG-based Biped Locomotion with a Policy Gradient Method: Application to a Humanoid Robot. *Int. J. Robot. Res.* **2008**, *27*, 213–228. [CrossRef]
2. Maufroy, C.; Kimura, H.; Takase, K. Integration of Posture and Rhythmic Motion Controls in Quadrupedal Dynamic Walking Using Phase Modulations Based on Leg Loading/Unloading. *Auton. Robot.* **2010**, *28*, 331–353. [CrossRef]
3. Ijspeert, A.J. A connectionist central pattern generator for the aquatic and terrestrial gaits of a simulated salamander. *Boil. Cybern.* **2001**, *84*, 331–348. [CrossRef] [PubMed]
4. Von Twickel, A.; Pasemann, F. Reflex-Oscillations in Evolved Single Leg Neuron Controllers for Walking Machines. *Nat. Comput.* **2007**, *6*, 311–337. [CrossRef]
5. Noble, F.K.; Potgieter, J.; Xu, W.L. Modelling and simulations of a central pattern generator controlled, antagonistically actuated limb joint. In Proceedings of the 2011 IEEE International Conference on Systems, Man, and Cybernetics, Anchorage, AK, USA, 9–12 October 2011; pp. 2898–2903.
6. Manoonpong, P.; Pasemann, F.; Wörgötter, F. Sensor-Driven Neural Control for Omnidirectional Locomotion and Versatile Reactive Behaviors of Walking Machines. *Robot. Auton. Syst.* **2008**, *56*, 265–288. [CrossRef]
7. Suter, D.M.; Miller, K.E. The Emerging Role of Forces in Axonal Elongation. *Prog. Neurobiol.* **2011**, *94*, 91–101. [CrossRef]
8. Mortimer, D.; Pujic, Z.; Vaughan, T.; Thompson, A.W.; Feldner, J.; Vetter, I.; Goodhill, G.J. Axon Guidance by Growth-Rate Modulation. *Proc. Natl. Acad. Sci. USA* **2010**, *107*, 5202–5207. [CrossRef]
9. Pearson, Y.E.; Castronovo, E.; Lindsley, T.A.; Drew, D.A. Mathematical Modeling of Axonal Formation Part I: Geometry. *Bull. Math. Boil.* **2011**, *73*, 2837–2864. [CrossRef]
10. Nima, D.; Liu, Y. Modelling Axon Growth Using Driven Diffusion. *APS* **2019**, *2019*, L70–L305.
11. Marinov, T.; Sánchez, H.A.L.; Yuchi, L.; Adewole, D.O.; Cullen, D.K.; Kraft, R.H. A Computational Model of Bidirectional Axonal Growth in Micro-Tissue Engineered Neuronal Networks (micro-TENNs). *Silico Boil.* **2020**, *14*, 15–29. [CrossRef]
12. Kassraian-Fard, P.; Pfeiffer, M.; Bauer, R. A Generative Growth Model for Thalamocortical Axonal Branching in Primary Visual Cortex. *PLoS ONE Comput. Boil.* **2020**, *16*, e1007315. [CrossRef] [PubMed]
13. Gafarov, F. Neural electrical activity and neural network growth. *Neural Netw.* **2018**, *101*, 15–24. [CrossRef] [PubMed]
14. Kiehn, O. Development and Functional Organization of Spinal Locomotor Circuits. *Curr. Opin. Neurobiol.* **2011**, *21*, 100–109. [CrossRef] [PubMed]
15. Zjajo, A.; Hofmann, J.; Christiaanse, G.J.; Van Eijk, M.; Smaragdos, G.; Strydis, C.; De Graaf, A.; Galuzzi, C.; Van Leuken, R. A Real-Time Reconfigurable Multichip Architecture for Large-Scale Biophysically Accurate Neuron Simulation. *IEEE Trans. Biomed. Circuits Syst.* **2018**, *12*, 326–337. [CrossRef] [PubMed]
16. O'Donnell, C. Simplicity, Flexibility, and Interpretability in a Model of Dendritic Protein Distributions. *NEURON.* **2019**, *103*, 950–952. [CrossRef]
17. Koene, R.A.; Tijms, B.M.; Van Hees, P.; Postma, F.; De Ridder, A.; Ramakers, G.J.A.; Van Pelt, J.; Van Ooyen, A. NETMORPH: A Framework for the Stochastic Generation of Large Scale Neuronal Networks With Realistic Neuron Morphologies. *Neuroinformatics* **2009**, *7*, 195–210. [CrossRef]
18. Moser, T. Low-Conductance Intercellular Coupling between Mouse Chromaffin Cells In Situ. *J. Physiol.* **1998**, *506*, 195–205. [CrossRef]
19. Mortimer, D.; Fothergill, T.; Pujic, Z.; Richards, L.J.; Goodhill, G.J. Growth Cone Chemotaxis. *Trends Neurosci.* **2008**, *31*, 90–98. [CrossRef]
20. Borisyuk, R.M.; Cooke, T.; Roberts, A. Stochasticity and Functionality of Neural Systems: Mathematical Modelling Of Axon Growth in the Spinal Cord of Tadpole. *Biosystems* **2008**, *93*, 101–114. [CrossRef]
21. Pearson, Y.E. *Discrete and Continuous Stochastic Models for Neuromorphological Data*; Rensselaer Polytechnic Institute: New York, NY, USA, 2009.
22. Isler, Y. A Software for Simulating Steady-State Properties of Passive Dendrites Based on the Cable Theory. *Comput. Methods Programs Biomed.* **2007**, *88*, 264–272. [CrossRef]
23. Bennett, M.R.; Fernandez, H.; Lavidis, N.A. Development of the Mature Distribution of Synapses on Fibres in the Frog Sartorius Muscle. *J. Neurocytol.* **1985**, *14*, 981–995. [CrossRef] [PubMed]

24. Matsuoka, K. Mechanisms of Frequency and Pattern Control in the Neural Rhythm Generators. *Boil. Cybern.* **1987**, *56*, 345–353. [CrossRef] [PubMed]

25. Kiehn, O. Locomotor Circuits in the Mammalian Spinal Cord. *Annu. Rev. Neurosci.* **2006**, *29*, 279–306. [CrossRef] [PubMed]

26. Van Heijst, J.; Vos, J. Self-Organizing Effects of Spontaneous Neural Activity on the Development of Spinal Locomotor Circuits in Vertebrates. *Boil. Cybern.* **1997**, *77*, 185–195. [CrossRef]

27. Van Heijst, J.J.; Vos, J.E.; Bullock, D. Development in a Biologically Inspired Spinal Neural Network for Movement Control. *Neural Netw.* **1998**, *11*, 1305–1316. [CrossRef]

4

Trajectory Optimization of Industrial Robot Arms using a Newly Elaborated "Whip-Lashing" Method

Rabab Benotsmane [1], László Dudás [1] and György Kovács [2,*]

[1] Department of Information Technology, University of Miskolc, 3515 Miskolc, Hungary;
iitrabab@uni-miskolc.hu (R.B.); iitdl@uni-miskolc.hu (L.D.)

[2] Institute of Logistics, University of Miskolc, 3515 Miskolc, Hungary

* Correspondence: altkovac@uni-miskolc.hu

Abstract: The application of the Industry 4.0's elements—e.g., industrial robots—has a key role in the efficiency improvement of manufacturing companies. In order to reduce cycle times and increase productivity, the trajectory optimization of robot arms is essential. The purpose of the study is the elaboration of a new "whip-lashing" method, which, based on the motion of a robot arm, is similar to the motion of a whip. It results in achieving the optimized trajectory of the robot arms in order to increase velocity of the robot arm's parts, thereby minimizing motion cycle times and to utilize the torque of the joints more effectively. The efficiency of the method was confirmed by a case study, which is relating to the trajectory planning of a five-degree-of-freedom RV-2AJ manipulator arm using SolidWorks and MATLAB software applications. The robot was modelled and two trajectories were created: the original path and path investigate the effects of using the whip-lashing induced robot motion. The application of the method's algorithm resulted in a cycle time saving of 33% compared to the original path of RV-2AJ robot arm. The main added value of the study is the elaboration and implementation of the newly elaborated "whip-lashing" method which results in minimization of torque consumed; furthermore, there was a reduction of cycle times of manipulator arms' motion, thus increasing the productivity significantly. The efficiency of the new "whip-lashing" method was confirmed by a simulation case study.

Keywords: manipulator arm; trajectory optimization; "whip-lashing" method; reduction of cycle time; trajectory planning; SolidWorks and MATLAB software applications

1. Introduction

Nowadays, the modern industry in all sectors is facing a new revolution known as Industry 4.0 [1,2], where many challenges and requirements are taken into consideration with the aim of building smart factories that combine flexibility and ability concepts [3,4] by developing a new paradigm based on the latest technologies, where automation and network systems present the efficient keys for realizing the new industrial revolution [5].

Recently, industrial robotics has become an important solution used in different sectors due to the advantages guaranteed by industrial robots [6] as manipulator arms and parallel robots represented with higher precision and higher productivity. This optimizes the lead time of the production process [7].

Especially with technological developments, the manipulator arm, for example, presents the most often used tool in the production sector [8], where it can cooperate with its environment and work safety [9,10]. In addition to the ability to pick huge products and control itself in a flexible and smart way, the physical structure of the manipulator arm regroups two essential parts [11]. These parts are the serial links articulated as an arm and the end-effector, which can be reconfigurable according to the task [12,13]. The motion planning of a manipulator arm is always based on the degree of freedom characterized by the joints placed in each link, where the number of the degree of freedom limits the workspace and defines the redundancy of the robot. The control of manipulator robots can be studied in two directions, depending on the requirements needed to achieve it [14]: (1) task execution, where the process is based on pick and place, welding, and painting; (2) path planning execution, where the process is based on the trajectory of the end effector, depending on each joint connected to the link of the robot arm.

In the area of robotics, the trajectory planning of manipulator arms represents an essential field for focus. The execution of a robot arm's defined task optimizes its trajectory, which can guarantee many benefits such as a reduced cycle time and energy consumption, as well as increased productivity. Basically, the main objective in the trajectory planning field is to compute the desired points that represent the reference input data for the controller of a robot using mathematical techniques [15,16]. The motion executed from the reference inputs always represents two categories known as forward and inverse kinematics: (1) in free space based on the joint angles, where the motion is limited by the structure constraints, i.e., velocity, torque, and workspace limits; or (2) in task space based on the position and the orientation of the end-effector, where it depends on precision and avoiding obstacles [17]. The approaches generally used are polynomial interpolation function, the bang-bang law, the trapezoid law, etc. [18].

Over the years, researchers studied this field deeply by proposing many methods and solutions to solve trajectory problems for industrial robots [19–22]. The definition of the optimality concept is divided in many directions. Some scientists focus on a time-optimal trajectory to increase productivity [23,24], while others work on the smoothness of trajectories [25,26], taking into account reducing cycle time by implementing fast trajectories combined with optimal jerk values in order to reduce the excitation of the resonant frequencies and limit the vibrations of the mechanical system [27–29]. From the literature, a basic approach is known for generating a trajectory using splines [30,31], where the virtual points are required to ensure the continuity of the trajectory from the starting point to the endpoint. The development of this approach motivated the authors of this study to apply an improved technique in the aspect of motion optimality using B-spline interpolation, based on the calculation of inverse of Jacobian matrix.

Regarding time optimality, an approach was proposed for a hyper-redundant robot taking into account the obstacles located in a 3D workspace [32,33]. It aims to minimize the cycle time during the execution of required tasks, regarding trajectory optimization for robots in terms of energy consumption and minimizing joint torque. Other researchers described a new scheme to determine the trajectory of a redundant robot arm with the purpose of minimizing the total energy consumption [34]. In order to optimize both the energy consumption and the time required for executing a trajectory, many researchers have elaborated new methods based on a fuzzy logic algorithm, a genetic algorithm, or an ant colony algorithm [34,35]. By using a genetic algorithm, a contribution was proposed to optimize the torque applied at the joints of the robot [36,37]. We can also cite the second contribution for the same target, which uses a unified quadratic-programming-based dynamic system [38], as well as the role of neural networks for the optimized dynamics of redundant robots [39]. Most of the literature in motion planning features deals with point-to-point applications in free space without any obstacles, where the starting and ending points of the end-effector are predefined.

The main purpose of this literature analysis was to guarantee a deeper understanding of the path planning field so that researchers could find an optimal solution without any constraints. Further, time and energy consumption presents the most important factors for evaluation [40].

The purpose of this study is to elaborate upon a new "whip-lashing" method that aims to realize an optimized trajectory for the five-degree-of-freedom RV-2AJ robot arm, i.e., to generate a path for a manipulator arm without any design constraints. This newly elaborated method seeks to minimize the cycle time of the trajectory with constrained torque values applied in joints for executing smooth motions. This new method originates from the motion of a whip analogue applied for a RV-2AJ robot arm. First, we introduce the methodology of this newly elaborated method, identifying all the steps required using SolidWorks and MATLAB software applications. After, the main features of whip-lashing are introduced and the Section 4 presents a concrete simulation executed for both paths in order to compare the normal motion of the arm to a whip lashing motion, based on cycle time. The final section presents the results of the joints torque variation according to the cycle time calculated in the previous section in order to show the reader the effect of the whip-lashing motion on the cycle time and the torque consumption.

The main value of the research is that a manipulator arm can be treated as a whip in certain conditions, which can guarantee running an improved path that results in reduced cycle time. The proof of this novelty is presented by applying the real parameters of an RV-2AJ arm to simulation tools.

2. Materials and Methods: Modelling the Robot Arm and the Original and Improved Trajectories

The idea of trajectory improvement is taken from the motion of a whip. When applying a huge force at the handle, it will propagate along the whip's length, with a wave running alongside the whip transferring the energy from the handle to the tip. In order to realize this idea and check its real effect on the cycle time of robot arm motion, SolidWorks [41] and MATLAB [42] software applications were used in the dynamic motion simulations. For comparison, an "original" path with simple angle interpolation at the joints was also simulated. The two will be compared.

2.1. Dynamic Analysis of RV-2AJ Robot Arm

The dynamic analysis of RV-2AJ arm requires the computation of the inertia matrix for each joint. This computation is performed using SolidWorks software by drawing the robot model and applying its real geometric and mass measurements. SolidWorks offers the possibility to transform all of the robot parameters in a URDF (Unified Robotic Description Format) file; this file can be used for the simulation in MATLAB software using Robotic Toolbox [43]. Figure 1 presents the CAD model of RV-2AJ arm in SolidWorks environment and the creating of its URDF file for MATLAB software. The URDF file allows us to import the dynamic parameters of RV-2AJ arm into a MATLAB environment, where we can visualize the robot arm and calculate its forward and inverse kinematics using the Robotic Toolbox that provides the following:

- RigidBodyTree (RBT) object,
- Home configuration function,
- Inverse Kinematic solver.

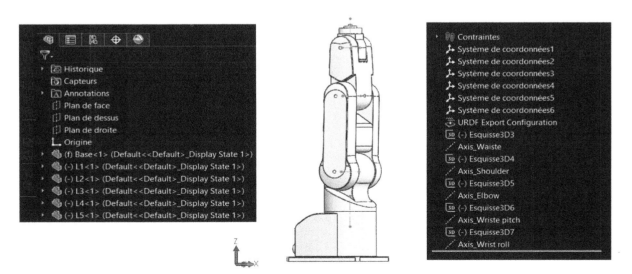

Figure 1. RV-2AJ model in SolidWorks and the created URDF (Unified Robotic Description Format) file for the CAD model of the RV-2AJ arm.

Figure 2 visualizes the imported structure of RV-2AJ arm in the MATLAB environment and the script code used for it. The visualization of the robot arm uses the RV-2AJ.URDF definition file and the "show (robot, Qhome)" command performs the visualization, as can be followed in the MATLAB code.

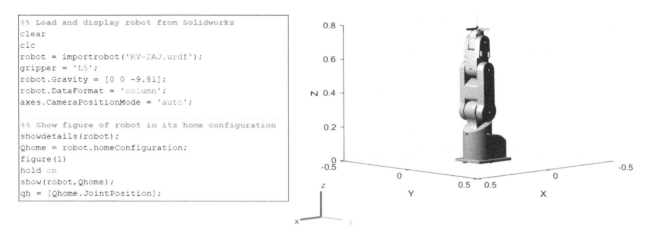

Figure 2. Visualization of RV-2AJ arm in the MATLAB environment.

For finding better or "quasi-optimal" solution and proving the effectiveness of the newly elaborated whip-lashing method, two paths were generated with the same starting and ending points but with different internal motions. During the application of the method both paths require the rotation of second–third–fourth joints, whereas the first and the fifth articulation are not used.

2.2. The Original Path

The original trajectory as a usual interpolated path is presented in Figure 3 as a continuous red arc starting from start point S and ending in final point E. In this path the RV-2AJ robot arm executes its motion by computing the angle steps for each internal point dividing the start-end angle of every joint with the number of points minus one step.

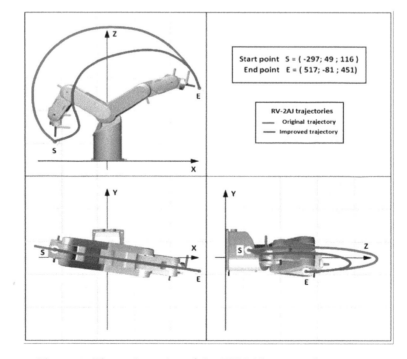

Figure 3. The trajectories of the RV-2AJ arm in three views.

2.3. The Improved Path

The suggested trajectory—named "improved path" given in blue in three views (front, top, and left side views) with the robot arm in Figure 3—is based on the newly elaborated special method that imitates the natural motion of a whip. The improved path aims to decrease the cycle time for RV-2AJ arm's movement from the starting point S to the final point E. This method is based on principle of the whip-lashing motion that determines the torques applied in closing and opening of joints of the arm, which will be discussed in the further sections.

3. Results: Newly Elaborated "Whip-Lashing" Method

For centuries whips have been considered instruments used to direct animals or torture slaves. Nowadays, this instrument has become an artform for some nations. In the last few years, researchers have been interested in the phenomenon executed by whips, which is characterized by the crack sound and the velocity that can reach supersonic speed [44,45]. Figure 4 presents the basic elements of a simple whip.

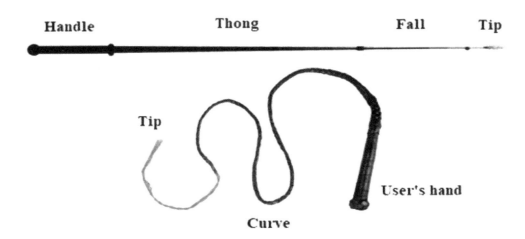

Figure 4. Basic elements of a whip.

When we give a force for a whip with the "handle" to move up to down and then stop, that produces kinetic energy. This energy will be transferred to the end of whip "tip" due to $p = m \times v$, where the mass (m) decreases and the velocity (v) increases.

The result of this is a sonic boom produced by a crack that is caused when some section of the whip moves faster than the speed of sound. The motion of a whip can include three types: a half wave, a full wave and a loop. Figures 5 and 6 present the transfer direction of momentum along the segments of whip and the diagrams describe the change of velocity, mass, kinetic energy and torque as a function of time. The input values used in the diagrams in Figure 6 were taken from Krehl et al. [46]. The subfigures show the following functions:

(1). The velocity diagram shows the changing of the velocity of the tip during a whip-lashing cycle.

(2). The mass diagram shows the changing of the mass of that part of the whip that has the most the kinetic energy during the whip-lashing as the wave impulse goes along the whip length.

(3). The kinetic energy diagram corresponds to the work of the whip user who moves the whip and conveys motion energy to the whip with his hand.

(4). The torque diagram shows the torque value changing in the wrist joint during the whip-lashing.

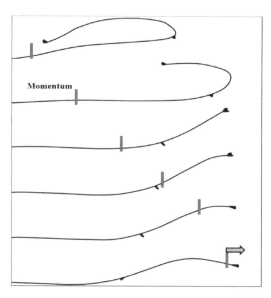

Figure 5. The motion of a whip.

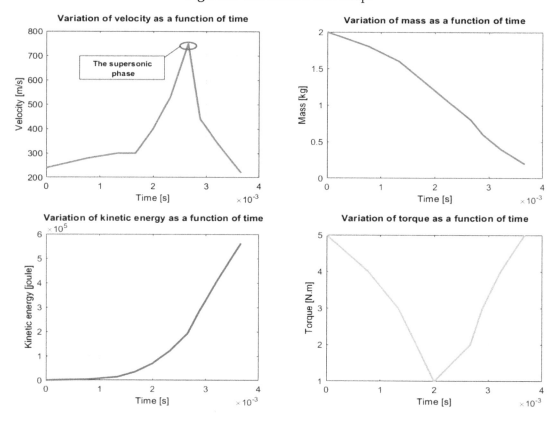

Figure 6. The variation of velocity, mass, kinetic energy, and torque as a function of time.

As we mentioned earlier, the motion of a whip is based on the momentum conservation from the beginning of the whip to the arrival to the tip, where—before cracking—many parameters vary, such as the moving mass along the whip which will be decreased and produce high velocity, presenting the supersonic boom. This phenomenon results in diminution in the torque and an increase in kinetic energy, as Figure 5 shows. A similar effect of conserving momentum can be seen in case of a figure skater doing a spin, when the skater closes his/her hands to his/her body, thus decreasing the rotational inertia and increasing the rotational velocity. Borrowing this characteristic of the motion from whip and considering the robot arm as a similar shape, the increase in the velocity of the robot parts will decrease the cycle time, so the motion of the robot arm can be used like in the mentioned whip or skater examples. The whip analogue fits the robot arm better because the arm is similar to a whip. It has many conjoined arm elements that become slimmer from the basement to the gripper.

3.1. Modelling of RV-2AJ Arm Motion as Whip-Lashing in MATLAB Software

Based on the brief analysis on the motion of the whip and considering the structural similarity to the manipulator arm, the suggestion of a whip-analogous motion and the resulted in trajectory of the gripper seems to be rational. In this article, we applied the same principle to the RV-2AJ arm, where the robot acted as a whip and the gripper executed the trajectory between two points faster than it otherwise would. The whip-lashing motion of the robot arm can be followed through the sub-figures of Figure 7. When the second joint rotation stopped, the momentum WAS transferred to the third and fourth joints. As the moving mass became smaller, the rotational velocity increased at the last links of the robot arm.

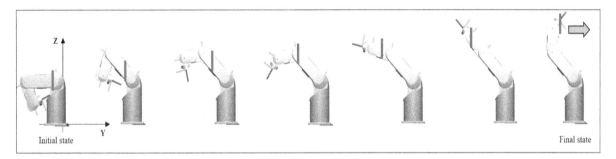

Figure 7. The motion of RV-2AJ arm.

The first step of operation of RV-2AJ arm started with the heavier segment, where the movement was realized by the torque applied at the second joint, at the "closer part to the base". In addition, the rotational inertia decreased when the motion of the links reached the smallest segment "end effector". As the motion proceeded the speedy "closer part to the base" decelerated and the outer joints opened. Finally, the "closer part to the base" stopped and the other joints finish the motion speeding up the motion of the "end effector".

In order to prove the efficiency of whip-lashing method, an iterative algorithm was created to execute the two trajectories (original and improved, see Sections 2.2 and 2.3) and calculated the minimal cycle time of each, trying to apply the maximum allowed torques for the joints.

The core of the algorithm inputted a given larger expected cycle time and determined the joint torque functions along the trajectory. If no torque maximum reaches the allowed torque limit for the given joint, the expected cycle time was decreased by a time step and the core process was repeated. If the expected cycle time was too small and the robot could only execute the trajectory with one or more joint torques exceeding the torque limit, then the decreasing time step was halved and the process applied a back step. This was repeated with a smaller time decrease. At the end of this successive approximation, the minimum cycle time that utilized the torque limit—usually this happens for one joint only—was obtained.

3.2. Elaboration of the Cycle Time Minimization (CTM) Algorithm

The organigram presented in Figure 8 contains different blocs (A–E) that define the following calculations:

- **A**: Filling up the **TM** torques matrix with the **T** torque vectors for every i-th trajectory point.
- **B**: Copying the i-th **T** torque vector in the i-th column of the **TM** torques matrix.
- **C**: Determining the $mT[j]$ maximum torque for the j-th joint.
- **D**: Checking if any joint torque maximum $mT[j]$ exceeds the allowed torque $aT[j]$ for the j-th joint.
- **E**: Refinement of time step ts if necessary and continuing iteration, or finishing if time step ts goes below ending time step ets.

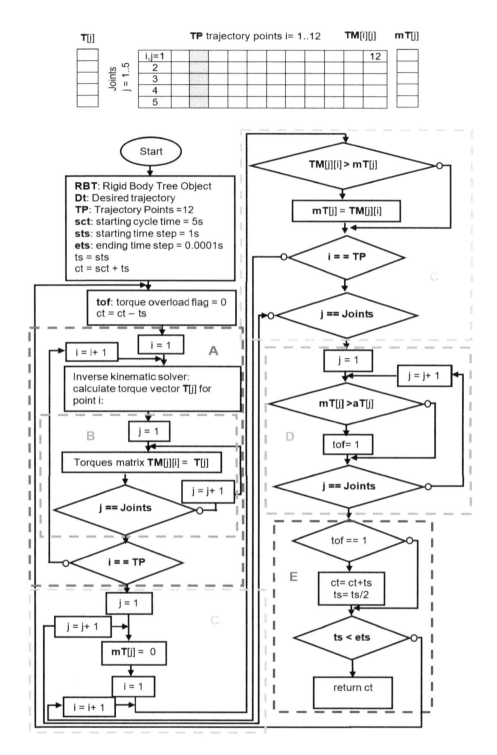

Figure 8. Cycle time minimization algorithm. A–E stand for different blocs that define different calculations.

By running the algorithm for the original and improved trajectories, the two minimum cycle times were determined and compared. In Figure 8, the elaborated cycle time minimization algorithm is introduced in detail.

The algorithm aimed to calculate the cycle time ct of the trajectory. It was made using a successive approximation algorithm as mentioned before. The concrete parameters of the algorithm are the next. The number of trajectory points $TP = 12$, where the index of a specified point is $i = 1, \ldots, 12$. The number of RV-2AJ arm joints is 5, where the index of a specified joint is $j = 1, \ldots, 5$. In the algorithm two conditions should be fulfilled:

- $ts \leq ets$ (*ts*—time step; *ets*—ending time step),
- $mT[j] \leq aT[j]$ (**mT**[*j*]—the joints' torque maximums; **aT**[*j*]—allowed torque for every *j*-th joint).

The algorithm aimed to determine the following data:

The **T**[*j*] torque vector is calculated for every *i*-th trajectory point and copied into the **TM**[*j*][*i*] torque matrix into the *i*-th column.

The **T** torque vector is determined by the "inverse Dynamics MATLAB Robotics System Toolbox function" of the MATLAB Robotic Toolbox. Then, the maximum of every *j*-th row of **TM** torque matrix is determined to the *j*-th cell of the **mT**[*j*] maximum torque vector. Then the **mT**[*j*] $\leq$ **aT**[*j*] condition setting the *tof* torque overload flag selects between cycle time decreasing or time step refinement and back stepping.

4. Discussion: Trajectory Optimization's Results of RV-2AJ Arm Using the Application of CTM Algorithm

The two trajectories are defined by positioning the same starting and ending points and 10 desired inner points that differ for each path using the spline function of MATLAB script to define the continuous path between the points. Figures 9–12 visualize the code script results for both trajectories (the original and the improved). The static Figures 10 and 12 cannot make the motion of the robot arms felt.

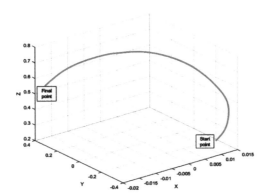

Figure 9. Original trajectory visualization.

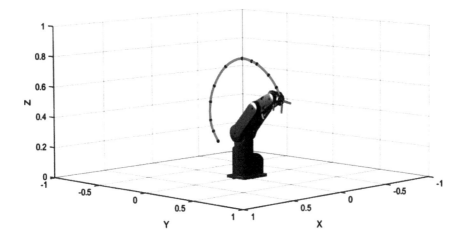

Figure 10. Visualization of the execution of the original trajectory by the RV-2AJ arm.

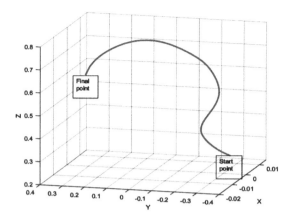

Figure 11. Improved trajectory visualization.

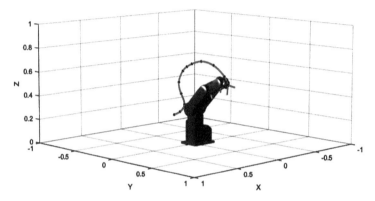

Figure 12. Visualization of the execution of the improved trajectory by RV-2AJ arm.

The execution of script code for both trajectories resulted in the value of cycle time ct for each path, as presented in Figure 13, uniting the two results with drawing manipulation to make comparable the relation between the two trajectories and robot arm motions alike. First, a starting large cycle time $ct = 5$ [s] was entered as input, then the algorithm was run till we obtained a new cycle time described as the searched cycle time sct when the torque limiting condition became unsatisfied, the algorithm continued to iterate the new values for cycle time ct and time step ts around the searched cycle time value till the value of tuned time step became smaller than the value of ending time step ets. For the original path, the minimum cycle time was $ct = 2.82$ [s], while for the improved path, the minimum cycle time was $ct = 1.86$ [s]. Comparing the two results, we proved that the improved trajectory, which is described as a whip motion analogue, utilized the maximum allowed torques with shorter cycle time than the original trajectory.

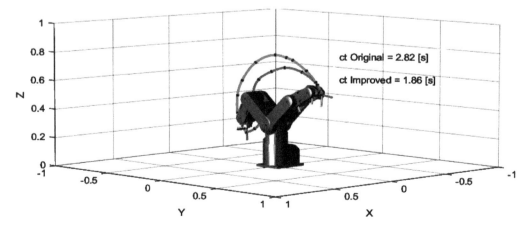

Figure 13. The original and the improved scenarios of RV-2AJ robot arm.

Modelling of RV-2AJ Arm Motion as Whip-Lashing in MATLAB Software

After studying the cycle time variation, the analysis of torque change effect regarding the both trajectories are discussed in this section. To perform the simulation in the Simulink environment, we imported the RV-2AJ body structure XML file into MATLAB. Then, we configured each link and joint in the structure to receive the vector of positions as input blocks that represent the original and the improved path, as well as to calculate the torques applied for providing the series of such positions.

Figures 14 and 15 present the block system needed to determine the finishing cycle time *ct* and torque vectors **T** calculated during the *ct* minimizing process, where the blocks "Improved Trajectory" and "Original Trajectory" contain spline function for each joint specified by angles vector that should RV-2AJ arm execute according to each position. "RV-2AJ arm block" contains the mechanical properties of the robot arm.

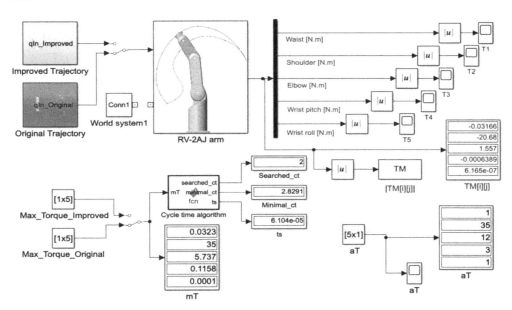

Figure 14. The block system scheme of RV-2AJ arm of the original trajectory.

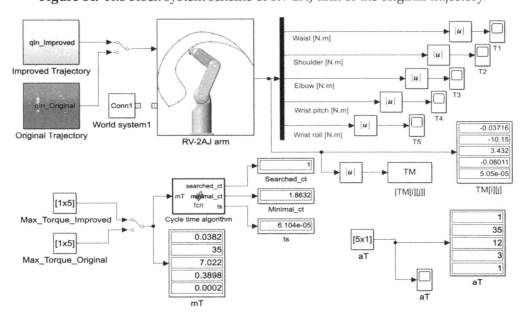

Figure 15. The block system scheme of RV-2AJ arm of the improved trajectory.

Using Scope Box, we visualized the torque diagram of each working joint (second–third–fourth, *j* = 2, 3, 4) for the original and improved paths (Figure 16).

In this section, we investigated the variation effect of the second joint (shoulder) and third joint (elbow) regarding the allowable torque values **aT** for those joints. As can be observed in Figure 16, the variation of torque values for the two paths for a starting cycle time $ct = 5$ [s] and $ts = 1$ [s] describes the behavior of the RV-2AJ arm.

In Figure 16, it can be seen that the torques of the fourth joint are close to zero due to the weight in this segment, which is very small compared to the other segments, which hold other segments. The comparison between the two paths was based on shoulder, which presents the second joint. It was clear that in the first iteration where $ct = 5$ [s], the torque curve for the original path reached higher peak values than the improved path, where the maximum torque value for the original path was **mT** = 31 Nm and for the improved path was **mT** = 27 Nm. We also observed an overshoot in the beginning stage for both curves, which is explained as the gravity effect on the mechanical structure.

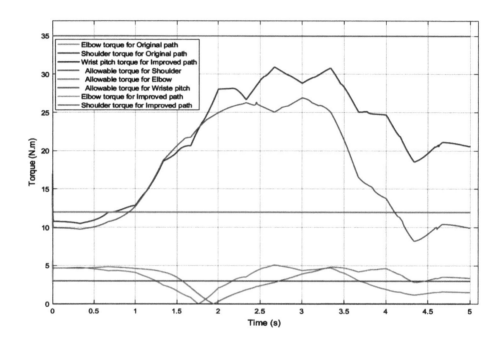

Figure 16. Variation of joint torques according to the two paths, $ct = 5$ [s].

As mentioned before, we started within a large cycle time $ct = 5$ [s] and within $ts = 1$ as a first iteration. The algorithm of minimization of cycle time started to calculate maximum torque for such a cycle time. If the maximum torque was always below the allowable torque for every joint, then the cycle time ct decreased by the time step ts.

Figure 17 presents the variation of joints torques according to the original path for different cycle times $ct = 5, 4, 3, 2$ [s] to see the behavior of torque curves and compare their peak values in each iteration, in order to observe when the torque peak of an iteration exceeds the allowable torque value **aT**.

Regarding the second joint (shoulder), in the interval 0–5 [s] we observed that the decrease of cycle time resulted in an increase of torque value for the original trajectory, where in $ct = 2$ [s] the maximum torque value of the original trajectory exceeded the allowable torque value with **mT** = 42 [N·m]. Therefore, this value $ct = 2$ [s] was stored as the searched cycle time value sct for the original trajectory and according to this value we calculated a new time step which was smaller than the first one, so the algorithm iterated the cycle time value more precisely.

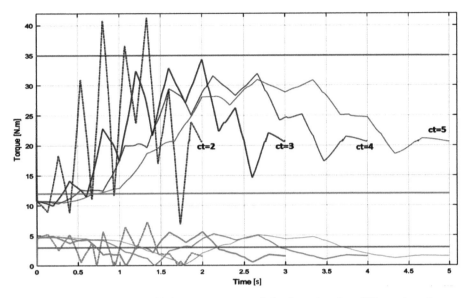

Figure 17. Variation of joint torques—original path—for different cycles.

As presented in Figure 18, the searched cycle time sct = 2 [s] obtained for the original path minimized the possible range to find the optimal maximum torque with optimal new cycle time. As a result, the optimal curve for shoulder was the red curve with ct = 2.82 [s]; this curve was obtained after the necessary iterations, starting from the first one, then minimizing the torque till obtaining the optimal curve, the "fourth iteration" m and because we had a condition to stop iteration when the time step ts was smaller than the ending time step ets; therefore, the algorithm completed the execution and found a better solution with ct = 2.82 [s], as presented in the first section.

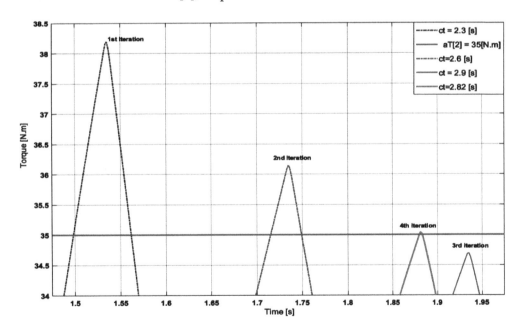

Figure 18. Different iterations of cycle time minimization algorithm for the original path around the searched cycle time sct = 2 [s].

Figure 19 presents the variation of joint torques for the improved path for different cycle times ct = 5, 4, 3, 2, 1 [s] by executing different iterations of cycle times ct in order to observe when the torque peak of an iteration exceeded the allowable torque value **aT**. For the improved path at ct = 1 [s], the maximum torque **mT** = 58 [N·m] exceeded the allowable torque value **aT**. Therefore, the searched cycle time for the improved path was sct = 1 [s].

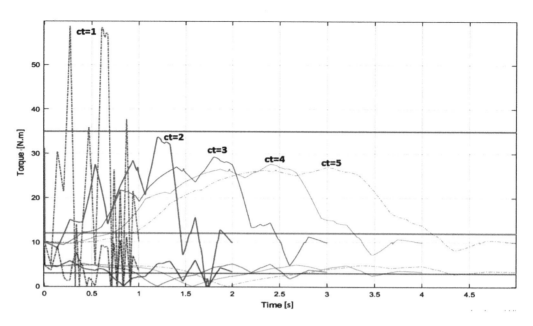

Figure 19. Variation of joints torques according to Improved path for different cycle times $ct = 5, 4, 3, 2, 1$ [s].

Figure 20 shows the execution of different iterations around the searched cycle time value obtained $sct = 1$ [s] from the results of Figure 19 to find the optimal curve. From the diagram, after 4 iterations around the searched cycle time sct = 1 [s], the algorithm achieved the optimal curve with $ct = 1.86$ [s].

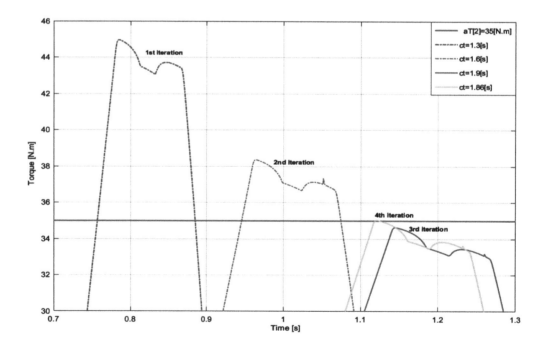

Figure 20. Different iterations of the cycle time minimization algorithm for improved path around the searched cycle time $sct = 1$ [s].

Based on the demonstration of diagrams and the comparison between the multi graphs of each path for the second joint, it was clear that the original path exceeded the allowable torque **aT** optimally with $ct = 2.82$ [s], unlike the improved path, which exceeded optimally **aT** with $ct = 1.86$ [s], as presented in Figure 21. Consequently, we proved that the improved path consumed 33% less time than the original path, which verified the concept of optimization.

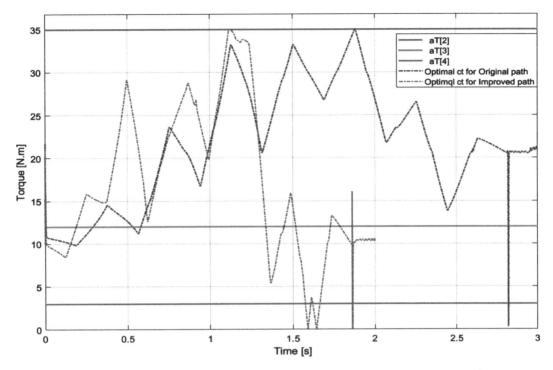

Figure 21. Optimal cycle time values for the original and improved paths.

5. Conclusions

The trajectory improvement of industrial robot arms plays an important role in the Industry 4.0 concept, since the trajectory optimization of robot arms reduces cycle times and energy consumption; furthermore, it increases productivity.

At first, a newly elaborated "whip-lashing" method was introduced for the trajectory planning of a robot arm. The idea of the "whip-lashing" method is that the motion and the shape of a whip are similar to the motion and shape of a robot arm, and it results in increasing the velocity of the robot arm's parts during operation; therefore, it reduces the cycle time. With the "whip-lashing" method, the optimized trajectory of the robot arm can be achieved in order to minimize the cycle times of manipulator arms' motion and utilize the torque of joints more effectively.

In the second part of the article, a case study was introduced to confirm the efficiency of the practical applicability of the "whip-lashing" method. In order to realize the idea of the "whip-lashing" method and check its real effect on the cycle time of robot arm motion, SolidWorks and MATLAB software applications were used in the dynamic motion simulations.

In the case study, the trajectory planning of a five-degrees-of-freedom RV-2AJ manipulator arm was described using SolidWorks and MATLAB software applications. At first, the RV-2AJ robot was modelled by the application of SolidWorks software. Two trajectories of the investigated manipulator arm were created using the application of MATLAB software and via the newly elaborated cycle time minimization algorithm. This was done in order to compare the original and the improved paths. It was found that application of the cycle time minimization algorithm resulted in a 33% shorter cycle time compared to the original path of RV-2AJ robot arm.

The main added value of the study is the elaboration and implementation of the newly elaborated "whip-lashing" method, which resulted in reduction of cycle times of manipulator arms' motion, thus increasing productivity significantly. The efficiency of the new "whip-lashing" method was confirmed by a simulation case study.

The application of this newly elaborated method can provide many advantages in industrial applications in the immediate future, where the robot arms will be designed as the shape of a whip; furthermore, the robot arms will be manufactured by usage of lightweight materials instead of recently used traditional metals. These innovative solutions (application of "whip-lashing" method and lightweight materials) will result in more flexible robot arms that can achieve motion with higher speeds without consuming higher energy, by the application of the momentum conservation law. This law can also be used by the existing rotation joint of robot arms where the motion can be achieved in a plain to increase the speed of the robot arm and decrease the motion time. This conception requires the application of new robot controllers and robot simulation software. It can be concluded that the application of the newly elaborated "whip-lashing" method results in achieving the optimized trajectory of the robot arms in order to increase velocity of the robot arm's parts, thereby minimizing motion cycle times and to utilize the torque of the joints more effectively. Consequently, the productivity will be increased significantly.

In the following research, the aim was to find a quasi-optimal trajectory between two given points. The searching for the quasi-optimum solution used the Tabu search method.

Author Contributions: Conceptualization, R.B., L.D., and G.K.; literature review and data collection, R.B.; methodology and software, R.B. and L.D.; formal analysis, R.B., L.D., and G.K.; writing—original draft preparation, R.B.; writing—review and editing, R.B., L.D., and G.K.; visualization, R.B.; supervision, L.D. and G.K.; project administration, L.D. and G.K. All authors have read and agreed to the published version of the manuscript.

References

1. Benotsmane, R.; Dudás, L.; Kovács, G. Collaborating robots in Industry 4.0 conception. In Proceedings of the XXIII International Conference on Manufacturing, IOP Conference Series: Materials Science and Engineering, Kecskemét, Hungary, 7–8 June 2018; pp. 1–9.
2. Benotsmane, R.; Dudás, L.; Kovács, G. Survey on new trends of robotic tools in the automotive industry. In *Vehicle and Automotive Engineering 3*; VAE 2020. Lecture Notes in Mechanical Engineering; Springer: Singapore, 2021; pp. 443–457.
3. Dima, I.C.; Kot, S. Capacity of production. In *Industrial Production Management in Flexible Manufacturing Systems*, 1st ed.; Dima, I.C., Ed.; Book News Inc.: Portland, OR, USA, 2013; pp. 40–67.
4. Kovács, G. Combination of Lean value-oriented conception and facility layout design for even more significant efficiency improvement and cost reduction. *Int. J. Prod. Res.* **2020**, *58*, 2916–2936. [CrossRef]
5. Benotsmane, R.; Kovács, G.; Dudás, L. Economic, social impacts and operation of smart factories in Industry 4.0 focusing on simulation and artificial intelligence of collaborating robots. *Soc. Sci.* **2019**, *8*, 143. [CrossRef]
6. Delgado, S.D.R.; Kostal, P.; Cagánová, D.; Cambál, M. On the possibilities of intelligence implementation in manufacturing: The role of simulation. *Appl. Mech. Mater.* **2013**, *309*, 96–104. [CrossRef]
7. Yildirim, C.; Sevil Oflaç, B.; Yurt, O. The doer effect of failure and recovery in multi-agent cases: Service supply chain perspective. *J. Serv. Theory Pract.* **2018**, *28*, 274–297. [CrossRef]
8. Gilchrist, A. *Industry 4.0: The Industrial Internet of Things*; Apress: Bangkok, Thailand, 2014; pp. 1–12.
9. El Zoghby, N.; Loscri, V.; Natalizio, E.; Cherfaoui, V. Robot cooperation and swarm intelligence. In *Wireless Sensor and Robot Networks: From Topology Control to Communication Aspects*, 1st ed.; Mitton, N., Simplot-Ryl, D., Eds.; World Scientific Publishing Company: Lille, France, 2014; pp. 168–201.
10. Alessio, C.; Maratea, M.; Mastrogiovanni, N.; Vallati, M. On the manipulation of articulated objects in human-robot cooperation scenarios. *Robot. Auton. Syst.* **2018**, *109*, 139–155.
11. Benotsmane, R.; Dudás, L.; Kovács, G. Trial—and—error optimization method of pick and place task for RV-2AJ robot arm. In *Vehicle and Automotive Engineering 3*; VAE 2020. Lecture Notes in Mechanical Engineering; Springer: Singapore, 2021; pp. 458–467.
12. Yim, M.; Shen, W.; Salemi, B.; Rus, D.; Moll, M.; Lipson, H.; Klavins, E.; Chirikjian, G.S. Modular self-reconfigurable robot systems. *IEEE Robot. Autom. Mag.* **2007**, *14*, 43–52. [CrossRef]
13. Koren, Y.; Heisel, U.; Jovane, F.; Moriwaki, T.; Pritschow, G.; Ulsoy, G.; Van Brussel, H. Reconfigurable manufacturing systems. *CIRP Ann. Manuf. Technol.* **1999**, *48*, 527–540. [CrossRef]
14. Lewis, F.L.; Abdallah, C.T.; Dawson, D.M.; Lewis, F.L. *Robot Manipulator Control: Theory and Practice*, 2nd ed.; Marcel Dekker: New York, NY, USA, 2004.
15. Wissama, K.; Etienne, D. *Modélisation Identification et Commande des Robots*, 2nd ed.; Harmes: Lavoisier, France, 1999.
16. Liu, X.; Qiu, C.; Zeng, Q.; Li, A. Kinematics analysis and trajectory planning of collaborative welding robot with multiple manipulators. *Procedia CIRP* **2019**, *81*, 1034–1039. [CrossRef]

17. Benotsmane, R.; Kacemi, S.; Benachenhou, M.R. Calculation methodology for trajectory planning of a 6 axis manipulator arm. *Ann. Fac. Eng. Hunedoara Int. J. Eng.* **2018**, *3*, 27–32.
18. Coiffet, P. *Les robots: Modélisation et Commande*, 1st ed.; Hermes Science Publications: Paris, France, 1986.
19. Kim, H.; Hong, J.; Ko, K. Optimal design of industrial manipulator trajectory for minimal time operation. *KSME J.* **1990**, *4*. [CrossRef]
20. Straka, M.; Žatkovič, E.; Schréter, R. Simulation as a means of activity streamlining of continuously and discrete production in specific enterprise. *Acta Logist.* **2014**, *1*, 11–16. [CrossRef]
21. Doan, Q.V.; Vo, A.T.; Le, T.D.; Kang, H.-J.; Nguyen, N.H.A. A novel fast terminal sliding mode tracking control methodology for robot manipulators. *Appl. Sci.* **2020**, *10*, 3010. [CrossRef]
22. Joo, S.-H.; Manzoor, S.; Rocha, Y.G.; Bae, S.-H.; Lee, K.-H.; Kuc, T.-Y.; Kim, M. Autonomous navigation framework for intelligent robots based on a semantic environment modeling. *Appl. Sci.* **2020**, *10*, 3219. [CrossRef]
23. Kim, J.; Kim, S.-R.; Kim, S.-J.; Kim, D.-H. A practical approach for minimum-time trajectory planning for industrial robots. *Ind. Robot Int. J.* **2010**, *37*, 51–61. [CrossRef]
24. Perumaala, S.; Jawahar, N. Synchronized trigonometric S-curve trajectory for jerk-bounded time-optimal pick and place operation. *Int. J. Robot. Autom.* **2012**, *27*, 385–395. [CrossRef]
25. Avram, O.; Valente, A. Trajectory planning for reconfigurable industrial robots designed to operate in a high precision manufacturing industry. *Procedia CIRP* **2016**, *57*, 461–466. [CrossRef]
26. Macfarlane, S.; Croft, E.A. Jerk-bounded manipulator trajectory planning: Design for real-time applications. *IEEE Trans. Robot. Autom.* **2003**, *19*, 42–52. [CrossRef]
27. Gasparetto, A.; Lanzutti, A.; Vidoni, R.; Zanotto, V. Experimental validation and comparative analysis of optimal time-jerk algorithms for trajectory planning. *Robot. Comput. Integr. Manuf.* **2012**, *28*, 164–181. [CrossRef]
28. Liu, H.; Xiaobo, L.; Wenxiang, W. Time-optimal and jerk-continuous trajectory planning for robot manipulators with kinematic constraints. *Robot. Comput. Integr. Manuf.* **2013**, *29*, 309–317. [CrossRef]
29. Martínez, J.R.G.; Reséndiz, J.R.; Prado, M.Á.M.; Miguel, E.E.C. Assessment of jerk performance s-curve and trapezoidal velocity profiles. In Proceedings of the XIII International Engineering Congress, Universidad Autónoma de Queretaro, Santiago de Queretaro, Mexico, 15–19 May 2017; pp. 1–7.
30. Fang, Y.; Hu, J.; Liu, W.; Shaw, Q.; Qi, J.; Peng, Y. Smooth and time-optimal S-curve trajectory planning for automated robots and machines. *Mech. Mach. Theory* **2019**, *137*, 127–153. [CrossRef]
31. Zheng, K.; Hu, Y.; Wu, B. Trajectory planning of multi-degree-of-freedom robot with coupling effect. *Mech. Sci. Technol.* **2019**, *33*, 413–421. [CrossRef]
32. Gasparetto, A.; Zanotto, V. A new method for smooth trajectory planning of robot manipulators. *Mech. Mach. Theory* **2007**, *42*, 455–471. [CrossRef]
33. Xidias, E.K. Time-optimal trajectory planning for hyper-redundant manipulators in 3D workspaces. *Robot. Comput. Integr. Manuf.* **2018**, *50*, 286–298. [CrossRef]
34. Hirakawa, A.; Kawamura, A. Trajectory planning of redundant manipulators for minimum energy consumption without matrix inversion. In Proceedings of the International Conference on Robotics and Automation, Albuquerque, NM, USA, 25 April 1997; Volume 3, pp. 2415–2420.
35. Baghli, F.Z.; El Bakkali, L.; Yassine, L. Optimization of arm manipulator trajectory planning in the presence of obstacles by ant colony algorithm. *Procedia Eng.* **2017**, *181*, 560–567. [CrossRef]
36. Saramago, S.F.P.; Steffen, V. Optimization of the trajectory planning of robot manipulators taking into account the dynamics of the system. *Mech. Mach. Theory* **1998**, *33*, 883–894. [CrossRef]
37. Devendra, G.; Manish, K. Optimization techniques applied to multiple manipulators for path planning and torque minimization. *Eng. Appl. Artif. Intell.* **2002**, *15*, 241–252.
38. Zhang, Y.; Shuzhi, S.G.; Tong, H.L. A unified quadratic-programming-based dynamical system approach to joint torque optimization of physically constrained redundant manipulators. *IEEE Trans. Syst.* **2014**, *34*, 2126–2132. [CrossRef]
39. Ding, H.; Li, Y.F.; Tso, S.K. Dynamic optimization of redundant manipulators in worst case using recurrent neural networks. *Mech. Mach. Theory* **2000**, *35*, 55–70. [CrossRef]
40. Saxena, P.; Stavropoulos, P.; Kechagias, J.; Salonitis, K. Sustainability assessment for manufacturing operations. *Energies* **2020**, *13*, 2730. [CrossRef]
41. Onwubolu, G. *A Comprehensive Introduction to SolidWorks*; SDC Publications: Mission, KS, USA, 2013.

42. Perutka, K. *MATLAB for Engineers—Applications in Control, Electrical Engineering, IT and Robotics*; Intech: Rijeka, Croatia, 2011.

43. Corke, P. *Robotics, Vision & Control: Fundamental Algorithms in MATLAB*, 2nd ed.; Springer: Victoria, Australia, 2017.

44. Goriely, A.; McMillen, T. Shape of a Cracking Whip. *Phys. Rev. Lett.* **2002**, *88*, 244301. [CrossRef]

45. Henrot, C. Characterization of Whip Targeting Kinematics in Discrete and Rhythmic Tasks. Bachelor's Thesis, MIT, Cambridge, MA, USA, 23 June 2016.

46. Krehl, P.; Engemann, S.; Schwenkel, D. The puzzle of whip cracking—Uncovered by a correlation of whip-tip kinematics with shock wave emission. *Shock Waves* **1998**, *8*, 1–9. [CrossRef]

5

Safe pHRI via the Variable Stiffness Safety-Oriented Mechanism (V2SOM): Simulation and Experimental Validations [†]

Younsse Ayoubi , Med Amine Laribi *, Marc Arsicault and Saïd Zeghloul

Dept. GMSC, Pprime Institute, CNRS, University of Poitiers, ENSMA, UPR 3346 Poitiers, France;
you.ayoubi@gmail.com (Y.A.); marc.arsicault@univ-poitiers.fr (M.A.); said.zeghloul@univ-poitiers.fr (S.Z.)
* Correspondence: med.amine.laribi@univ-poitiers.fr
† This paper is an extended version of the paper entitled "New Variable Stiffness Safety Oriented Mechanism for Cobots' Rotary Joints", presented at International Conference on Robotics in Alpe-Adria Danube Region, Patras, Greece, 6–8 June 2018.

Abstract: Robots are gaining a foothold day-by-day in different areas of people's lives. Collaborative robots (cobots) need to display human-like dynamic performance. Thus, the question of safety during physical human–robot interaction (pHRI) arises. Herein, we propose making serial cobots intrinsically compliant to guarantee safe pHRI via our novel designed device, V2SOM (variable stiffness safety-oriented mechanism). Integrating this new device at each rotary joint of the serial cobot ensures a safe pHRI and reduces the drawbacks of making robots compliant. Thanks to its two continuously linked functional modes—high and low stiffness—V2SOM presents a high inertia decoupling capacity, which is a necessary condition for safe pHRI. The high stiffness mode eases the control without disturbing the safety aspect. Once a human–robot (HR) collision occurs, a spontaneous and smooth shift to low stiffness mode is passively triggered to safely absorb the impact. To highlight V2SOM's effect in safety terms, we consider two complementary safety criteria: impact force (ImpF) criterion and head injury criterion (HIC) for external and internal damage evaluation of blunt shocks, respectively. A pre-established HR collision model is built in Matlab/Simulink (v2018, MathWorks, France) in order to evaluate the latter criterion. This paper presents the first V2SOM prototype, with quasi-static and dynamic experimental evaluations.

Keywords: pHRI; variable stiffness actuator; V2SOM; friendly cobots; safety criteria; human–robot collisions

1. Introduction

The currently emerging manufacturing paradigm, known as Industry 4.0, is behind the rethinking of how industrial processes are designed in order to increase their efficiency and flexibility, together with higher levels of automatization [1]. To this end, Industry 4.0 integrates multiple technologies such as the Internet of Things (IoT) [1], artificial intelligence (AI) [1], and cyber-physical systems. Accordingly, robotics researchers are proposing new solutions [2,3] whereby collaborative robots, known as cobots, physically collaborate with well-qualified operators to achieve the goals of this revolution. Hence, the problem of the human subject's safety vs. the robot's high dynamic performances arises. This means that the next generation of widely used cobots should manifest human-friendly attributes. In the literature, two main approaches were proposed to tackle this problem: active impedance control (AIC) and passive compliance (PC). In the former, the impedance [4] is controlled to display safe behavior vis-à-vis the robot's environment, including the humans within it. In the case of a fast human–robot (HR) collision, this approach, because it cannot respond as quickly as within 200 ms [5,6], can allow severe damage to the impacted human [7]. As a result, the PC approach attracted interest due to its instantaneous reaction to any potential impact. This latter, potential impact is defined as a quantification of the maximum impact force a robot can exert in a collision with a stationary object [8]. To emphasize, it is the combination of high mobile inertia and high velocity (i.e., the high kinetic energy) that makes robots dangerous [7,9]. With this in mind, achieving safety without compromising the desired dynamics boils down to reducing the reflected inertia, which is the key feature of the PC approach. Indeed, from a dynamic perspective, integrating passive mechanisms in robot joints decouples a certain colliding inertia from the rest of the robot. This reduces the overall kinetic energy absorbed by the impacted human. In line with this, the series elastic actuator [10] is among the first solutions that has a constant stiffness profile. A passive compliance system composed of purely mechanical elements often provides faster and more reliable responses to dynamic collisions [11]. To enhance the latter's design capacity to react to load variation, a series parallel elastic actuator was introduced by Mathijssen and co-workers [12]. To improve the safety of physical human–robot interaction (pHRI), Zinn et al. presented [13] a distributed macro-mini (DM [2]) actuation system that puts forward low-inertia actuators to interact with the human subject. This allows for both safety and a fast control reaction via the low inertia actuated part. In order to deal with any unsupervised collision while handling variable loads, the concept of the variable stiffness actuator (VSA) emerged. Through the years, several VSAs have been proposed, as discussed in previous studies [3,14]. Their design concepts resulted in different structural paradigms (e.g., serial or antagonistic), different stiffness profiles, and a wide range of power to mass ratios. Herein, a great emphasis is placed on a VSA's stiffness profile vs. the safety aspect. The proposed approach, leading to prototype V2SOM (variable stiffness safety-oriented mechanism) [1], presents the following novelties compared to the literature:

- The stiffness behavior, in the vicinity of zero deflection, is smoothened via a cam follower mechanism.

- The stiffness sharply sinks to maintain, theoretically, as discussed in Section 2, a constant torque threshold in case of a collision.

- The torque threshold, T_{max}, is tunable according to the load variation.

This paper presents the V2SOM design as well as a simulation and experimental validations. V2SOM is primarily conceived to enhance safety in normal working routines, as well as in the case of uncontrolled HR collisions. In Section 2, this aspect is discussed in light of a comparative study between several stiffness profiles. Then, V2SOM's working principle and mechanical structure are illustrated. In Section 3, the impact of V2SOM on human safety is studied in terms of two complementary safety criteria: the head injury criterion (HIC) and impact force criterion (ImpF). The study is carried out via a HR collision model simulation found in previous works [15,16]. The experimental validation of the V2SOM is presented in Section 4. This section includes the quasi-static characterizations as well as the HR collision tests. Section 5 summarizes the outcomes of the present study and gives some future perspectives.

2. Materials and Methods

The VSA's design concept aims to make load-adjustable compliant robots by implementing a variable stiffness mechanism (VSM) in series with the actuation system, as depicted in Figure 1. However, a VSM can simply be described as a tunable spring with a basic nonlinear stiffness profile. With this in mind, we will discuss the properties of the different existing basic stiffness curves and highlight the V2SOM profile that we propose in this work.

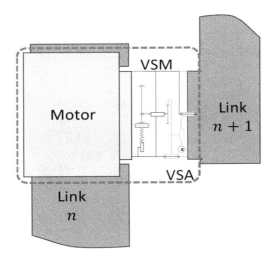

Figure 1. Variable stiffness actuator (VSA) including the variable stiffness safety-oriented mechanism (V2SOM).

2.1. Different VSMs' Basic Stiffness Curves

Previous works on VSAs resulted in design concepts that differ in several aspects, such as mass to volume ratio, elastic energy to mass or volume ratios, working principal, etc. (see [13,17–20] for more details). Herein, we focus on the VSA's stiffness profile. The VSAs mentioned in other studies (see [14,17–20], representative of the current state-of-the-art) can be classified into one of the three categories illustrated in Table 1.

The stiffness profiles of VSAs might be viewed singularly as a tunable basic stiffness profile. These basic stiffness profiles are shown in Figure 2:

- Constant stiffness: this basic stiffness profile is shown in Figure 2a and exemplified in the second column of Table 1 by an actuator with adjustable stiffness (AwAS-II) [19]. At each curve of the AwAS-II characteristic, the stiffness remains practically constant. To adapt this mechanism to a variable load, the torque's slope is tuned via a small motor.
- Biomimetically inspired stiffness, shown in Figure 2b, represents different VSAs, such as the ones developed at the German Aerospace Center (DLR) Institute of Robotics and Mechatronics, the floating spring joint (FSJ) [20], QA-joint (Quasi-Antagonistic Joint) [21], FAS (Flexible Antagonistic Spring) [22] or the one presented by Ayoubi et al. [23]. The third column of Table 1 shows the floating spring joint (FJS) prototype with a stiffness that increases along with the deflection, the same as in biological muscles [23].
- Torque limiter: Park introduced [24,25] the safe joint mechanism (SJM) (see fourth column of Table 1), which is based on a slider-crank mechanism. Its basic stiffness curve corresponds to the one in Figure 2c. This mechanism is supposed to remain stiff under a certain torque level T_1. Upon exceeding it, the stiffness rapidly drops to maintain the torque at T_{max} level, which represents the safety threshold. As the SJM is a slider-crank-based mechanism that is well known for its sensitivity to friction in the vicinity of zero deflection, the safety threshold may be easily exceeded in the case of blunt shocks, as shown by Park and colleagues [26]. For this reason, we introduced the V2SOM basic curve (see Figure 2c,d). In the next section, the full V2SOM characteristic, which allows for coping with load variation, unlike the SJM, is shown. The latter curve presents a finite value stiffness near zero deflection, which continuously drops to display a constant torque threshold T_{max}. The following equation depicts the desired behavior:

$$T_\gamma(\gamma) = T_{max}(1 - e^{-s\gamma}),\qquad(1)$$

where s is a positive constant and the γ elastic deflection angle is in the range of 0 to $\pi/2$.

Table 1. The three main categories of VSAs according to their stiffness characteristic.

Stiffness Profile	Variable-Constant Stiffness	Variable-Biomimetically Inspired Stiffness	Torque Limiter
Illustrative prototype	Actuator with Adjustable Stiffness (AwAS-II) 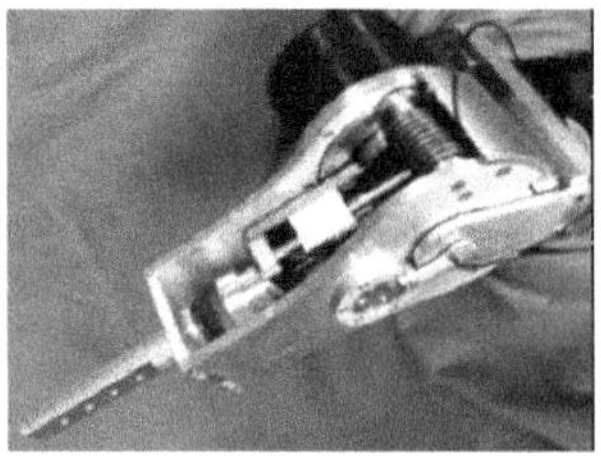	Floating Spring Joint (FSJ) 	Safe Joint Mechanism (SJM) 
Prototype's full characteristic	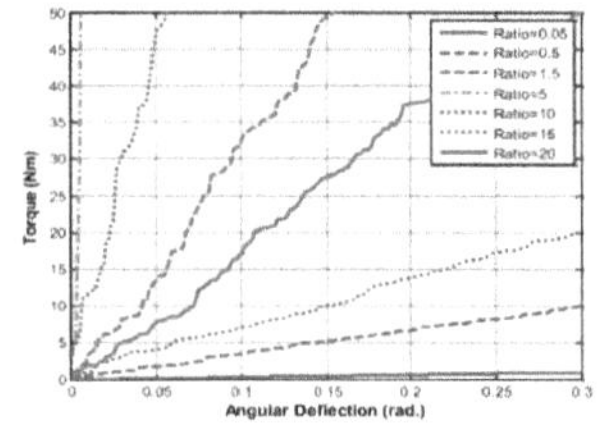	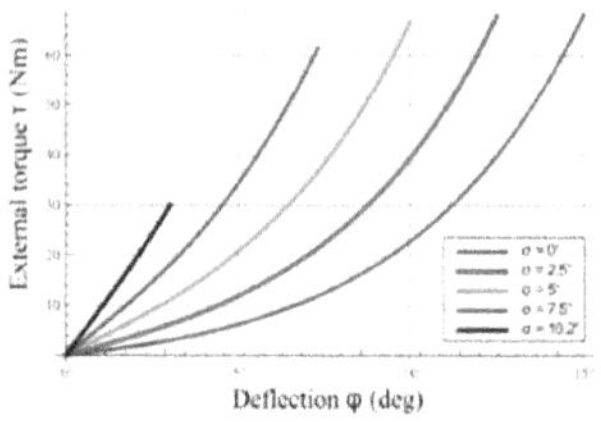	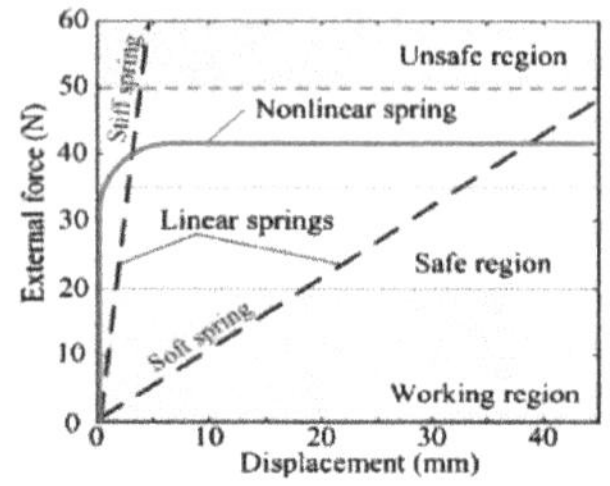

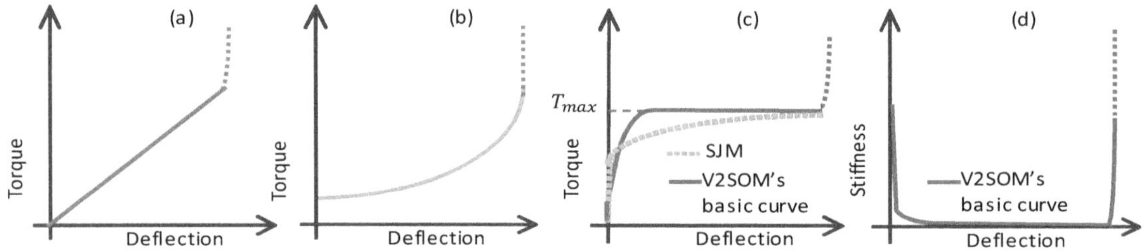

Figure 2. Different basic torque curves and V2SOM stiffness curves: (**a**) Constant stiffness, (**b**) Biomimetically inspired stiffness, (**c,d**) V2SOM basic curves.

To compare the three stiffness profiles, we adopted the following factors:

- Factor 1. The maximal stored elastic energy in "normal working conditions," (i.e., not during collision scenarios). This quantifies how much elastic energy is stored in the VSM that can be unleashed as collision kinetic energy, hence increasing the damaging effect of a potential collision.
- Factor 2. Passive torque limitation: in the case of a fast HR collision, it is more convenient to instantaneously contain the exerted torque with the VSM rather than as a control-based reaction [5,6].
- Factor 3. Gravity-induced elastic deflection: this criterion quantifies the VSM's ability to passively limit this elastic deflection, which reduces the controller's compensation action.

Based on the three factors, we ranked the stiffness profiles in Figure 2 accordingly, where the number of '+' reflects the factor's qualitative value. The results are given in Table 2, which shows that the V2SOM basic curve displays better features in normal working conditions (i.e., factors 1 and 3), and in a collision scenario based on factor 2.

Table 2. Results of a comparative analysis of different VSAs' basic torque curves.

Factors	Profile Figure 2a	Profile Figure 2b	V2SOM Basic Profile
Maximal stored elastic energy	++	+++	+
Passive torque limitation	+ +	+	+ + +
Gravity-induced elastic deflection	+ +	+	+ + +

2.2. Working Principle of V2SOM

V2SOM contains two functional blocks, as depicted in Figure 3a: a nonlinear stiffness generator block (SGB) and a stiffness adjusting block (SAB). To simplify the understanding of the V2SOM kinematic scheme, Figure 3b represents its cross section, which is symmetrical to the rotation axis L_1. The SGB is based on a cam follower mechanism, whereby the cam's rotation γ about the L_5 axis, between $-90°$ and $90°$, induces the translation of its follower according to the slider L_6. Then the follower extends its attached spring. At this level, a deflection angle γ corresponds to a torque value T_γ exerted on the cam. The wide range of this elastic deflection needs to be reduced to a lower range of $-20 \leq \theta \leq 20$, as is widely considered in most VSAs [21,27].

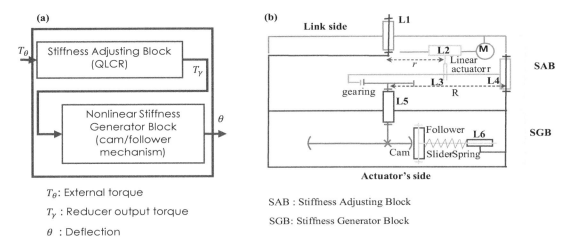

T_θ: External torque

T_γ : Reducer output torque

θ : Deflection

SAB : Stiffness Adjusting Block

SGB: Stiffness Generator Block

Figure 3. Block representation of the V2SOM, (**a**) Two functional blocks of V2SOM (**b**) V2SOM kinematic scheme.

Figure 4 shows various simplified diagrams necessary to understand the functioning principle of this block. In Figure 4a, the cam follower system is in the resting position, meaning that the springs are relaxed and the cam's applied torque about its rotation axis is $T_\gamma = 0$. Applying a torque $T_\gamma \neq 0$ generates a deflection angle γ and the translation of the sliders supporting the followers, which results in the compression of the springs (Figure 3b).

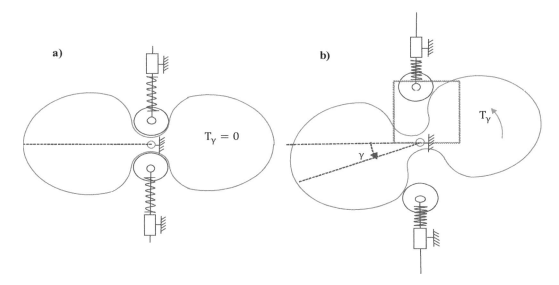

Figure 4. Stiffness generator block: (**a**) at rest $\gamma = 0$ $(T_\gamma = 0)$; (**b**) at deflection $\gamma \neq 0$ (T_γ).

The stiffness adjusting block (SAB) acts as a reducer by using a gear ring system, which is considered to be made up of gear mechanisms commonly used to transform the rotary motion into either rotary or linear motion [28]. Furthermore, the SAB serves as a variable reducer thanks to the linear actuator M, which controls the distance r while driving the gears in a lever-like configuration. The reduction ratio of SAB is continuously tunable, allowing V2SOM to cope with the external load T_θ, where the link side makes a deflection angle θ relative to the actuator side.

Figure 5 shows the symmetrical two ring gears (in blue and green in Figure 5a) geared to a central spur gear (in red). These two ring gears are driven by a symmetrical double lever arm system via two rods (in yellow). These rods, which are part of the lever arm L_2, as illustrated in Figure 4, slide freely along the pocket of the ring gear, creating the prismatic joint shown in Figure 5b. Adjusting the position r of the rods along the lever arm via linear actuator M changes the transmission ratio of the lever arm system, and hence the reduction ratio of the SAB. The gears' rotation induces a variation of the rods' position along the pockets, which is characterized by the distance x in Figure 5b.

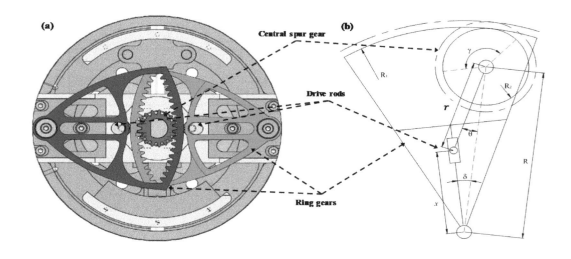

Figure 5. Stiffness adjusting block (SAB) illustration of (**a**) cross section of CAD (computer aided design) model and (**b**) half scheme with a single ring gear.

The V2SOM CAD (computer aided design) model is presented in Figure 6. More details are given in the V2SOM patent [29,30]. Figure 6a shows the two blocks, the stiffness adjustable block and the generator block, with a zoomed-in view. Each block is depicted by additional views of their mechanical parts (Figure 6b–d) with correspondence to the kinematic sketch in Figure 3b. One can identify the linear actuators, the gearing, and cam follower system with its actuation side.

The V2SOM blocks, as shown in Figure 4, are connected rigidly to fulfill separate dedicated tasks:

- The SGB is characterized by the curve of the torque T_γ vs. deflection angle γ. This curve is obtained through the cam profile, the followers, and other design parameters. The basic torque curve leading to the torque characteristic of the V2SOM is depicted in Figure 6a. This basic curve is elaborated with a torque threshold equal to $T_{max} = 2.05$ Nm.

- The SAB is considered a quasi-linear continuous reducer (QLCR) and defined by its ratio expression given in Figure 6b. The ratio is a function of the NL (Nonlinear) factor, the deflection angle γ, and the reducer's tuning parameter r. The NL (Nonlinear) factor is linked to the SAB's internal parameters and can be approximated with a constant when the deflection θ range is between −20 and 20; this process is discussed in the next section.

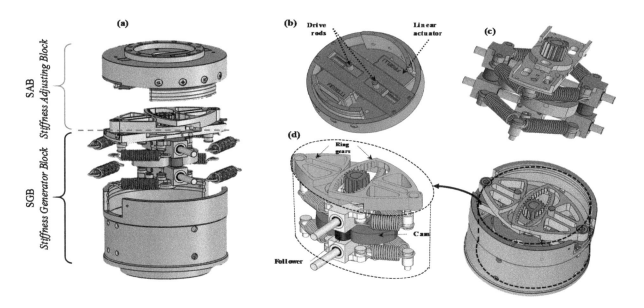

Figure 6. The V2SOM prototype (**a**) Exploded view of V2SOM (**b**) CAD model of SAB (**c**) CAD model of the cam follower mechanism with springs (**d**) CAD model of SGB.

Figure 7c shows the V2SOM characteristics resulting from Figure 7a, with seven increasing reduction ratio settings (seven values of torque tuning). Thanks to the QLCR behavior of the SAB, the curves in Figure 7c follow a formula similar to Equation (1), with the specific tunable constant s and a deflection range.

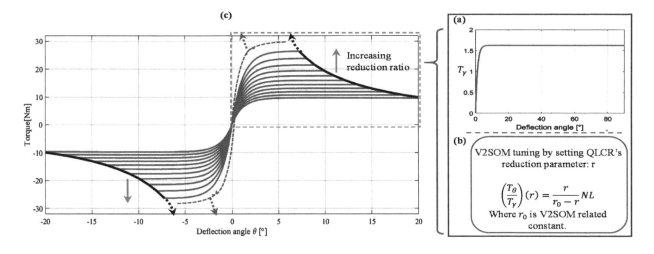

Figure 7. (**a**) Example of V2SOM basic torque curve with (**b**) quasi-linear continuous reducer (QLCR); (**c**) illustration of the V2SOM torque characteristic with seven QLCR settings.

In general, V2SOM has two working modes, between which a transition smoothly takes place in the case of blunt shock, as illustrated in Figure 8. The normal working mode of the V2SOM is the one with linear region, mode (I), which allows the system to avoid a loss of control if the load exceeds the maximum. In case of a collision, mode (II), with the quasi-linear region, is activated. A high stiffness mode (I) is defined within the deflection range $[0, \theta_1]$ and the torque range $[0, T_1]$. The T_1 value defines the torque in normal working conditions. Exceeding this torque value means that the shock absorbing mode is triggered, characterized by a low stiffness that leads to the torque threshold T_{max}. The T_1 torque value, which limits the normal working mode of V2SOM, is an online tunable value. Adjusting this value allows us to cope with possible load variations due to the robot's dynamics.

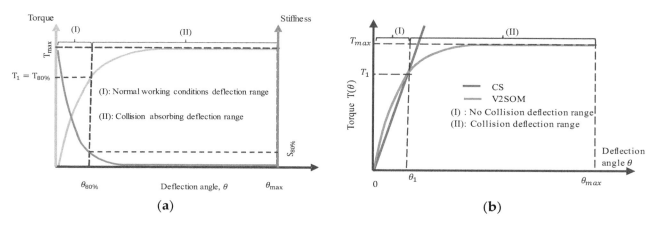

Figure 8. (a) V2SOM working modes; (b) statically equivalent V2SOM and constant stiffness (CS) profiles.

2.3. Understanding Stiffness and Adjusting the Block's Behavior

The QLCR behavior is exemplified in Figure 9, where two kinds of curves are shown:

- Solid curves $\frac{CT_\theta}{T_\gamma}(\gamma, r)$ represent the QLCR reduction ratio, multiplied by a constant C related to the mechanism's parameters.
- Dotted line curves $IR(\gamma, r)$ represent an ideally equivalent reducer (IR), where the reduction ratio is constant while the deflection angle γ changes. The following equation describes this ideal approximation:

$$IR(\gamma, \ r) = \frac{r}{C(r_0 - r)}. \tag{2}$$

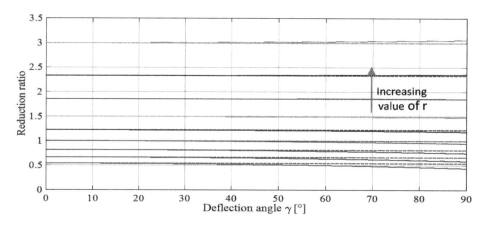

Figure 9. Reduction ratio: QLCR in solid lines: $\frac{CT_\theta}{T_\gamma}(\gamma, r)$ and ideally equivalent reducer (IR) in dotted lines: $IR(\gamma, r)$.

Both curves present the same reduction ratio in the vicinity of zero deflection. This means that QLCR can be considered as an ideal tunable reducer near zero deflection (i.e., not during a collision scenario). Thereupon, QLCR can be tuned using Equation (2). When a collision takes place, the QLCR reduction ratio starts to diverge slightly from the corresponding IR curves. This slight change is not problematic because the function of V2SOM in this phase is to absorb shock energy rather than to precisely set a safety threshold.

At each torque vs. deflection curve of Figure 7 (i.e., for a given reduction ratio r), the stiffness passively varies from a high stiffness value near zero deflection to a practically null stiffness for which the torque attains its threshold; for example, see Figure 10b at the setting $\frac{r}{r_0} = 0.475$.

As for the stiffness modulation in the vicinity of zero deflection angle $\frac{dT_\theta}{d\theta}\Big|_{\theta=0}$, Figure 10a shows the variation range of the presented prototype. As can be seen, the V2SOM's stiffness near zero deflection allows for a wide modulation range that varies from 670 to 13560 Nm/rad for $\frac{r}{r_0} = 0.4 \rightarrow 0.75$. From a theoretical viewpoint, the setting at $r \sim r_0$ gives an infinite stiffness value. However, practically, V2SOM's stiffness is limited by the stiffness of its components. It should be mentioned that the main goal of the V2SOM stiffness profile is to provide a high stiffness in the smaller range (I) to handle a robot's dynamics upon collision (i.e., when it exceeds the tunable threshold, the stiffness drops rapidly and passively to guarantee safe collision absorption).

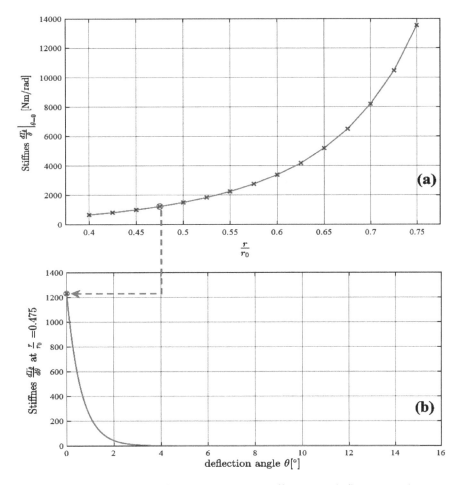

Figure 10. (**a**) Stiffness variation range of the V2SOM; (**b**) stiffness vs. deflection at the setting $\frac{r}{r_0} = 0.475$.

3. Safety Criteria: V2SOM vs. Constant Stiffness

The safety of pHRI is quite problematic in terms of quantification as well as its application to the whole body. However, the abbreviated injury scale (AIS) [31] presents a simple mapping system, from a medical perspective, of different safety criteria, with a unified scale with values ranging from 0 (no injury) to 6 (severe injury or death).

3.1. Safety Criteria

To achieve safe HR collaboration, a level 1 in AIS must be respected. In this work, we consider 0 in AIS as an ergonomic threshold for making human-friendly cobots. ISO/TS-15066 and Newman [32] adopted the most widely considered safety criteria, namely:

- G: The generalized model for brain injury threshold was introduced by Newman (see [33,34]). This index considers both the direction of the impact and the angular accelerations. The G index is valid for 50% of probability of AIS ≥ 3, which does not help to evaluate safe and human-friendly HR collisions, where AIS ≤ 1.

- NIR (new safety index): This index is quite similar to the HIC formulation. While HIC is generic, NIR is specific to the robot's technical data, which are provided in the manufacturer catalogs [34,35]. In our case, HIC is used to include the cover's effect on safety.

- HIC quantifies high accelerations of the brain (concussion) during blunt shocks even for a short amount of time; for example, HIC_{15} less than 15 ms can cause severe, irreversible health effects [35].

- ImpF (also known as contact force) is quite interesting as it can be applied to the whole body. This value is considered for a specific contact surface with a minimum 2.70 cm^2 area.

- Compression criterion (CC) reflects the damaging effect of human–robot (HR) collisions by means of deformation depth, mainly adopted for the naturally compliant chest and belly regions.

Regarding the data in Table 3, the head region is the most critical part of the human body compared to the trunk region, which is more naturally resilient, as indicated by the CC column. The CC criterion is not relevant to the head region as the skull is quite rigid. In contrast, HIC and ImpF are considered for their complementary aspect of HR shock evaluation. HIC is suitable for internal damage evaluation as it quantifies dangerous brain concussions. ImpF is suitable for external damage evaluation. Note that the HIC is only valid in the case of nonclamped head collision scenarios. On the other hand, a constrained head is a dangerous scenario, as shown by Heinzmann and Zelinsky [8]. Thus, a collaborative workspace should be designed, as note ISO/TS15066 permits, in such a way that free head motion is not compromised; this is the first step to guaranteeing safe pHRI.

Table 3. Safety criteria thresholds for most critical body regions from ISO/TS15066 and Payne [31]. ImpF = impact force; AIS = abbreviated injury scale; CC = compression criterion; HIC = head injury criterion.

Body Region		ImpF (N) for AIS ≤ 1	CC (N/mm)	HIC_{15ms} for AIS = 0
Head/Neck	Face	90	75	150
	Neck/sides	190	50	
Trunk	Belly	160	10	

3.2. Human–Robot Collision Model

The human head is the most critical body region when dealing with the safety problems of pHRI. Indeed, some previous works [13,14] have investigated this issue. Furthermore, they proceeded with theoretical modeling of dummy head hardware in crash tests. The resulting model (see Figure 11) has been well-tuned experimentally and validated.

The model in Figure 11 is parameterized according to Wolf et al. [14], with:

- Neck viscoelastic parameters $d_N = 12$ (N·s/m) , $k_N = 3300$ [N/m].
- Head's mass $M_{head} = 5.09$ (Kg) and linear displacement x.
- The contact surface viscoelastic parameters $d_c = 10$ (N·s/m), $k_c = 1500$ (N/m).
- Robot arm contact position l and inertia I_{arm}.
- Rotor inertia I_{rotor}, torque τ_{rotor}, and angular position θ_1.
- VSM's stiffness K and angular deflection $\theta = \theta_1 - \theta_2$.

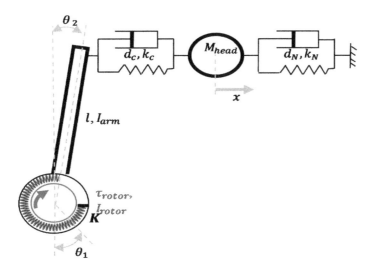

Figure 11. Mechanical model of dummy head hardware collision against a robot arm. Simulation results of human-robot collision.

The collision model allows us to evaluate both ImpF and HIC. The first criterion is directly deduced from the simulation data as the maximum value of the contact force. HIC is evaluated by solving the following optimization problem:

$$
HIC_{15} = \max_{t_1,t_2} \left[\left(\frac{1}{(t_2-t_1)} \int_{t_1}^{t_2} \ddot{x}(t)dt \right)^{2.5} (t_2 - t_1) \right],
$$

$$
\text{Subject to } t_2 - t_1 \leq 15ms,
$$

(3)

where $\ddot{x}(t)$ is the head acceleration value at instant t.

In the next sections, to compare between V2SOM and constant stiffness (CS), a VSM is carried out via a simulation of the HR collision model in the Matlab/Simulink platform. Here, the CS value is set so that the two profiles match at a torque value of 0.8 T_{max}, as shown in Figure 8. This torque value defines the deflection range of the normal operational mode for the V2SOM, after which the shock-absorbing mode is considered to be triggered.

The following simulations are meant to highlight V2SOM's inertia and torque decoupling capacities in comparison to a statistically equivalent CS-based VSM. These capacities represent the robust passive tackling of a HR collision in the fast and critical phase before the collision detection and reaction take place (e.g., within a range of 15 ms), as quantified via the HIC. These simulations were carried out using parametrization of Table 4 to evaluate both safety criteria (HIC and ImpF).

Table 4. Simulation parameters.

Parameter	Figure 13	Figure 14
(I_{rotor}, I_{arm}) (kgm^2)	$(0.0875{\rightarrow}0.525, 0.14)$	$(0.175, 0.14)$
$(\tau_{rotor}, T_{max}, T_1)$ (Nm)	$(10, 15, 12)$	$(7.5{\rightarrow}30, 15, 12)$
c	37	37
$\dot{\theta}_1 \left(\text{rad} \cdot \text{s}^{-1} \right)$	π	π
k_c (N/m)	1500	1500
l (m)	0.6	0.6

3.2.1. Inertia Decoupling

Figure 12 shows that V2SOM presents more than 80% gains on an HIC basis compared to CS. On the other hand, a gain of 10%–40% is noticed for ImpF curves. HIC_{V2SOM} and $ImpF_{V2SOM}$ are steady for a large range of rotor inertia. This property leads us to conclude that V2SOM presents a high inertia decoupling capability compared to a CS-based VSM. Ideally, this characteristic means that the human body, in the case of a HR shock, is subject to only arm-side inertia rather than the heavy resulting arm and rotor inertia.

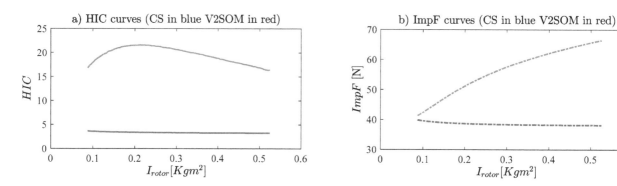

Figure 12. I_{rotor} impact on safety criteria, (**a**) HIC and (**b**) ImpF, for both V2SOM and CS characteristics.

As previously shown by Haddadin et al. [7], lower values of the mobile mass allow for higher velocities to maintain the same safety level. By considering the V2SOM inertia decoupling capability in addition to Haddadin's results, the proposed design allows for better dynamic performance of the cobot without overreaching the safety thresholds.

3.2.2. Torque Decoupling

V2SOM presents a quasi-constant response that can be observed in Figure 13. The large change in motor-applied torque, τ_{rotor}, leads to a significant variation in the HIC and ImpF values of CS-based VSM in comparison with V2SOM, which maintains the same values. The significant variation of 10%–40% in ImpF can be improved with a customized contact surface of the robot arm. However, HIC cannot be reduced and it will be difficult to attenuate the 80% gap between the two responses (Figure 13a). The benefit of V2SOM use, in terms of a reduction in concussions, is visible through HIC mitigation vs. CS use.

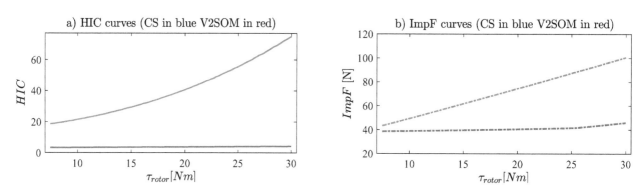

Figure 13. τ_{rotor} impact on safety criteria, (**a**) HIC and (**b**) ImpF, for both V2SOM and CS characteristics.

4. Experimental Validation of V2SOM

In this section a static characterization of the first V2SOM prototype and a preliminary HR collision testing are presented. Figure 14 shows the developed prototype, which is a cylinder 92 mm in diameter and 78 mm in height that weighs about 970 g. A lighter and more compact version is under development, along with a safety-oriented control strategy using V2SOM.

Figure 14. V2SOM first prototype.

4.1. Quasi-Static Characterization of V2SOM

Figure 15 shows the V2SOM characteristics in terms of torque vs. deflection for different settings of the tuning parameter r ranging from 16.6 to 26.6 mm. Quasi-static characterization means the load is applied at a slow rate like a static load [11]. The solid curves represent the theoretical curves, while the experimental curves are represented by dotted lines. The former displays relatively constant torque thresholds, proving that the cam correction brings V2SOM behavior near to its ideal form. The experimental curves deviate slightly from the theoretical ones. This error will be mitigated in upcoming versions by measures such as friction sources analysis and in-depth study of the mechanical parts' deformation. In addition, some parts will be enhanced, such as improving the stiffness profile time change by opting for faster linear actuators. For more details about this prototype, see the table in Appendix A of the mechanical and electrical specifications, which are written according to the recommendations of Grioli et al. [3].

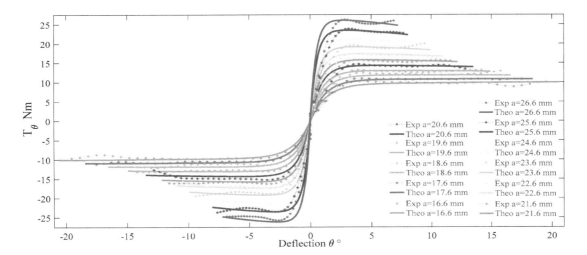

Figure 15. V2SOM characteristics: theoretical (in solid lines) and experimental (in dotted lines) with the 11 different SAB settings.

4.2. Preliminary HR Collision Tests

Currently, a safety-oriented control strategy is in ongoing development, wherein a V2SOM-based cobot functions under preemptive safety conditions in terms of joints' angular velocities. For this purpose, an optimization problem that maximizes the safely reachable angular velocities was formulated by taking into account the actual robot configuration. These constraints result from the simulation and are stored as lookup tables for real-time access.

In this paper, a simple example of tackling a collision with a human subject using V2SOM is demonstrated. The experimental setup is shown in Figure 16.

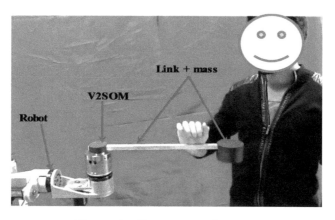

Figure 16. Experimental setup.

In Figure 17, the results of two collision tests are presented. In the first test (Figure 17a,b), the link's inertia is $I_{link} = 0.050$ Kgm^2 with an impact velocity of 90°/s. In the second test, (Figure 17c,d), the link's inertia is $I_{link} = 0.179$ Kgm^2 with an impact velocity of 32°/s. In both cases, V2SOM presents small deflections corresponding to the normal working conditions range (range I; see Figure 8). Upon collision, the safety threshold is exceeded and V2SOM warns of the collision by sending a collision detection message via a CAN (Controller Area Network) bus to the robot's main controller. Hence the robot, through its controller, stops the joint rotation. In the in-between phase (range I and range II in Figure 8), V2SOM passively absorbs the shock's kinetic energy. The elastic end stroke of the V2SOM should not be reached either by a V2SOM maximum storable elastic energy or by a possible slow control reaction. For this reason, both previous features are considered in developing the safety-oriented control strategy.

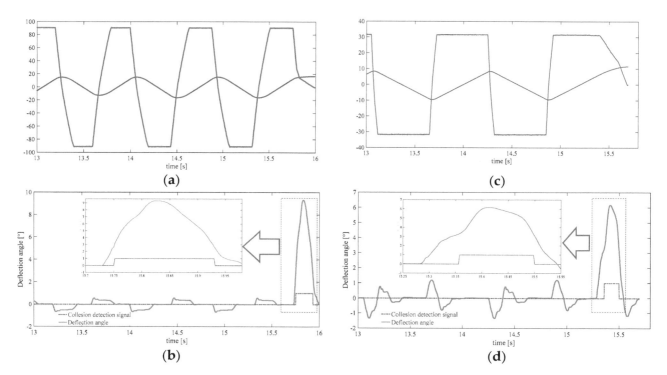

Figure 17. Results of experimental collision tests: (**a**) angular velocity and position of the link in the first test; (**b**) V2SOM deflection angle in the first test; (**c**) angular velocity and position of the link in the second test; (**d**) V2SOM deflection angle in the second test.

The time elapsed between the collision detection (exiting deflection range I) and returning is 180 ms in the first test and 140 ms in the second test.

These illustrative tests are meant first to show the inertia decoupling and torque decoupling capability of the V2SOM and then to shed light on the problem of a collision control strategy that is specifically designed to benefit from the V2SOM properties (e.g., inertia decoupling capacity). This problem is the subject of our upcoming work (for a general review of HR collision control see [4]).

5. Discussion

The simulation allowed us to cover a large range of inertia as well as torque. Based on the obtained results, we can conclude that V2SOM presents a high inertia decoupling capability, which is correlated with the results obtained during the experimentation. Safe physical human–robot interaction is ensured in the case of collision due to the presence of two continuously linked functional modes of V2SOM: high and low stiffness modes. Future work will be focused on the evaluation of the dynamic performance of robots using V2SOM as a way to cope with possible limitations.

6. Conclusions and Future Work

In this paper, we presented V2SOM as the basis of a human-friendly cobot in compliance with safety level 1 on the AIS. V2SOM displays two complementary working modes:

- In normal working conditions, characterized by high stiffness, the proposed mechanism limits the storable elastic energy that increases the absorbed kinetic energy in case of a sudden HR shock. Moreover, this high stiffness limits the elastic deflection to near zero, which eases the robot control.
- In the case of a blunt HR impact, V2SOM stiffness promptly decreases, allowing fast absorption of the collision energy.

From a safety perspective, as shown via the collision simulations, two complementary safety criteria were adopted for internal and external damage evaluation of the human body. Based on these two criteria, HIC and ImpF, important improvements were observed for the V2SOM profile in contrast to an equivalent CS profile. Especially in terms of brain concussion quantification, HIC was reduced by over 70% relative to the results of the CS profile. Moreover, V2SOM presented a high inertia decoupling capacity (i.e., large variation in rotor-side inertia slightly impacts the evaluation of safety on a HIC basis).

The experimental implementation, through the HR collision tests, shows the decoupling capacities of V2SOM. In case of collision, the deflection is detected, and the kinetic energy absorbed, which guarantees safe physical human–robot interaction.

Thus, the present work is focused on an illustrative case study of a one degree-of-freedom robot arm. In future work, different HR collision scenarios will be investigated by considering a generic multi-degrees of freedom (DoFs) robot with V2SOM joints, its dynamics, and the impact location on the robot arm.

7. Patents

WO 2019/043068 A1—7 March 2019: "Mechanical Device with Passive Compliance for Transmitting Rotational Movement," M. Arsicault, Y. Ayoubi, M.A Laribi, S. Zeghloul, and F. Courrèges.

Author Contributions: Conceptualization, Y.A. and M.A.L.; validation, Y.A. and M.A.; designed the experiments and wrote the paper, Y.A., M.A.L., and M.A.; supervised the research work, M.A.L. and S.Z. All authors have read and agreed to the published version of the manuscript.

Acknowledgments: This work was supported by the French National Research Agency, convention ANR-14-CE27-0016. This work was sponsored by the French government research program "Investissements d'avenir" through the Robotex Equipment of Excellence (ANR-10-EQPX-44).

Appendix A

Table A1. V2SOM mechanical and electrical specifications.

	Mechanical			
1	Lowest Safety Threshold Torque		(Nm)	9.7
2	Safety Threshold Variation Time (from nominal level to completely stiff state)	With/Without Load (prone improvement)	(s)	1.6
3	Maximum Stiffness		(Nm/rad)	∞
4	Minimum Stiffness		(Nm/rad)	~0
5	Maximum Elastic Energy		(J)	2.98
6	Maximum Deflection	With Maximum Safety Threshold	(°)	0
		With Minimum Safety Threshold	(°)	20
7	Active Rotation Angle		(°)	±∞
8	Angular Resolution		(°)	0.0313
9	Weight		(Kg)	0.970
	Electrical			
10	Nominal Voltage		(V)	12
11	Nominal Current		(A)	0.010
12	Maximum Current		(A)	0.500
	Control			
13	Voltage Supply		(V)	12
14	Nominal Current		(A)	0.105
15	I/O Protocol		CAN (Controller Area Network) (1 Mbit/s)	

References

1. Lu, Y. Industry 4.0: A survey on technologies, applications and open research issues. *J. Ind. Inf. Integr.* **2017**, *6*, 1–10. [CrossRef]
2. Tobe, F. Why Co-Bots Will Be a Huge Innovation and Growth Driver for Robotics Industry. Available online: https://www.aitrends.com/robotics/why-co-bots-will-be-a-huge-innovation-and-growth-driver-for-robotics-industry/ (accessed on 1 January 2016).
3. Grioli, G.; Wolf, S.; Garabini, M.; Catalano, M.G.; Burdet, E.; Caldwell, D.; Carloni, R.; Friedl, W.; Grebenstein, M.; Laffranchi, M.; et al. Variable stiffness actuators: The user's point of view. *Int. J. Robot. Res.* **2015**, *34*, 727–743. [CrossRef]
4. Haddadin, S.; De Luca, A.; Albu-Schaffer, A. Robot Collisions: A Survey on Detection, Isolation, and Identification. *IEEE Trans. Robot.* **2017**, *33*, 1292–1312. [CrossRef]
5. Zinn, M.; Khatib, O.; Roth, B.; Salisbury, J. Playing it safe. *IEEE Robot. Autom. Mag.* **2004**, *11*, 12–21. [CrossRef]
6. Jianbin, H.; Zongwu, X.; Minghe, J.; Zainan, J.; Hong, L. Adaptive Impedance-controlled Manipulator Based on Collision Detection. *Chin. J. Aeronaut.* **2009**, *22*, 105–112. [CrossRef]
7. Haddadin, S.; Albu-Schaffer, A.; Hirzinger, G. The role of the robot mass and velocity in physical human-robot interaction—Part I: Non-constrained blunt impacts. In Proceedings of the IEEE International Conference on Robotics and Automation, Nice, France, 22–26 September 2008.
8. Heinzmann, J.; Zelinsky, A. Quantitative Safety Guarantees for Physical Human-Robot Interaction. *Int. J. Robot. Res.* **2003**, *22*, 479–504. [CrossRef]
9. Haddadin, S.; Albu-Schaffer, A.; Frommberger, M.; Hirzinger, G. The role of the robot mass and velocity in physical human-robot interaction—Part II: Constrained blunt impacts. In Proceedings of the IEEE International Conference on Robotics and Automation, Nice, France, 19 May 2008.

10. Pratt, G.A.; Williamson, M.M. Series elastic actuators. In Proceedings of the IEEE/RSJ International Conference on Intelligent Robots and Systems. Human Robot Interaction and Cooperative Robots, Pittsburgh, PA, USA, 5–9 August 1995.

11. Corral, E.; García, M.; Castejon, C.; Meneses, J.; Gismeros, R. Dynamic Modeling of the Dissipative Contact and Friction Forces of a Passive Biped-Walking Robot. *Appl. Sci.* **2020**, *10*, 2342. [CrossRef]

12. Mathijssen, G.; Cherelle, P.; Lefeber, D.; VanderBorght, B. Concept of a Series-Parallel Elastic Actuator for a Powered Transtibial Prosthesis. *Actuators* **2013**, *2*, 59–73. [CrossRef]

13. Zinn, M.; Roth, B.; Khatib, O.; Salisbury, J.K. A New Actuation Approach for Human Friendly Robot Design. *Int. J. Rob. Res.* **2004**, *23*, 379–398. [CrossRef]

14. Wolf, S.; Grioli, G.; Eiberger, O.; Friedl, W.; Grebenstein, M.; Hoppner, H.; Burdet, E.; Caldwell, D.G.; Carloni, R.; Catalano, M.G.; et al. Variable Stiffness Actuators: Review on Design and Components. *IEEE/ASME Trans. Mechatron.* **2015**, *21*, 2418–2430. [CrossRef]

15. López-Martínez, J.; García-Vallejo, D.; Giménez-Fernández, A.; Torres-Moreno, J.L. A Flexible Multibody Model of a Safety Robot Arm for Experimental Validation and Analysis of Design Parameters. *J. Comput. Nonlinear Dyn.* **2013**, *9*, 011003. [CrossRef]

16. Hyun, D.; Yang, H.S.; Park, J.; Shim, Y. Variable stiffness mechanism for human-friendly robots. *Mech. Mach. Theory* **2010**, *45*, 880–897. [CrossRef]

17. Tagliamonte, N.L.; Sergi, F.; Accoto, D.; Carpino, G.; Guglielmelli, E. Double actuation architectures for rendering variable impedance in compliant robots: A review. *Mechatronics* **2012**, *22*, 1187–1203. [CrossRef]

18. VanderBorght, B.; Albu-Schaeffer, A.; Bicchi, A.; Burdet, E.; Caldwell, D.; Carloni, R.; Catalano, M.G.; Eiberger, O.; Friedl, W.; Ganesh, G.; et al. Variable impedance actuators: A review. *Robot. Auton. Syst.* **2013**, *61*, 1601–1614. [CrossRef]

19. Jafari, A.; Tsagarakis, N.G.; Caldwell, D. AwAS-II: A new Actuator with Adjustable Stiffness based on the novel principle of adaptable pivot point and variable lever ratio. In Proceedings of the IEEE International Conference on Robotics and Automation, Shanghai, China, 9–13 May 2011.

20. Wolf, S.; Eiberger, O.; Hirzinger, G. The DLR FSJ: Energy based design of a variable stiffness joint. In Proceedings of the 2011 IEEE International Conference on Robotics and Automation, Shanghai, China, 9–13 May 2011.

21. Eiberger, O.; Haddadin, S.; Weiß, M.; Albu-Schaffer, A.; Hirzinger, G. On joint design with intrinsic variable compliance: Derivation of the DLR QA-Joint. In Proceedings of the 2010 IEEE International Conference on Robotics and Automation, Anchorage, Alaska, 3 May 2010.

22. Friedl, W.; Chalon, M.; Reinecke, J.; Grebenstein, M. FAS A flexible antagonistic spring element for a high performance over. In Proceedings of the 2011 IEEE/RSJ International Conference on Intelligent Robots and Systems, San Francisco, CA, USA, 25–30 September 2011.

23. Ayoubi, Y.; Laribi, M.A.; Courreges, F.; Zeghloul, S.; Arsicault, M. A complete methodology to design a safety mechanism for prismatic joint implementation. In Proceedings of the 2016 IEEE/RSJ International Conference on Intelligent Robots and Systems (IROS), Daejeon, Korea, 9–14 October 2016.

24. Lan, N.; Crago, P.E. Optimal control of antagonistic muscle stiffness during voluntary movements. *Boil. Cybern.* **1994**, *71*, 123–135. [CrossRef] [PubMed]

25. Park, J.-J.; Song, J.-B. A Nonlinear Stiffness Safe Joint Mechanism Design for Human Robot Interaction. *J. Mech. Des.* **2010**, *132*, 061005. [CrossRef]

26. Park, J.-J.; Lee, Y.-J.; Song, J.-B.; Kim, H.-S. Safe joint mechanism based on nonlinear stiffness for safe human-robot collision. In Proceedings of the 2008 IEEE International Conference on Robotics and Automation, Pasadena, CA, USA, 19–23 May 2008.

27. Petit, F.; Friedl, W.; Höppner, H.; Grebenstein, M. Analysis and Synthesis of the Bidirectional Antagonistic Variable Stiffness Mechanism. *IEEE/ASME Trans. Mechatron.* **2014**, *20*, 684–695. [CrossRef]

28. Meneses, J.; Garcia-Prada, J.C.; Castejon, C.; Rubio, H.; Corral, E. The kinematics of the rotary into helical gear transmission. *Mech. Mach. Theory* **2017**, *108*, 110–122. [CrossRef]

29. Ayoubi, Y.; Laribi, M.A.; Arsicault, M.; Zeghloul, S.; Courreges, F. Mechanical Device with Variable Compliance for Rotary Motion Transmission. WO 2019/043068 A1, 7 March 2019.

30. Ayoubi, Y.; Laribi, M.A.; Zeghloul, S.; Arsicault, M. Design of V2SOM: The safety mechanism for cobot's rotary joints. In *IFToMM Symposium on Mechanism Design for Robotics*; Springer: Cham, Switzerland, 2018; pp. 147–157.

31. Payne, D.A.R.; Patel, S. Levels Of Consciousness In Relation To Head Injury Criteria. 2001. Available online: http://www.eurailsafe.net/subsites/operas/HTML/appendix/Table13.htm (accessed on 1 January 2017).

32. HIC Tolerance Levels Correlated To Brain Injury. Available online: http://www.eurailsafe.net/subsites/operas/HTML/appendix/Table14.htm (accessed on 3 June 2015).

33. Newman, J.A. A generalized acceleration model for brain injury threshold (GAMBIT). In Proceedings of the International IRCOBI Conference, Zurich, Switzerland, 2–4 September 1986.

34. Alén-Cordero, C.; Carbone, G.; Ceccarelli, M.; Echávarri, J.; Muñoz, J.L. Experimental tests in human–robot collision evaluation and characterization of a new safety index for robot operation. *Mech. Mach. Theory* **2014**, *80*, 184–199. [CrossRef]

35. Gao, D.; Wampler, C.W. Assessing the Danger of Robot Impact. *IEEE Robot. Autom. Mag.* **2009**, *16*, 71–74. [CrossRef]

6

A Novel Fast Terminal Sliding Mode Tracking Control Methodology for Robot Manipulators

Quang Vinh Doan [1], Anh Tuan Vo [2], Tien Dung Le [1,*], Hee-Jun Kang [3] and Ngoc Hoai An Nguyen [2]

[1] The University of Danang—University of Science and Technology, 54 Nguyen Luong Bang street, Danang 550000, Vietnam; dqvinh@dut.udn.vn
[2] Electrical and Electronic Engineering Department, The University of Danang—University of Technology and Education, Danang 550000, Vietnam; voanhtuan2204@gmail.com (A.T.V.); hoaian2206@gmail.com (N.H.A.N.)
[3] School of Electrical Engineering, University of Ulsan, Ulsan 44610, Korea; hjkang@ulsan.ac.kr
*   Correspondence: ltdung@dut.udn.vn

Featured Application: Featured Application: This paper proposed the control synthesis, which can be performed in the trajectory tracking control for various robot manipulators as well as in other mechanical systems, the control of the higher-order system, several uncertain nonlinear systems, or chaotic systems.

Abstract: This paper comes up with a novel Fast Terminal Sliding Mode Control (FTSMC) for robot manipulators. First, to enhance the response, fast convergence time, against uncertainties, and accuracy of the tracking position, the novel Fast Terminal Sliding Mode Manifold (FTSMM) is developed. Then, a Supper-Twisting Control Law (STCL) is applied to combat the unknown nonlinear functions in the control system. By using this technique, the exterior disturbances and uncertain dynamics are compensated more rapidly and more correctly with the smooth control torque. Finally, the proposed controller is launched from the proposed sliding mode manifold and the STCL to provide the desired performance. Consequently, the stabilization and robustness criteria are guaranteed in the designed system with high-performance and limited chattering. The proposed controller runs without a precise dynamic model, even in the presence of uncertain components. The numerical examples are simulated to evaluate the effectiveness of the proposed control method for trajectory tracking control of a 3-Degrees of Freedom (DOF) robotic manipulator.

Keywords: super-twisting control law; robot manipulators; fast terminal sliding mode control

1. Introduction

Robots are gradually replacing people in the fields of social life, manufacturing, exploring, and performing complex tasks. In order to improve productivity, product quality, system reliability, electronics, measurement, and mechanical systems of robotic systems, more advanced designs are required. Therefore, this leads to an increase in the complexity of the structural and mathematical model when there is an additional occurrence of uncertain components.

Sliding Mode Control (SMC) [1–14] is capable of handling high non-linearity and external noise when it possesses outstanding features such as fast response, and robustness towards the existing uncertainties. However, the chattering problem in the SMC causes oscillations in the control input system leading to vibrations in the mechanical system, heat, and even causing instability. Furthermore, the SMC does not yield convergence in a defined period and provides a slow convergence time when it uses a linear sliding manifold. As a result of the linear sliding manifold, the SMC only ensures asymptotic convergence [15]. Therefore, in case that the pressure of a large control force is unavailable, the asymptotic stability status is less likely to converge fast with high-precision control. Terminal Sliding Mode Control (TSMC) [16,17] is proposed to solve convergence problems in finite time, enhancing the transient performance. However, in several situations, TSMC does not offer the desired performance with initial state variables far from the equilibrium point. Additionally, it has not solved the chattering and slow convergence, as well as creating a new problem that is the singularity phenomenon.

To resolve the issues of SMC and TSMC in a synchronized manner, Nonsingular Fast Terminal Sliding Mode Control (NFTSMC) has been evolved successfully in robotic systems [10,12,18–24]. It thoroughly solves the problems, including singularity, convergence in finite-time, and slow convergence. Unfortunately, the chattering has not yet been addressed as these controllers still use a robust reaching control law to deal with uncertain components, and also require the upper limit value of unknown components. For chattering, many solutions have been proposed that can be mentioned, such as the Boundary Layer Method (BLM) [6,25,26], the Adaptive Super-Twisting Method (ASTM) [27], the Second-Order SMC (SOSMC) or the Third-Order SMC (TOSMC) [28–31], the Full-Order SMC (FOSMC) [32–35], and the Fuzzy-SMC (F-SMC) [5,36–38]. For the requirement of the upper limit value of unknown components, many control methods were based on a combination of Neural Networks (NNs), Fuzzy Logic Systems (FLSs), or Adaptive Control Laws (ALCs) with NFTSMC. Although the behavior of unknown components can be well learned by NNs or FLSs. However, it increases the complexity of the control method design because there are more parameters to be adjusted. ACLs are more applicable than the other two methods because they use simple and effective updating rules.

Based on the mentioned analysis, our paper attempts to propose an advanced Fast Terminal Sliding Mode Control (FTSMC) with contributions for robot manipulators: (1) inherits the benefits of the FTSMC and Supper-Twisting Control Law (STCL) in the characteristics of robustness towards the existing uncertainties, finite-time convergence, singularity elimination, estimation capability, and good transient performance; (2) proposes a new Fast Terminal Sliding Mode Manifold (FTSMM) and provides sufficient evidence of finite-time convergence; (3) further improves the precision in the trajectory tracking control; (4) the control torque commands are smooth with less oscillation.

The rest of our paper has the following arrangement. The issue statements are outlined in Section 2. Section 3 presents a synthesis of the designed controller. Continued after Section 3, simulation examples are conducted to assess the influence of the designed controller for a 3-Degrees of Freedom (DOF) robot manipulator in Section 4. Its control performance was then evaluated along with the performance of different control algorithms, including SMC and NFTSMC. Section 5 presents some noteworthy conclusions.

2. Issue Description

Consider the dynamic equation of robot manipulators without the loss of generality:

$$M(\theta)\ddot{\theta} + Q(\theta,\dot{\theta})\dot{\theta} + G(\theta) + f_r(\dot{\theta}) + \delta_d(t) = \tau \qquad (1)$$

where $\theta = \begin{bmatrix} \theta_1, & \ldots, & \theta_n \end{bmatrix}^T \in R^{n\times1}$, $\dot{\theta} = \begin{bmatrix} \dot{\theta}_1, & \ldots, & \dot{\theta}_n \end{bmatrix}^T \in R^{n\times1}$, and $\ddot{\theta} = \begin{bmatrix} \ddot{\theta}_1, & \ldots, & \ddot{\theta}_n \end{bmatrix}^T \in R^{n\times1}$ declare the position angle vector, the velocity angle vector, and the acceleration angle vector, respectively. $M(\theta) = \hat{M}(\theta) + \Delta M(\theta) \in R^{n\times n}$ declares the real inertia matrix, $Q(\theta,\dot{\theta}) = \hat{Q}(\theta,\dot{\theta}) + \Delta Q(\theta,\dot{\theta}) \in R^{n\times n}$ declares the real Coriolis and centrifugal force matrix, and $G(\theta) = \hat{G}(\theta) + \Delta G(\theta) \in R^{n\times n}$ represents the real gravitational matrix. $\hat{M}(\theta) \in R^{n\times n}$ declares the estimated inertia matrix, $\hat{Q}(\theta,\dot{\theta}) \in R^{n\times n}$ represents the estimated Coriolis and centrifugal force matrix, and $\hat{G}(\theta) \in R^{n\times n}$ declares the estimated gravitational matrix. $f_r(\dot{\theta}) \in R^{n\times1}$ and $\delta_d(t) \in R^{n\times1}$ declare the friction vector and disturbance vector, respectively. $\Delta M(\theta) \in R^{n \times n}$ and $\Delta Q(\theta, \dot{\theta}) \in R^{n \times n}$ are the errors of the real dynamic model.

Consequently, the real dynamic equation of robot manipulators is achieved as:

$$\hat{M}(\theta)\ddot{\theta} + \hat{Q}(\theta, \dot{\theta})\dot{\theta} + \hat{G}(\theta) + \Delta U = \tau \tag{2}$$

The lumped uncertain component Δu in Equation (2) is given as:

$$\Delta u = \Delta M(\theta)\ddot{\theta} + \Delta Q(\theta, \dot{\theta})\dot{\theta} + \Delta G(\theta) + f_r + \delta_d(t) \tag{3}$$

Accordingly, the robotic dynamic in Equation (1) is reorganized as:

$$\ddot{\theta} = \hat{M}^{-1}(\theta)\tau - \hat{M}^{-1}(\theta)\left[\hat{Q}(\theta, \dot{\theta})\dot{\theta} + \hat{G}(\theta)\right] - \hat{M}^{-1}(\theta)\Delta u \tag{4}$$

Then, the dynamic Equation (4) can be transferred into the following state-space form as:

$$\begin{cases} \dot{x}_1 = x_2 \\ \dot{x}_2 = \Pi(x,t) + \Phi(x,t)u + D(\theta, \Delta u) \end{cases} \tag{5}$$

where, we set $u = \tau$ as the control input and $x = [x_1, x_2]^T$ as the state vector in which x_1, x_2 correspond to θ, $\dot{\theta} \in R^{n \times 1}$. $\Pi(x,t) = -\hat{M}^{-1}(\theta)\left[\hat{Q}(\theta, \dot{\theta})\dot{\theta} + \hat{G}(\theta)\right]$, $\Phi(x,t) = \hat{M}^{-1}(\theta)$, and $D(\theta, \Delta U) = -\hat{M}^{-1}(\theta)\Delta u$.

The control target of the system is to further increase the response speed and accuracy of the trajectory tracking control for robot manipulators, even if the effects of uncertain dynamics and external perturbations are valid. First, to enhance the response, fast convergence time against uncertainties, and accuracy of the tracking position, the novel FTSMM is developed. Then, STCL is applied to combat the unknown nonlinear functions in the control system. By using this combined technique, the exterior disturbances and uncertain dynamics will be compensated more rapidly and more correctly with the smooth control torque. Finally, the designed controller is launched from the proposed sliding mode manifold and the STCL to obtain the control efficiency.

3. Main Results

The position control error and the velocity control error on each joint are, respectively, defined as follows:

$$x_{ei} = x_{1i} - x_{di} \tag{6}$$

$$x_{dei} = x_{2i} - \dot{x}_{di}; i = 1, \ldots n \tag{7}$$

where $x_d \in R^{n \times 1}$ represents the angle of the expected position.

3.1. The Designed FTSMM

To enhance the response, fast convergence time, and accuracy of the tracking position, the novel FTSMM is developed as follows:

$$s_i = x_{dei} + \frac{2\gamma_1}{1 + e^{-\mu_1(|x_{ei}|-\phi)}}x_{ei} + \frac{2\gamma_2}{1 + e^{\mu_2(|x_{ei}|-\phi)}}|x_{ei}|^\alpha \mathrm{sgn}(x_{ei}) \tag{8}$$

where s_i is the proposed sliding mode manifold, $\gamma_1, \gamma_2, \mu_1, \mu_2$ are the positive constants, $0 < \alpha < 1$, and $\phi = \left(\frac{\gamma_2}{\gamma_1}\right)^{1/(1-\alpha)}$.

Based on the SMC, when the control errors operate in the sliding mode, the following constrain is satisfied [1]:

$$\begin{aligned} s_i &= 0; \\ \dot{s}_i &= 0 \end{aligned} \tag{9}$$

From condition in Equation (9), it is pointed out that:

$$x_{dei} = -\frac{2\gamma_1}{1+e^{-\mu_1(|x_{ei}|-\phi)}}x_{ei} - \frac{2\gamma_2}{1+e^{\mu_2(|x_{ei}|-\phi)}}|x_{ei}|^{\alpha}\mathrm{sgn}(x_{ei}) \tag{10}$$

Remark 1. *When the control error of $|x_{ei}|$ is much greater than ϕ, the first component of Equation (10) offers the role of providing a quick convergence rate and the second component has a smaller role. Contrariwise, when the control error of $|x_{ei}|$ is much smaller than ϕ, the second component of Equation (10) offers a greater role than the first one.*

The following theorem is launched to guarantee that convergence takes place within the defined time.

Theorem 1. *Let us consider dynamic of Equation (10). $x_{ei} = 0$ is defined as the equilibrium point and the state variables of the dynamic of Equation (10), including x_{ei} and x_{dei} stabilize to zero in finite-time.*

Proof. To validate the correctness of Theorem 1, the Lyapunov function candidate is proposed as follows:

$$L_1 = 0.5x_{ei}^2, \tag{11}$$

and its time derivative is

$$\begin{aligned}\dot{L}_1 &= x_{ei}x_{dei} \\ &= -\frac{2\gamma_1}{1+e^{-\mu_1(|x_{ei}|-\phi)}}x_{ei}^2 - \frac{2\gamma_2}{1+e^{\mu_2(|x_{ei}|-\phi)}}|x_{ei}|^{\alpha+1}\mathrm{sgn}(x_{ei}) \\ &< 0\end{aligned} \tag{12}$$

It is shown that $\dot{L}_1 < 0$, hence, x_{ei} and x_{dei} concentrate on the equilibrium state in finite time.
When $|x_{ei}(0)| > \phi$, the sliding motion consists of two phases:
The first phase: $x_{ei}(0) \to |x_{ei}| = \phi$, the first component of Equation (10) offers the role of providing a quick convergence rate and the second component has a smaller role.

$$\begin{aligned}\int_0^{t_1} dt &= \int_{\phi}^{x_{ei}(0)} \frac{1}{\frac{2\gamma_1}{1+e^{-\mu_1(|x_{ei}|-\phi)}}x_{ei}+\frac{2\gamma_2}{1+e^{\mu_2(|x_{ei}|-\phi)}}|x_{ei}|^{\alpha}}d(|x_{ei}|) \\ &< \int_{\phi}^{x_{ei}(0)} \frac{1}{\gamma_1|x_{ei}|}d(|x_{ei}|) = \frac{\ln(|x_{ei}(0)|)-\ln(\phi)}{\gamma_1}\end{aligned} \tag{13}$$

The second phase: $|x_{ei}| = \phi \to x_{ei} = 0$, the second component of Equation (10) offers a role greater than the first one.

$$\begin{aligned}\int_0^{t_2} dt &= \int_0^{\phi} \frac{1}{\frac{2\gamma_1}{1+e^{-\mu_1(|x_{ei}|-\phi)}}x_{ei}+\frac{2\gamma_2}{1+e^{\mu_2(|x_{ei}|-\phi)}}|x_{ei}|^{\alpha}}d(|x_{ei}|) \\ &< \int_0^{\phi} \frac{1}{\gamma_2|x_{ei}|^{\alpha}}d(|x_{ei}|) = \frac{1}{\gamma_2(1-\alpha)}|\phi|^{1-\alpha}\end{aligned} \tag{14}$$

The total time of the sliding motion phase is defined as:

$$T_s = t_1 + t_2 < \frac{\ln(|x_{ei}(0)|)-\ln(\phi)}{\gamma_1} + \frac{1}{\gamma_2(1-\alpha)}|\phi|^{1-\alpha} \tag{15}$$

The state variable of the dynamic in Equation (10) converges to sliding manifold ($s(0) \rightarrow 0$) within the defined time T_r, which was pointed out in [8]. Therefore, the total time for stability on the sliding manifold is computed as:

$$T \leq T_r + T_s \tag{16}$$

3.2. The Designed Control Methodology

Let us take the time derivative of Equation (8):

$$\dot{s} = \dot{x}_{de} + \frac{2\gamma_1}{1+e^{-\mu_1(|x_e|-\phi)}}x_{de} + \frac{2\gamma_1\mu_1 x_{de}\text{sgn}(x_e)e^{-\mu_1(|x_e|-\phi)}}{\left(1+e^{-\mu_1(|x_e|-\phi)}\right)^2}x_e$$
$$+\frac{2\gamma_2\alpha}{1+e^{\mu_2(|x_e|-\phi)}}|x_e|^{\alpha-1}x_{de} - \frac{2\gamma_2\mu_2 x_{de}e^{\mu_2(|x_e|-\phi)}}{\left(1+e^{\mu_2(|x_e|-\phi)}\right)^2}|x_e|^{\alpha} \tag{17}$$

With $\dot{x}_{de} = \ddot{x}_2 - \ddot{x}_d$, the time derivation of Equation (17) gets along with the system in Equation (5) as follows:

$$\dot{s} = \Pi(x,t) + \Phi(x,t)u + D(\theta, \Delta u) - \ddot{x}_d + \frac{2\gamma_1}{1+e^{-\mu_1(|x_e|-\phi)}}x_{de} + \frac{2\gamma_1\mu_1 x_{de}\text{sgn}(x_e)e^{-\mu_1(|x_e|-\phi)}}{\left(1+e^{-\mu_1(|x_e|-\phi)}\right)^2}x_e$$
$$+\frac{2\gamma_2\alpha}{1+e^{\mu_2(|x_e|-\phi)}}|x_e|^{\alpha-1}x_{de} - \frac{2\gamma_2\mu_2 x_{de}e^{\mu_2(|x_e|-\phi)}}{\left(1+e^{\mu_2(|x_e|-\phi)}\right)^2}|x_e|^{\alpha} \tag{18}$$

In order to facilitate controller design, there is the following assumption:

Assumption 1. *The lumped uncertain terms, $D(\theta, \Delta u) = [D_1(\theta, \Delta u), \ldots, D_n(\theta, \Delta u)]$, need to satisfy the following standard condition:*

$$\|D_i(\theta, \Delta u)\| \leq K_i|s_i|^{\frac{1}{2}}; i = 1, \ldots, n \tag{19}$$

where $K_i > 0$.

In order to achieve the stabilization target of the robot system, the following control action is proposed:

$$u = -\Phi^{-1}(x,t)\left(u_{eq} + u_r\right). \tag{20}$$

Here, it should be noted that the u_{eq} is designed as:

$$u_{eq} = \Pi(x,t) - \ddot{x}_d + \frac{2\gamma_1}{1+e^{-\mu_1(|x_e|-\phi)}}x_{de} + \frac{2\gamma_1\mu_1 x_{de}\text{sgn}(x_e)e^{-\mu_1(|x_e|-\phi)}}{\left(1+e^{-\mu_1(|x_e|-\phi)}\right)^2}x_e$$
$$+\frac{2\gamma_2\alpha}{1+e^{\mu_2(|x_e|-\phi)}}|x_e|^{\alpha-1}x_{de} - \frac{2\gamma_2\mu_2 x_{de}e^{\mu_2(|x_e|-\phi)}}{\left(1+e^{\mu_2(|x_e|-\phi)}\right)^2}|x_e|^{\alpha} \tag{21}$$

and u_r is designed as:

$$u_r = \Sigma_1|s|^{\frac{1}{2}}\text{sgn}(s) + \eta$$
$$\dot{\eta} = -\Sigma_2\text{sgn}(s) \tag{22}$$

where $\Sigma_1 = diag(\Sigma_{11}, \ldots, \Sigma_{1n})$ and $\Sigma_2 = diag(\Sigma_{21}, \ldots, \Sigma_{2n})$. Σ_{1i} and Σ_{2i} are assigned to satisfy the following relationship [8]:

$$\begin{cases} \Sigma_{1i} > 2K_i \\ \Sigma_{2i} > \Sigma_{1i}\frac{5K_i\Sigma_{1i}+4K_i^2}{2(\Sigma_{1i}-2K_i)} \end{cases} ; i = 1, 2, \ldots, n \tag{23}$$

Based on those above statements, the following theorems are written to prove the stability problem.

Theorem 2. *Consider the robot system in Equation (1). If the designed torque actions are proposed for system in Equation (1) as Equations (20)–(22), then x_{ei} and x_{dei} stabilize to zero in finite time. That means that robot system in Equation (1) runs in a stable mode.*

Proof. Applying control torque in Equation (20)–(22) to Equation (19) gains:

$$\begin{cases} \dot{s} = D(\theta, \Delta u) - \Sigma_1 |s|^{1/2} \text{sgn}(s) - \eta \\ \dot{\eta} = -\Sigma_2 \text{sgn}(s) \end{cases} \tag{24}$$

Based on the assumption in Equation (19), and the selection condition of the sliding gains in Equation (23), it can be verified that the sliding manifold and its time derivative will converge to zero in finite time. Now, considering one of the elements in Equation (24) as follows:

$$\begin{cases} \dot{s}_i = D_i(\theta, \Delta u) - \Sigma_{1i} |s_i|^{1/2} \text{sgn}(s_i) - \eta_i \\ \dot{\eta}_i = -\Sigma_{2i} \text{sgn}(s_i) \end{cases} \tag{25}$$

The following Lyapunov function is defined for the system in Equation (25):

$$L_2 = \sigma^T \Gamma \sigma \tag{26}$$

Here, $\sigma = \left[s_i^{1/2}, \lambda_i \right]^T$, $\Gamma = \frac{1}{2} \begin{bmatrix} 4\Sigma_{2i} + \Sigma_{1i}^2 & -\Sigma_{1i} \\ -\Sigma_{1i} & 2 \end{bmatrix}$. If $\Sigma_{2i} > 0$, so, according to Rayleigh's inequality:

$$\lambda_{\min}(\Gamma) \|\sigma\|^2 \leq L_2 \leq \lambda_{\max}(\Gamma) \|\sigma\|^2 \tag{27}$$

with $\|\sigma\|^2 = |s_i| + \eta_i^2$.

The time derivation of Equation (26) is:

$$\dot{L}_2 = -\frac{1}{|s_i|^{1/2}} \sigma^T \Phi \sigma + \frac{1}{|s_i|^{1/2}} [D_i(\theta, \Delta u), 0] \Gamma \sigma \tag{28}$$

with $\Phi = \frac{\Sigma_{1i}}{2} \begin{bmatrix} 2\Sigma_{2i} + \Sigma_{1i}^2 & -\Sigma_{1i} \\ -\Sigma_{1i} & 1 \end{bmatrix}$.

Based on the assumption in Equation (19), we can gain:

$$\begin{aligned} \dot{L}_2 &\leq -\frac{1}{|s_i|^{1/2}} \sigma^T \widetilde{\Phi} \sigma \\ &\leq -\frac{1}{|s_i|^{1/2}} \lambda_{\min}\left(\widetilde{\Phi}\right) \|\sigma\|^2 \end{aligned} \tag{29}$$

where $\widetilde{\Phi} = \frac{\Sigma_{1i}}{2} \begin{bmatrix} 2\Sigma_{2i} + \Sigma_{1i}^2 - (4\Sigma_{2i} + \Sigma_{1i})K_i & -(\Sigma_{1i} + 2K_i) \\ -(\Sigma_{1i} + 2K_i) & 1 \end{bmatrix}$.

$\widetilde{\Phi}$ is selected to be greater than zero. Consequently, $\dot{L}_2 < 0$.

Applying the Equation (27) gives:

$$|s_i|^{1/2} \leq \|\sigma\| \tag{30}$$

It follows that:

$$\dot{L}_2 \leq \upsilon L_2^{1/2} \tag{31}$$

with $\upsilon = \frac{\lambda_{\min}\left(\widetilde{\Phi}\right)}{\lambda_{\max}^{1/2}(\Gamma)}$.

According to [8], s_i and $\dot{s}_i$ are equal to zero in finite-time ($t_{ri} = 2L_2^{1/2}(t = 0)/\upsilon$). Therefore, s and $\dot{s}$ equal to zero in finite time ($T_r = \max_{i=1,\ldots,n}\{t_{ri}\}$) and both x_{ei} and x_{dei} also stabilize to equilibrium in finite time ($T \leq T_r + T_s$) under the control action in Equation (20)–(22).

4. Numerical Simulation Studies

In this numerical example, a 3-DOF PUMA560 robot manipulator (with the first three joints and the last three joints blocked) is adopted. The MATLAB/SIMULINK software (2019a MATLAB Version of The MathWorks, Inc. 3 Apple Hill Drive Natick, MA 01760 USA) was used for all computation, the sampling time was set to 10^{-3} s, and the solver ode3 was used. The kinematic description for the robot system is displayed in Figure 1. The design parameters and dynamic models of the robot system are referenced from the document [39]. There are many essential parameters of a robot that need to be presented. Therefore, to present briefly, the design parameters and dynamic models of the robot system are reported in [39].

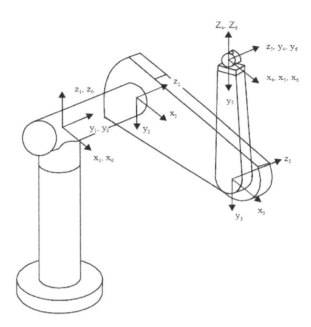

Figure 1. The kinematic description of 3-Degrees of Freedom (DOF) PUMA560 robot manipulator.

To explore the potential of our designed approach, the robot is controlled to follow the designated trajectory configuration at first. Later, its control performance is then evaluated and compared with the performance of different control algorithms, including SMC and NFTSMC. These control methods for comparison are briefly explained as follows:

The normal SMC [14] has the following control torque:

$$u = -\Phi(x,t)^{-1}\left[\Pi(x,t) + c\left(x_2 - \dot{x}_d\right) - \ddot{x}_d + (\Sigma + \xi)\mathrm{sgn}(s)\right] \tag{32}$$

where $s = x_{de} + cx_e$ is the linear sliding manifold, c is a positive constant.

Further, the NFTSMC [40] has the following control torque:

$$u(t) = -\Phi(x,t)^{-1}\left[\Pi(x,t) + \omega\frac{q}{l}x_{de}^{2-l/q} - \ddot{x}_d + (\Sigma + \xi)\mathrm{sgn}(s)\right] \tag{33}$$

where $s = x_e + \omega^{-1}x_{de}^{l/q}$ is a nonlinear sliding manifold.

The designed parameters of three control methodologies are given in Table 1.

Table 1. Parameter values for three different control methodologies.

Control Schemes	Parameters	Values
SMC	c	2
	ξ, Σ	$0.01, 20$
NFTSMC	l, q, ω	$5, 3, 2$
	ξ, Σ	$0.01, 20$
Proposed FTSMC	$\gamma_1, \gamma_2, \mu_1, \mu_2$	$1, 1, 1.2, 1.4$
	$\phi, \alpha, \Sigma_1, \Sigma_2$	$1, 0.6, 40, 50$

The designated trajectory configuration for position tracking when the robot manipulator operates:

$$\theta_d = \begin{cases} \theta_{d1} = 0.6 + \cos\left(\frac{t}{6\pi}\right) - 1 \\ \theta_{d2} = -0.6\sin\left(\frac{t}{6\pi} + 0.5\pi\right) - 1 \\ \theta_{d3} = 0.6 + \sin\left(0.2\frac{t}{\pi} + 0.5\pi\right) - 1 \end{cases} \tag{34}$$

Friction and disturbance models are hypothesized to analyze the strong capability of the designed FTSMC. It is not amenable to accurately calculate these friction and disturbance terms; therefore, the physical values of frictions and disturbances are not measured. Therefore, the following friction forces and disturbances were modeled, respectively:

$$f_r(\dot{\theta}) = \begin{cases} f_{r1} = 2.3\dot{\theta}_1 + 2.14\mathrm{sgn}(3\dot{\theta}_1) \\ f_{r2} = 4.5\dot{\theta}_2 + 2.35\mathrm{sgn}(2\dot{\theta}_2) \\ f_{r3} = 1.7\dot{\theta}_3 + 1.24\mathrm{sgn}(2\dot{\theta}_3) \end{cases} \tag{35}$$

$$\delta_d(t) = \begin{cases} \delta_{d1}(t) = 7.6\sin(\dot{\theta}_1) \\ \delta_{d2}(t) = 6.6\sin(\dot{\theta}_2) \\ \delta_{d3}(t) = 4.23\sin(\dot{\theta}_3) \end{cases} \tag{36}$$

To clearly present the results within the simulation period and to facilitate easier comparison, the averaged tracking error i:

$$E_j = \sqrt{\frac{1}{N}\sum_{i=1}^{N}\left(\|e_j(k)\|^2\right)}; \; j = 1, 2, 3 \tag{37}$$

where Z is the number of simulation steps.

To demonstrate the superiority of the designed controller, the average control error is calculated over two different simulation periods (10 s and 30 s).

The averaged tracking errors are reported in Table 2.

Table 2. The averaged tracking errors under the input of the controllers.

Error Control System	E_1 (10 s)	E_2 (10 s)	E_3 (10 s)	E_1 (30 s)	E_2 (30 s)	E_3 (30 s)
SMC	0.03102	0.01946	0.03113	0.01077	0.00671	0.01082
NFTSMC	0.02322	0.01284	0.02208	0.00774	0.00428	0.00737
Proposed Controller	0.01939	0.01037	0.01793	0.00646	0.00345	0.00597

The designated trajectory configuration and real trajectory under three control methods at the first three joints are displayed in Figure 2. It can be seen from Figure 2 that all three controllers appear to have a similar tracking control performance. However, they have different convergence times in the following order: the designed controller has the fastest convergence time among all three control methods, and NFTSMC has faster convergence time than the normal SMC.

Figures 3 and 4 show the position control errors and the velocity control errors, respectively. It can be seen from Figure 3 and Table 2, the position control errors of the designed control scheme are relatively small compared to those of the other control methods, in the order of $10^{-7}rad$. The position control errors of the NFTSMC are in the order of $10^{-6}rad$. SMC provides the largest position control errors of the three control methods, in the order of $10^{-4}rad$.

From Figure 4, it is seen that the designed control method also has the smallest velocity control errors among all the three control methods.

The control torque for all three control manners, including SMC, NFTSMC, and the designed FTSMC, are displayed in Figure 5. It can be recognized from Figure 5, SMC and NFTSMC have discontinuous control torque because of using the high-frequency control law. Meanwhile, the designed system has smooth control torque with a significant elimination of the chattering phenomena. To achieve this goal, the suggested controller applies STCL to substitute the high-frequency control law in removing chattering behavior.

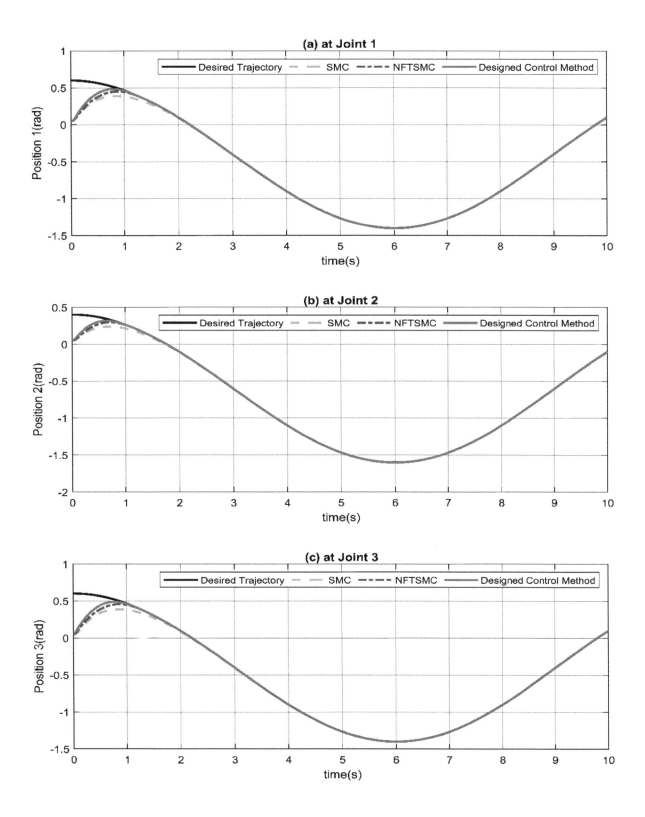

Figure 2. The designated trajectory configuration and real trajectory under three control methods: (**a**) at Joint 1, (**b**) at Joint 2, and (**c**) at Joint 3.

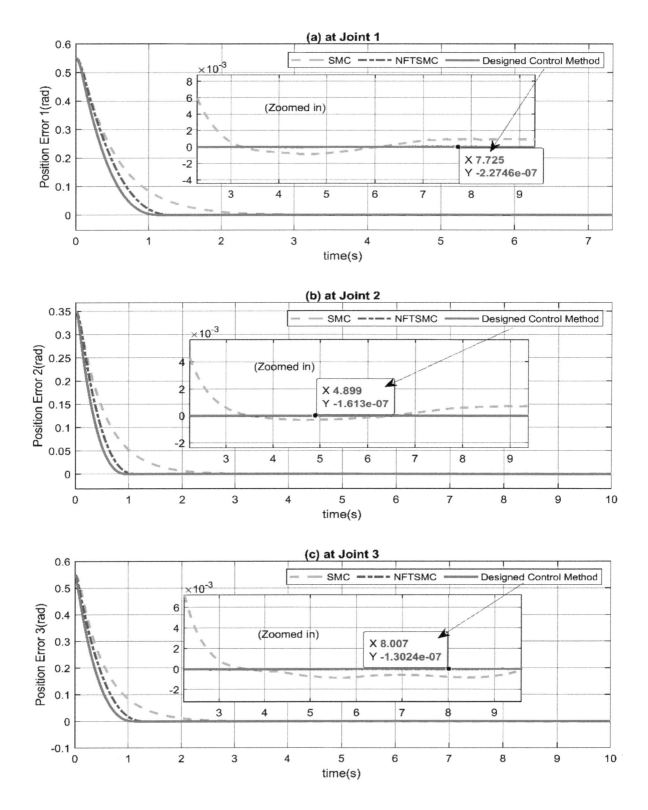

Figure 3. The position control errors: (**a**) at Joint 1, (**b**) at Joint 2, and (**c**) at Joint 3.

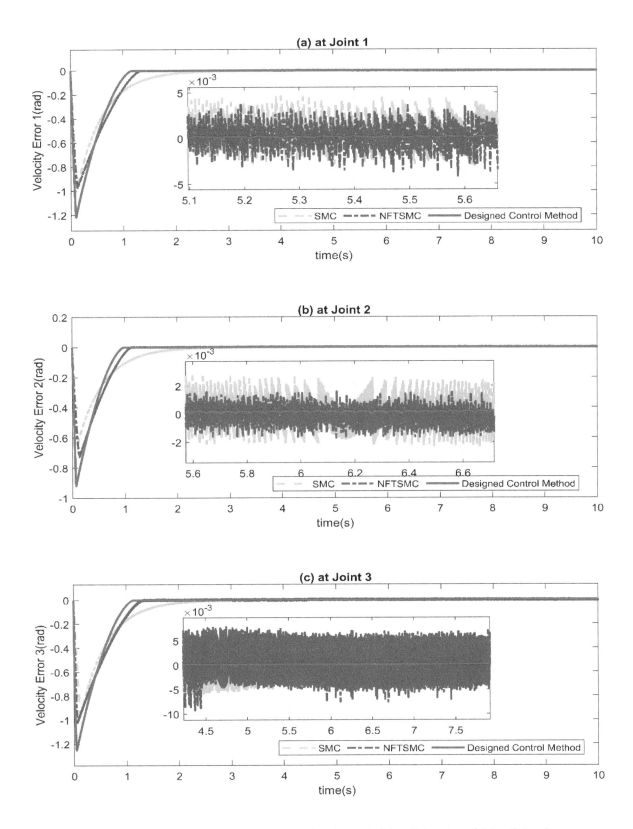

Figure 4. The velocity control errors: (**a**) at Joint 1, (**b**) at Joint 2, and (**c**) at Joint 3.

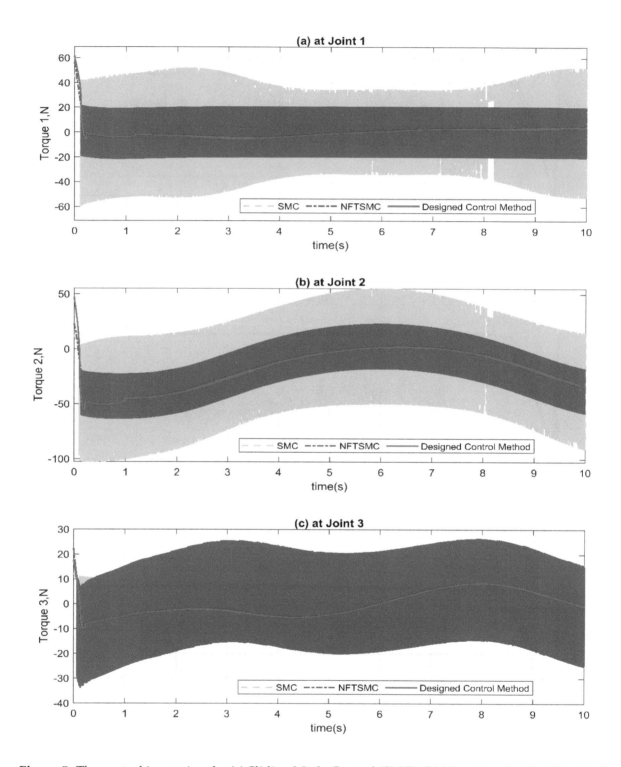

Figure 5. The control input signals: (**a**) Sliding Mode Control (SMC), (**b**) Non-singular Fast Terminal Sliding Mode Control (NFTSMC), and (**c**) designed Fast Terminal Sliding Mode Control (FTSMC).

Response time of the sliding mode manifolds, including SMC, NFTSMC, and designed FTSMC, are shown in Figure 6.

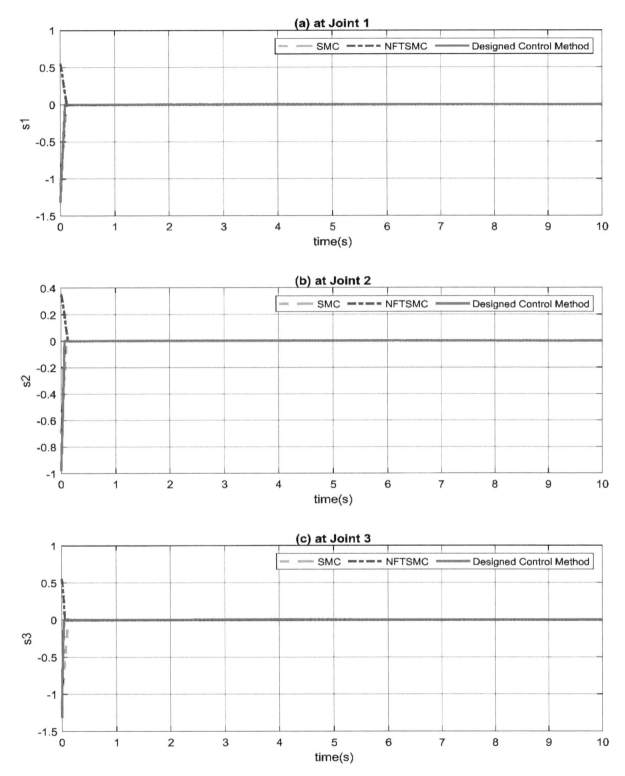

Figure 6. Response time of the sliding mode manifolds: (**a**) at Joint 1, (**b**) at Joint 2, and (**c**) at Joint 3.

5. Conclusions

This paper focuses on designing a novel FTSMC for robot manipulators. In the first step, the novel FTSMM is developed to enhance response capability, fast convergence time, uncertainties opposition, and especially, improve the accuracy of the tracking position. To alleviate unknown nonlinear parameters in the control system, STCL is then applied. Thanks to this valuable technique, exterior disturbances and nonlinear elements are compensated more rapidly and more correctly with the smooth control torque. Finally, combining STCL and our proposed sliding mode manifold, under the flexible controller, the stability and robustness of the control system are guaranteed with high-performance and limited chattering. To evaluate the efficiency, a simulation example is performed for the trajectory tracking control of a 3-DOF robotic manipulator.

From theoretical evidence, simulation results, and a comparison with SMC and NFTSMC, our proposed controller has some of the following contributions: (1) the proposed controller provides finite-time convergence and faster transient performance without singularity problem in controlling; (2) the proposed controller inherits the benefits of the FTSMC and CRCL in the characteristics of robustness towards the existing uncertainties; (3) a new FTSMM was proposed, and evidence of finite-time convergence was sufficiently proved; (4) the precision of the proposed controller was further improved in the trajectory tracking control; (5) the proposed controller shows the smoother control torque commands with lesser oscillation.

Author Contributions: Conceptualization, Q.V.D., A.T.V., and T.D.L.; methodology, Q.V.D. and T.D.L.; software, A.T.V.; validation, Q.V.D., A.T.V., and N.H.A.N.; formal analysis, Q.V.D., A.T.V., and H.-J.K.; investigation, T.D.L. and N.H.A.N.; resources, A.T.V. and T.D.L.; data curation, H.-J.K. and T.D.L.; writing—original draft preparation, Q.V.D., A.T.V., and N.H.A.N.; writing—review and editing, Q.V.D., H.-J.K., and N.H.A.N.; visualization, A.T.V.; supervision, T.D.L.; project administration, T.D.L. and Q.V.D.; funding acquisition, T.D.L. and Q.V.D. All authors have read and agreed to the published version of the manuscript.

References

1. Shtessel, Y.; Edwards, C.; Fridman, L.; Levant, A. *Sliding Mode Control and Observation*; Springer: Berlin/Heidelberg, Germany, 2014.
2. Young, K.D.; Utkin, V.I.; Ozguner, U. A control engineer's guide to sliding mode control. *IEEE Trans. Control Syst. Technol.* **1999**, *7*, 328–342. [CrossRef]
3. Islam, S.; Liu, X.P. Robust sliding mode control for robot manipulators. *IEEE Trans. Ind. Electron.* **2011**, *58*, 2444–2453. [CrossRef]
4. Ferrara, A.; Incremona, G.P. Design of an integral suboptimal second-order sliding mode controller for the robust motion control of robot manipulators. *IEEE Trans. Control Syst. Technol.* **2015**, *23*, 2316–2325. [CrossRef]
5. Roopaei, M.; Jahromi, M.Z. Chattering-free fuzzy sliding mode control in MIMO uncertain systems. *Nonlinear Anal. Theory Methods Appl.* **2009**, *71*, 4430–4437. [CrossRef]
6. Utkin, V. Discussion aspects of high-order sliding mode control. *IEEE Trans. Automat. Contr.* **2016**, *61*, 829–833. [CrossRef]
7. Perruquetti, W.; Barbot, J.-P. *Sliding Mode Control in Engineering*; CRC Press: Boca Raton, FL, USA, 2002.
8. Moreno, J.A.; Osorio, M. A Lyapunov approach to second-order sliding mode controllers and observers. In Proceedings of the 2008 47th IEEE Conference on Decision and Control, Cancun, Mexico, 9–11 December 2008; pp. 2856–2861.
9. Qi, Z.; McInroy, J.E.; Jafari, F. Trajectory tracking with parallel robots using low chattering, fuzzy sliding mode controller. *J. Intell. Robot. Syst.* **2007**, *48*, 333–356. [CrossRef]
10. Vo, A.T.; Kang, H.-J. An Adaptive Neural Non-Singular Fast-Terminal Sliding-Mode Control for Industrial Robotic Manipulators. *Appl. Sci.* **2018**, *8*, 2562. [CrossRef]
11. Vo, A.T.; Kang, H.-J.; Le, T.D. An Adaptive Fuzzy Terminal Sliding Mode Control Methodology for Uncertain Nonlinear Second-Order Systems. In Proceedings of the International Conference on Intelligent Computing, Chennai, India, 2–3 February 2018; pp. 123–135.
12. Kamal, S.; Moreno, J.A.; Chalanga, A.; Bandyopadhyay, B.; Fridman, L.M. Continuous terminal sliding-mode controller. *Automatica* **2016**, *69*, 308–314. [CrossRef]
13. Moreno, J.A.; Negrete, D.Y.; Torres-González, V.; Fridman, L. Adaptive continuous twisting algorithm. *Int. J. Control* **2016**, *89*, 1798–1806. [CrossRef]
14. Edwards, C.; Spurgeon, S. *Sliding Mode Control: Theory and Applications*; CRC Press: Boca Raton, FL, USA, 1998.
15. Lee, H.; Kim, E.; Kang, H.-J.; Park, M. A new sliding-mode control with fuzzy boundary layer. *Fuzzy Sets Syst.* **2001**, *120*, 135–143. [CrossRef]

16. Mu, C.; Xu, W.; Sun, C. On switching manifold design for terminal sliding mode control. *J. Franklin Inst.* **2016**, *353*, 1553–1572. [CrossRef]
17. Tan, C.P.; Yu, X.; Man, Z. Terminal sliding mode observers for a class of nonlinear systems. *Automatica* **2010**, *46*, 1401–1404. [CrossRef]
18. Zhang, F. High-speed nonsingular terminal switched sliding mode control of robot manipulators. *IEEE/CAA J. Autom. Sin.* **2017**, *4*, 775–781. [CrossRef]
19. Vo, A.T.; Kang, H. An Adaptive Terminal Sliding Mode Control for Robot Manipulators with Non-singular Terminal Sliding Surface Variables. *IEEE Access* **2018**, *7*, 7801–8712. [CrossRef]
20. Vo, A.T.; Kang, H. A Chattering-Free, Adaptive, Robust Tracking Control Scheme for Nonlinear Systems with Uncertain Dynamics. *IEEE Access* **2019**, *7*, 10457–10466. [CrossRef]
21. Tuan, V.A.; Kang, H.-J. A New Finite-time Control Solution to The Robotic Manipulators Based on the Nonsingular Fast Terminal Sliding Variables and Adaptive Super-Twisting Scheme. *J. Comput. Nonlinear Dyn.* **2019**, *14*, 031002. [CrossRef]
22. Nojavanzadeh, D.; Badamchizadeh, M. Adaptive fractional-order non-singular fast terminal sliding mode control for robot manipulators. *IET Control Theory Appl.* **2016**, *10*, 1565–1572. [CrossRef]
23. Mobayen, S.; Baleanu, D.; Tchier, F. Second-order fast terminal sliding mode control design based on LMI for a class of non-linear uncertain systems and its application to chaotic systems. *J. Vib. Control* **2017**, *23*, 2912–2925. [CrossRef]
24. Boukattaya, M.; Mezghani, N.; Damak, T. Adaptive nonsingular fast terminal sliding-mode control for the tracking problem of uncertain dynamical systems. *ISA Trans.* **2018**, *77*, 1–19. [CrossRef]
25. Edelbaher, G.; Jezernik, K.; Urlep, E. Low-speed sensorless control of induction machine. *IEEE Trans. Ind. Electron.* **2006**, *53*, 120–129. [CrossRef]
26. Li, H.; Dou, L.; Su, Z. Adaptive nonsingular fast terminal sliding mode control for electromechanical actuator. *Int. J. Syst. Sci.* **2013**, *44*, 401–415. [CrossRef]
27. Wang, Y.; Chen, J.; Yan, F.; Zhu, K.; Chen, B. Adaptive super-twisting fractional-order nonsingular terminal sliding mode control of cable-driven manipulators. *ISA Trans.* **2019**, *86*, 163–180. [CrossRef] [PubMed]
28. Van, M.; Franciosa, P.; Ceglarek, D. Fault diagnosis and fault-tolerant control of uncertain robot manipulators using high-order sliding mode. *Math. Probl. Eng.* **2016**, *2016*, 7926280. [CrossRef]
29. Van, M.; Ge, S.S.; Ren, H. Robust fault-tolerant control for a class of second-order nonlinear systems using an adaptive third-order sliding mode control. *IEEE Trans. Syst. Man, Cybern. Syst.* **2017**, *47*, 221–228. [CrossRef]
30. Levant, A. Higher-order sliding modes, differentiation and output-feedback control. *Int. J. Control* **2003**, *76*, 924–941. [CrossRef]
31. Rubio-Astorga, G.; Sánchez-Torres, J.D.; Cañedo, J.; Loukianov, A.G. High-order sliding mode block control of single-phase induction motor. *IEEE Trans. Control Syst. Technol.* **2014**, *22*, 1828–1836. [CrossRef]
32. Feng, Y.; Han, F.; Yu, X. Chattering free full-order sliding-mode control. *Automatica* **2014**, *50*, 1310–1314. [CrossRef]
33. Feng, Y.; Zhou, M.; Zheng, X.; Han, F.; Yu, X. Full-order terminal sliding-mode control of MIMO systems with unmatched uncertainties. *J. Franklin Inst.* **2018**, *355*, 653–674. [CrossRef]
34. Song, Z.; Duan, C.; Wang, J.; Wu, Q. Chattering-free full-order recursive sliding mode control for finite-time attitude synchronization of rigid spacecraft. *J. Franklin Inst.* **2018**, *356*, 998–1020. [CrossRef]
35. Xiang, X.; Liu, C.; Su, H.; Zhang, Q. On decentralized adaptive full-order sliding mode control of multiple UAVs. *ISA Trans.* **2017**, *71*, 196–205. [CrossRef]
36. Li, H.; Wang, J.; Wu, L.; Lam, H.-K.; Gao, Y. Optimal Guaranteed Cost Sliding-Mode Control of Interval Type-2 Fuzzy Time-Delay Systems. *IEEE Trans. Fuzzy Syst.* **2018**, *26*, 246–257. [CrossRef]
37. Van, M. An Enhanced Robust Fault Tolerant Control Based on an Adaptive Fuzzy PID-Nonsingular Fast Terminal Sliding Mode Control for Uncertain Nonlinear Systems. *IEEE/ASME Trans. Mechatron.* **2018**, *23*, 1362–1371. [CrossRef]
38. Duc, T.M.; Van Hoa, N.; Dao, T.-P. Adaptive fuzzy fractional-order nonsingular terminal sliding mode control for a class of second-order nonlinear systems. *J. Comput. Nonlinear Dyn.* **2018**, *13*, 31004. [CrossRef]
39. Armstrong, B.; Khatib, O.; Burdick, J. The explicit dynamic model and inertial parameters of the PUMA 560 arm. In Proceedings of the 1986 IEEE International Conference on Robotics and Automation, San Francisco, CA, USA, 7–10 April 1986; Volume 3, pp. 510–518.
40. Feng, Y.; Yu, X.; Man, Z. Non-singular terminal sliding mode control of rigid manipulators. *Automatica* **2002**, *38*, 2159–2167. [CrossRef]

7

Energy-Efficiency Improvement and Processing Performance Optimization of Forging Hydraulic Presses Based on an Energy-Saving Buffer System

Xiaopeng Yan and Baijin Chen *

State Key Laboratory of Materials Processing and Die & Mould Technology, School of Materials Science and Engineering, Huazhong University of Science and Technology, Wuhan 430074, China; yanxp@hust.edu.cn
* Correspondence: chenbaijin@hust.edu.cn

Abstract: This paper proposes an energy-saving system based on a prefill system and a buffer system to improve the energy efficiency and the processing performance of hydraulic presses. Saving energy by integrating such systems into the cooling system of a hydraulic press has not been previously reported. A prefill system, powered by the power unit of the cooling system, is used to supply power simultaneously with the traditional power unit during the pressurization stage, thus reducing the usage of pumps and installed power of the hydraulic press. In contrast to the traditional prefill system, the proposed energy-saving system is controlled by a servo valve to adjust flow according to the load profile. In addition, a buffer system is employed to the cooling system to absorb the hydraulic shock generated at the unloading stage, store those shares of hydraulic energy as a recovery accumulator, and then release this energy to power the prefill system and the hydraulic actuator in the subsequent productive process. Finally, through a series of comparative experiments, it was preliminarily validated that the proposed system could reduce the installed power and pressure shock by up to 22.85% and 41%, respectively, increase energy efficiency by up to 26.71%, and provide the same processing characteristics and properties as the traditional hydraulic press.

Keywords: hydraulic press; energy saving; energy efficiency; installed power; processing performance

1. Introduction

Hydraulic presses are commonly used for forging, molding, blanking, punching, deep drawing, and other metal forming operations because of their high load capacity, high power-to-mass ratio, and large force/torque output capacity. However, they are also known for their high energy consumption, low energy efficiency, and poor processing characteristics. As a result of the significant difference between the installed power and the demanded power, 70% of the total energy consumption is attributed to power dissipation and actuation of the hydraulic system. Therefore, improving the energy efficiency of hydraulic presses is an urgent issue that manufacturing industries must resolve [1–5].

Energy dissipation generates from each part of the hydraulic press system when motion or power are transmitted, as shown in Figure 1. It is estimated that only 9.32% of the input energy is transmitted into the forming energy. To better study the energy-saving methods for hydraulic presses, it is necessary to be clear on the energy consumption characteristics of hydraulic system. For this purpose, Zhao [6] established the basic energy flow model of the hydraulic press system, revealed the energy dissipation mechanism of each component, and indicated that the imbalance between installed power and demanded power is the main cause of low energy efficiency. Installed power is designed to meet the maximum power requirements of PS. However, as the same power unit also serves other low-power operations, mismatch between installed power and demanded power occurs.

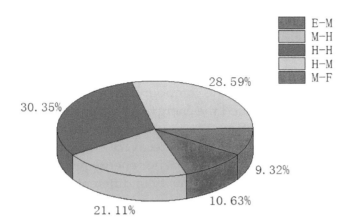

Figure 1. Energy dissipation of each energy conversion unit during a working cycle. (E-M, electrical–mechanical energy unit; M-H, mechanical–hydraulic energy unit; H-H, hydraulic–hydraulic energy unit; H-M, hydraulic–mechanical energy unit; M-F, mechanical-forming energy unit).

During the past decades, researchers are applying an increasing amount of effort to achieve the match between the installed power and the required power for hydraulic presses. As reported by Zhao [6] and Huang [7], energy-matching methods for hydraulic presses can be divided into two categories: energy-matching methods and energy-recovery methods.

A common approach based on an energy-matching mechanism is the volume control electrohydraulic system driven directly by various kinds of variable-speed motors and variable-displacement pumps. The control of pressure, flow, and direction of working liquid is achieved by changing the rotation speed or output displacement. Quan and Helduser [8] applied variable-speed motors and constant-displacement pumps in hydraulic drive units to match load variations. Su et al. [9] applied variable-frequency motors in the hydraulic press to reduce the mismatch between output power and load power. Camoirano and Dellepiane [10] employed variable-frequency drive (VFD) technology to achieve more efficient energy management as well as the precise control of torque and speed of AC motors. Wang et al. [11] adopted a calibration model with a genetic algorithm to adjust the variable-speed pump flow rate to their designated value and achieve an energy-saving ratio of at least 16.1%. Ge [12] adopted a variable-speed motor to drive a variable-displacement pump and employed a matching method based on segmented speed and continuous displacement control of the pump to reduce the throttle loss. The energy-saving ratio under partial load condition can be up to 33%. Since there is limited scope for further increasing the electrohydraulic unit's efficiency, researchers have also focused on the design of hydraulic control circuits to reduce the mismatch. Load sensing systems [13], hydraulic adaptive systems [14], fuzzy control systems [15], close-loop volume control systems [16], negative flow control systems [17], and secondary regulation systems are all useful methods to match the load by regulating operating parameters and system states. As these methods achieve matching by adjusting the output flow, speed, and pressure, the installed power remains unaltered.

Many papers on energy-recovery methods have been published in recent years. A lot of energy-recovery circuits have been applied in hydraulic press machines. Yan et al. [18] proposed a flywheel energy-saving system (FESS) that can store the redundant energy at no-load stages and low-load stages and then release the stored energy at high-load stages. Dai et al. [19] applied a hydraulic accumulator to a 20 MN fast forging hydraulic press to realize energy conversion by absorbing large flow–pressure pulses and hydraulic shock. The results show that the hydraulic accumulator has promising energy-saving effects. Triet and Ahn [20] utilized a hydraulic accumulator and a flywheel to realize energy recovery and presented a control strategy for this energy-saving method. Ven [21] developed a novel hydraulic accumulator that can keep the hydraulic system pressure constant by using a piston with an area that varies with stroke. Compared with conventional recovery accumulators, this new accumulator could significantly increase the energy storage density. Xia et al. [22] proposed an integrated drive and energy recuperation system for a hydraulic excavator, and the large gravitational potential energy of the boom was recovered by a three-chamber hydraulic cylinder. Lin et al. [23] combined the advantages of hydraulic accumulators and electric accumulators and presented a compound energy recovery system to improve the energy efficiency of hydraulic equipment. Through a series of experiments, they validated that the compound system could increase the energy efficiency by approximately 39%. Fu et al. [24] studied the energy-saving potential of the boom cylinder with an accumulator in the hybrid excavator system. The results of simulations and experiments showed that the closed hydraulic regeneration system had a high recovery efficiency. Examples of applying energy recovery systems also include generator–super capacitor energy recovery and the gas cylinder energy recovery system [25]. Figure 2 gives the energy density of different energy recovery circuits. However, Huang [26] pointed out that the low utilization efficiency of the recovery energy is also an important problem that cannot be ignored. In general terms, an energy recovery process is composed of two sub-processes: the recovery process and the reutilization process. The number of energy conversions increases with the integration of an energy recovery system, thereby increasing the system's structural complexity and causing low energy efficiency as well.

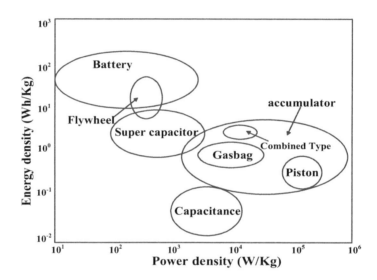

Figure 2. Power density and energy density of different energy-recovery systems.

In summary, energy efficiency and mismatch between the demanded power and the installed power can be significantly improved by using the aforementioned methods. However, most of those methods only consider energy efficiency and ignore the impact that they have on the hydraulic press itself. Therefore, most of the energy-saving presses present complicated structures and poor practicality. Furthermore, most of the existing studies in the literature only focus on the hydraulic power units and hydraulic control system; they ignore that the hydraulic press is a forming machine composed of various functional systems, such as the hydraulic actuating system, hydraulic cooling system, and other auxiliary systems. Each of these systems waste a large amount of energy during the forming process and has great potential to save energy.

Therefore, in contrast to existing energy-saving methods, the system in this paper focuses on the hydraulic cooling system as a novel starting point and proposes an energy-saving buffer system to improve the energy efficiency of a single press. A reduction of the installed power and improvement in energy efficiency can be achieved by integrating the prefill system into the hydraulic cooling system, whereas noise pollution, processing properties, and structural complexity can be improved by adding a buffer system. In addition, a servo valve is employed to adjust the supplied flow rate of the energy-saving buffer system according to the load profiles. Finally, the proposed energy-saving system was applied to a 13MN forging hydraulic press as a case study, and the results shows its significant economic and energy-saving potential.

2. Energy-Saving Method of the Hydraulic Press

2.1. Energy Characteristics of the Hydraulic Press

In Figure 3, a diagram showing the working cycle of a hydraulic press is presented. The operations performed by the hydraulic press include waiting within a working cycle (Stages 1 and 8, WT), fast falling (Stage 2, FF), pressing with slow falling (Stage 3, PS), pressure maintaining (Stage 4, PM), unloading (Stage 5, UL), fast returning (Stage 6, FR), and slow returning (Stage 7, SR). All these stages are also part of the forming process.

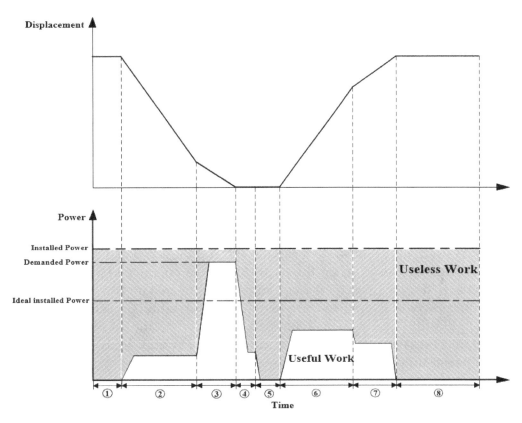

Figure 3. Hydraulic press working cycle diagram and comparison between installed and demanded power.

In traditional methods, installed power is designed to meet the maximum power requirements of Stage 3, which is much larger than the operating power of the other working stages. However, the duration of Stage 3 is only of a very small percentage of the total operating time, leading to high energy consumption and low energy efficiency. Furthermore, hydraulic presses operate performing periodic movements; the intermittent time between two adjacent cycles is almost equal to the duration of a cycle. Once hydraulic presses start running, they would not stop unless the hydraulic system reaches the unloaded state and the demanded power is zero; this leads to a large amount of power lost inadvertently. As shown in Figure 1, the white and blue areas represent useful work and useless work, respectively. Useless work is converted into heat or another useless form of energy during the operation of hydraulic presses; this is harmful to the processing properties.

When the hydraulic press is unloading, a large flow is unloaded during the processing cycle, leading to a considerable amount of energy being wasted by the traditional power unit. In addition, when the hydraulic cylinder rises rapidly with high working pressure, the high-pressure oil cannot be completely unloaded. In consequence, the residual high-pressure oil enters the hydraulic pipeline causing high vibration and huge noise pollution. The phenomena presented above constitute the main reasons for the low energy efficiency and poor processing performance in hydraulic presses.

Theoretically, if an external power system could supply energy to the hydraulic press simultaneously with the main power system during Stage 3, the installed power would be reduced and the energy efficiency would be increased.

2.2. Methodology

By analyzing in detail the reasons for the high installed power of hydraulic presses, this paper envisages the addition of an instantaneous power source during the PS stage (Stage 3). The power source can energize the hydraulic press simultaneously with the traditional power unit so as to reduce the power burden of the traditional power unit and thereby reduce the installed power of the hydraulic press. Based on this idea, an energy-saving method featuring a prefill system and a buffer system is proposed.

The proposed energy-saving system is developed by integrating a prefill system and a buffer system into the cooling system of the traditional hydraulic press, as shown in Figure 4. The pivot of the proposed system is the prefill system, which could supply high-pressure oil to the main cylinder similar to a motor-pump unit. In order to ensure the efficiency of the supplemental power and reduce pipeline losses, the prefill system is installed next to the hydraulic cylinder instead of tens of meters away from the hydraulic press, as in the traditional power unit. In addition, a servo valve is employed to adjust the output flow according to the load profiles, thereby achieving high-precision energy supplement. Based on this method, the installed power can be set at the ideal installed power as shown in Figure 1. During Stages 1, 2, 4, 5, 6, 7, and 8, the traditional power unit based on the ideal installed power can energize the press by itself, the prefill system is not used, and the servo valve outside the liquid-filling tank closes to reduce energy loss. When the hydraulic press operates at Stage 3, the power (or oil flow) required by the hydraulic press increases and exceeds the maximum value, it is difficult for the traditional hydraulic power unit alone to complete the pressurization process. At this time, the servo valve outside the liquid-filling tank opens with high accuracy, and the prefill system starts to supply power to the hydraulic press together with the traditional power unit. In this manner, power reliance on the traditional power unit is significantly reduced.

Furthermore, as depending on the power unit of the traditional cooling system, the addition of the prefill system does not require an increase in the number of motors and pumps to meet the flow and power requirements of Stage 3. Therefore, the usage of motor-pump units in the whole hydraulic system can be significantly reduced; the energy dissipation caused by the high installed power can be remarkably reduced in the low-load and no-load stage. Moreover, by reducing the number of pumps, the equipment footprint can be reduced, the structure of the equipment as a whole can be more compact, and the equipment utilization rate can be significantly improved.

When the hydraulic press operates at Stages 4 and 5, since the hydraulic press moves extremely quickly during the unloading stage, high-pressure oil cannot be completely unloaded. In consequence, the oil returning from the returning cylinder has high pressure and thus has a great impact on the prefill system and causes massive large noise pollution. To ensure the power supply of the prefill system successively improves the processing performance of the hydraulic press, the advantages of hydraulic accumulators are adopted in the method described in this paper by integrating a buffer system as an auxiliary facility. The buffer system, installed after the prefill system, can temporarily store the returning oil, absorb the high pressure in the oil to reduce the pressure shock in the system, and thus improve the processing performance of the hydraulic press. Furthermore, the high-pressure oil temporarily stored in the buffer system can also energize the hydraulic press together with the prefill system during Stage 3, allowing the energy-saving system to have a higher kinetic energy output and thus attain higher energy efficiency.

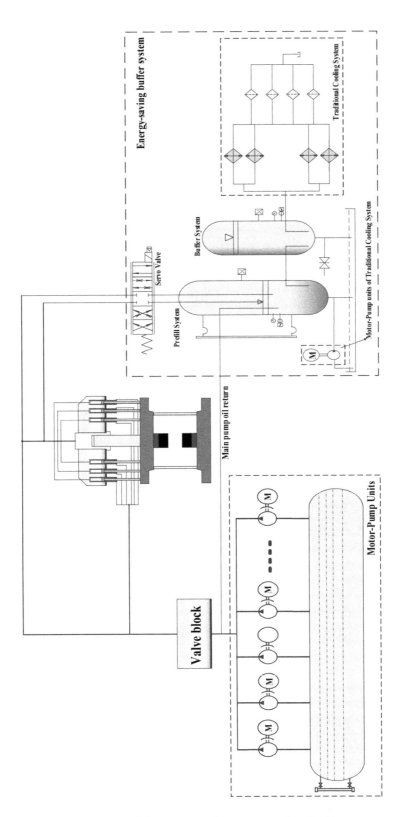

Figure 4. Diagram of the proposed energy-saving buffer system.

3. Energy Consumption Model and Theoretical Analysis

A hydraulic press is a closed energy conversion system composed of electric energy, mechanical energy, hydraulic energy, and forming energy, and the total energy consumption of a single hydraulic press is in the form of electrical energy.

The required electrical energy of a hydraulic press is determined by the flow rate q and the working pressure p of the actuator, which changes with the operation stage. Taking into account the energy conversion efficiency of the hydraulic system, the electrical energy demanded ($E_{electric-1}$) by the hydraulic machine to complete forming actions can be expressed as Equation (1):

$$E_{electric-1} = \sum_i \int_{T_0}^{T_0+T_1} p \cdot q \frac{1}{\eta_{Drive}} dt \tag{1}$$

where stage i includes the PS stage and the FF stage; η_{Drive} is the energy efficiency of motor-pump units; T_0 is the start time of stage i; and T_1 is the duration of stage i.

In terms of a hydraulic press with an energy-saving buffering system, Equation (1) can be divided into two parts,

$$\begin{aligned} E_{electric-1} &= E_1 + E_2 \\ &= \sum_i \left(\int_{T_0}^{T_0+T_1} p_{TMP} \cdot q_{TMP} \frac{1}{\eta_{Drive}} dt \right. \\ &\quad \left. + \int_{T_0}^{T_0+T_1} p_{PFS} \cdot q_{PFS} dt \right) \end{aligned} \tag{2}$$

where E_1 and E_2 are the energy provided by the traditional motor pumps and the prefill system, respectively; p_{TMP} and q_{TMP} are the pressure and flow rate provided by the traditional motor pumps, respectively; and p_{PFS} and q_{PFS} are the pressure and flow rate provided by the prefill system, respectively.

Apart from the electrical energy demanded by the hydraulic main circuit during the FF and PS stages, other electrical energy ($E_{electric-2}$) demanded by the overflow circuit and the unloading circuit during the PM, PR, and WT stages can be expressed as Equation (3):

$$E_{electric-2} = \sum_e \left(\int_{Tm} P_1 dt + \int_{Tn} P_2 dt + \int_{Ti} P_3 dt \right) \tag{3}$$

where P_1, P_2, and P_3 are the motor power of the UL stage, the FR stage, and the WT stage, respectively; and Tm, Tn, and Ti are the duration of the UL, FR, and WT stages.

Therefore, total energy consumption in a working cycle ($E_{electric-Total}$) can be obtained:

$$E_{electric-Total} = E_1 + E_{electric-2}. \tag{4}$$

Since the hydraulic actuator does not output mechanical energy until the forming process completes, the forming energy ($E_{Forming}$) of one working cycle can be expressed as:

$$E_{Forming} = \sum_i \int_i F_{NP} \cdot \dot{h} dt \tag{5}$$

where h is the height of the piston rod; and F_{NP} is the forming pressure of the hydraulic press.

Therefore, the energy efficiency (η_{E-F}, electrical-forming energy) of the proposed hydraulic press can be expressed as Equation (6):

$$\eta_{E-F} = \frac{E_{Forming}}{E_{electric-Total}}. \tag{6}$$

As shown in the above analysis, the integration of the prefill system and the buffer system can significantly reduce the power burden of traditional motor pumps during the high-load stage, and a reduction of installed power can be realized. In addition, both overflow losses during the PM stage and throttling losses during the idle state of the press could be greatly reduced because of the reduction of installed power, thereby improving the energy efficiency of the hydraulic press. Detailed energy-efficiency improvement is discussed in the following section.

4. Case Study

4.1. Experimental Scheme

To assess the energy-saving efficiency and processing performance of the proposed system, the system was tested on a 13 MN hydraulic press machine. The experimental setup (see Figure 5) consisted of a liquid-providing tank (volume and flow rate of 23 m^3 and 5800 L/min, respectively, Huawei, Yancheng, China), a buffering tank (volume of 6 m^3, Huawei, Yancheng, China), a hydraulic power unit containing an asynchronous motor (Y315S-4; rated power and speed of 110 kW and 1480 rpm, respectively, YKP, Botou, China), an oil pump (A2F250R2P2; rated speed and flow rate of 1480 rpm and 360 L/min, respectively, Rexroth, Shanghai, China), coolers, and filters. In addition, a DT-8850 noise meter is employed to measure the noise generated during the machining process, and body vibration is measured by the technical operator at the processing site.

Figure 5. Photograph of the energy-saving buffering system.

To test the operating stability of the system and the dynamic responsiveness during the rehydration process, two common forging methods, fast forging and constant forging, were selected and performed both on the proposed hydraulic press and a traditional hydraulic press. The forging frequency of constant forging and fast forging was set at 20 times per minute and 90 times per minute, respectively. The maximum forming pressure was 35 MPa. The installed power of the proposed forging press was set at 810 kW, while the traditional hydraulic press was 1050 kW.

Table 1 details the experimental parameters of this experiment. The pressure and flow rate of the prefill system (PFS) and the traditional power units (TPU) were recorded by a pressure gauge and a flow sensor and then uploaded to a computer.

Table 1. Parameters of the upsetting process.

Process Type	Forging Frequency	Load Pressure	Main Pump Unit Power	Displacement
Regular forging	20 times per minute	35 MPa	810 kW	450 mm
Fast forging	90 times per minute	35 MPa	810 kW	200 mm

4.2. Processing Performance Analysis

The forging frequency was set to 90 times per minute and 20 times per minute, respectively. The nominal pressure was set to 35 MPa. The relationships between TPU pressure and prefill pressure are shown in Figures 6 and 7.

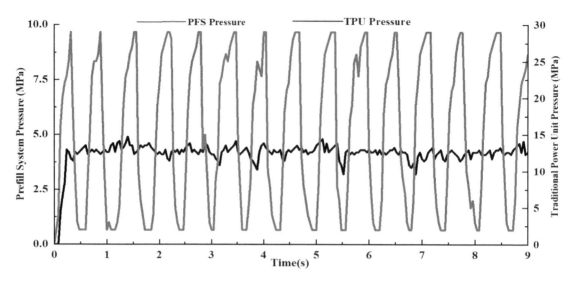

Figure 6. Output pressure of the prefill system and the main power unit during fast forging.

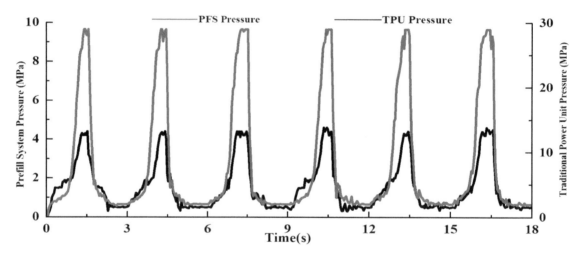

Figure 7. Output pressure of the prefill system and the main power unit during constant forging.

As it can be seen in Figures 6 and 7, the hydraulic press ran smoothly (without vibration) in the no-load and low-load stages. The TPU was used to supply energy to maintain the normal operation of the press, while the servo valve outside the liquid-providing tank was closed so that the output pressure and flow of the liquid-filling system was approximately zero. When the hammer of the hydraulic press contacted the forging workpiece, the load power rose and exceeded the nominal power supplied by the TPU; the servo valve outside the PFS opened with high accuracy, and the output power and flow of the PFS were used in conjunction with those of the TPU as the load pressure was reaching its maximum pressure. In the fast forging process, the servo valve remained open in order to ensure timely rehydration, since the timespan of one fast forging cycle was extremely short (see Figure 6).

When the press operated at the unloading stage, the traditional hydraulic system generated an impact pressure of up to 20 MPa in 0.04 s, causing a great impact on hydraulic pipes and beams. Many oil pipe ruptures and beam bending deformations are caused by this phenomenon, seriously affecting machining accuracy and the service life of the hydraulic press. Figures 8 and 9 show that the pressure-relief impact was well improved by integrating a buffer system. Since the PFS and the buffer system were installed close to the hydraulic cylinder, the high-pressure oil generated at the unloading stage directly entered the liquid-filling tank and the buffer tank without traveling a long distance along the hydraulic pipelines. Therefore, hydraulic impact and oil leakage could be reduced to a large extent, so as to improve the hydraulic working environment and reduce environmental pollution. According to experimental statistics, regardless of the forging method, adding the proposed system to a hydraulic press can reduce hydraulic impact by at least 9.2 MPa.

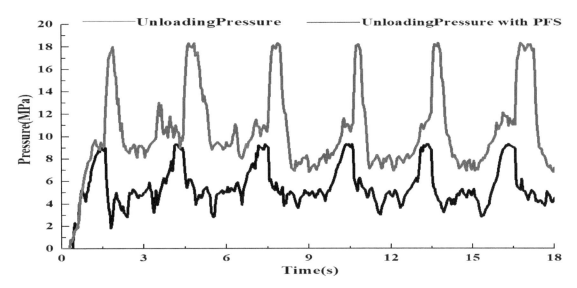

Figure 8. Unloading pressure of the hydraulic press with the energy-saving system and the traditional hydraulic press during fast forging.

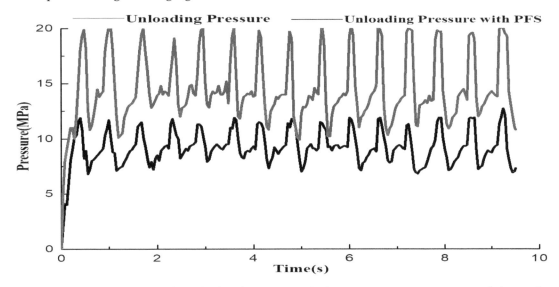

Figure 9. Unloading pressure of the hydraulic press with the energy-saving system and the traditional hydraulic press during fast forging.

Through several test experiments, the experimental data from Figures 6–9 are summarized in Tables 2 and 3, in which Tables 2 and 3 are the results of the experiments at the PS and unloading stage for different forging methods, respectively.

When the forging hydraulic press operates at the low-load or no-load stage, the energy supplied by the TPU is sufficient for the hydraulic press to operate normally, and the PFS does not work. As seen from Figures 6 and 7 and Table 3, when the hydraulic press operates at Stage 3, the load power rises and exceeds the maximum power the TPU could output, the servo valve outside the PFS opens, and then the PFS outputs high-pressure oil and energizes the main working cylinder together with the TPU; the maximum output pressure and flow rate of the TPU are 28.1 MPa and 2900 L/min, respectively, while the load pressure and flow rate of the press are 35 MPa and 4400 L/min. The difference between the load pressure and the output pressure of the TPU is the energy provided by the PFS, which are 5.4 MPa and 1480 L/min, respectively. The average pressure-saving rate for the two forging methods is

19.3% and 17.5%, respectively. Combined with the flow rate and the formula $P = p * q$, the maximum power saving rate of 20.1% could be obtained through calculations.

Table 2. Results of the experiments at the PS stage for different forging methods. PFS: prefill system, PS: pressing with slow falling, TPU: traditional power units.

Forging Method	Fast Forging	Constant Forging
Load pressure	35.0 MPa	35.0 MPa
Peak pressure of the TPU	28.1 MPa	28.0 MPa
Peak pressure of the PFS	5.4 MPa	4.9 MPa
Mean pressure of the TPU	27.1 MPa	26.9 MPa
Mean pressure of the PFS	4.9 MPa	4.5 MPa
Pressure saving rate	19.3%	17.5%
Peak flow rate of the TPU	2900 L/min	2870 L/min
Peak flow rate of the PFS	1480 L/min	1460 L/min
Power-saving rate	20.1%	19.6%

Table 3. Results of the experiments at the unloading stage for different forging methods.

Forging Method	Fast Forging	Constant Forging
Pipeline pressure	9.7 MPa	9.1 MPa
Traditional pipeline pressure	20.3 MPa	18.5 MPa
Buffering system pressure	6.3 MPa	5.2 MPa
Noise intensity	75 dB	72 dB
Traditional noise intensity	105 dB	100 dB
Oil temperature	35.7 °C	40.2 °C
Body vibration	Slight	Slight
Pressure impact reduction	10.6 MPa	9.2 MPa
Pressure decline rate	52.2%	49.73%
Noise decline rate	28.6%	28%

From Table 3, when the forging frequency was set at 90 times per minute (fast forging), the pipeline pressure dropped to 9.7 MPa, while the pressure shock and noise pollution decreased by 52.2% and 28.6%, respectively, compared with the traditional hydraulic press. When the forging frequency was set at 20 times per minute (constant forging), the pipeline pressure dropped to 9.2 MPa, the pressure shock and noise pollution decreased by 49.7% and 28%, respectively. In addition, although the cooling system was retrofitted, the system temperature remained within a low range, which was generally lower than that of the traditional hydraulic press.

4.3. Energy Consumption Analysis of the Two Presses

During the forging process, each motor pump in the drive unit changes the state of its access to the hydraulic system according to the energy demand of different operations, and one drive unit, which is useless in an operation, is set at an unloading state by switching the pressure-relief valve. Table 4 shows the compositions of the drive system of the two hydraulic presses and the number of motor pumps connected for each processing stage.

Table 4. Compositions of the drive system and the number of motor pumps used in different operations. WT: waiting within a working cycle, Stages 1 and 8, WT; FF: fast falling, Stage 2; UL: unloading, Stage 5; FR: fast returning, Stage 6; SR: slow returning, Stage 7.

	Total Number	WT	FF	PS	UL	FR	SR
Hydraulic press with PFS	7	0	3	7	1	7	2
Traditional hydraulic press	9	0	3	9	1	9	3

In this study, the energy consumption of a hydraulic press was scaled by measuring the active electricity consumption of motor pumps. A power meter and an data acquisition card were selected to measure and receive the voltage and current through the motors at any time, respectively. Then, the real-time measured data were transmitted to the PC after processing.

During the forming process, the energy efficiency of a hydraulic press can be expressed as follows,

$$\eta = \frac{E_{Forming}}{\sum_i E_{i-Input}} \cdot 100\% \tag{7}$$

where i is all the forming operations, including FF, PS, UL, FR, SR, and WT. $E_{i-Input}$ is the energy consumed by the hydraulic machine to complete the i-th operation, which can be obtained by Equation (8).

$$E_{i-Input} = \int_{t_{i-Start}}^{t_{i-End}} P_i(t)dt = \sum_{t_{i-Start}}^{t_{i-Start}} P_i \cdot \Delta t \tag{8}$$

where P_i is the active power of each operation, which can be obtained by the power meter, and $t_{i-Start}$ and $t_{i-Start}$ are the start time and the end time of the i-th operation, respectively.

According to Equation (8), the active electrical energy consumption of motors was obtained by experiments. The electricity consumption, useful energy, and energy efficiency of the two presses from experimental testing are shown in Tables 5 and 6.

Table 5. Energy dissipation of each operation in the hydraulic press with PFS (Constant Forging).

Operations (i)		FF	PS	UL	FR	SR	WT	Total
Energy (KJ)	Time (s)	0–0.5	0.5–1.4	1.2–1.3	1.3–1.9	1.9–2.2	2.2–3	3
Input energy ($E_{i-Input}$)		389.91	703.28	68.21	472.37	189.11	559.88	2382.76
Useful energy		165.41	692.87	9.28	439.78	68.19	0	1375.53
Energy efficiency		42.41%	98.42%	13.23%	93.00%	35.98%	0	57.72%
η						29.05%		

Table 6. Energy consumption of each operation of the traditional hydraulic press (Constant Forging).

Operations (i)		FF	PS	UL	FR	SR	WT	Total
Energy (KJ)	Time (s)	0–0.5	0.5–1.4	1.2–1.3	1.3–1.9	1.9–2.2	2.2–3	3
Input energy ($E_{i-Input}$)		492.28	901.89	99.38	604.22	297.09	832.80	3227.66
Energy efficiency		33.60%	72.82%	9.05%	74.00%	22.95%	0	37.04%
$\Delta E_{i-Input}$		+102.37	+198.61	+31.17	+121.85	+107.98	+242.92	+804.90
η						17.35%		

The traditional hydraulic press is equipped with 9 sets of motor pumps, of which the installed power is up to 1050 kW. According to Table 6, the total electric energy consumed to complete a working cycle is 3227.66 KJ, and only 31.01% of the input energy is converted into useful energy. Besides this, as a result of the pipe-valve energy loss, overflow energy loss, and other energy losses, only 17.35% of the input energy is used to form the workpiece. However, by integrating PFS into the press, the energy consumption of each stage significantly improved. The experimental data from Tables 5 and 6 compared with the traditional hydraulic press are summarized in Figure 10.

As shown above, since the installed power of the hydraulic press equipped with PFS was lower than that of the traditional press by 22.86%, the throttling loss and the no-load loss of the press were significantly reduced, and the average electrical energy consumption was minimized in each stage, as shown in Figure 10a. Total electrical energy consumption was reduced from 3227.66 KJ to 2382.76 KJ, so 804.90 KJ energy was saved per working cycle. During the PS stage, as the hammer contacted the forming workpiece, the load power rose and exceeded the nominal power supplied by the TPU, the servo valve outside the PFS opened, and the output power and flow of the PFS were used in conjunction with those of the TPU as the load pressure was reaching its maximum pressure. The difference between the input energy of the two presses is the energy provided by the PFS, which is about 198.61 KJ. After several experimental tests, compared with the traditional hydraulic press, the overall energy efficiency and the forming efficiency of the hydraulic press with PFS were improved by 26.71% and 11.70%, respectively.

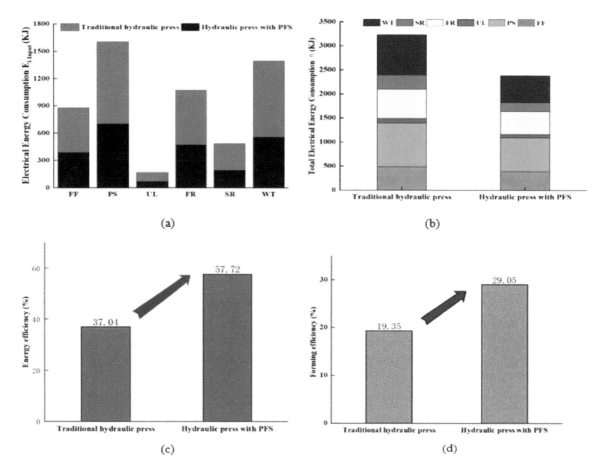

Figure 10. (**a**) Energy consumption under different processing stages; (**b**) Total energy consumption; (**c**) Total energy efficiency; (**d**) Forming efficiency.

5. Conclusions

Given the low energy efficiency and poor processing performance of the hydraulic press, this paper proposes a method featuring an energy-saving buffer system to reduce the installed power and improve processing performance. In this method, a liquid-filling system is used to supply power and an oil flow rate that the TPU cannot provide at the high-load stage. As it is not necessary to increase the number of motor pumps to meet the flow requirements, the motor-pump usage of the whole system is significantly reduced. In consequence, the installed power of the hydraulic press is reduced, and cost reduction and energy savings are achieved. Furthermore, to mitigate the high-pressure impact and noise pollution at the unloading stage, a buffer system is integrated into the system to absorb the high-pressure returning oil. In consequence, the pressure shock problem is addressed, and the service lifespan of the hydraulic press is increased. The proposed system was tested in a 13 MN hydraulic press in which an industrial press was used. Through a series of comparative experiments, it was preliminarily validated that the proposed system can reduce pump usage and pressure shock by up to 30% and 41%, respectively, increase energy efficiency by up to 26.71%, reduce noise pollution and installed power by 28% and 22.85%, respectively, and provide the same processing characteristics and properties as the traditional hydraulic press.

To better improve the energy efficiency of the hydraulic press, our future work will concentrate on the control strategies of the PFS to achieve higher precision output. Other research directions include the optimal design of the PFS and the buffering system for a given hydraulic press.

Author Contributions: Conceptualization, X.Y.; Methodology, X.Y. and B.C.; validation, X.Y. and B.C.; formal analysis, X.Y. and B.C.; data curation, X.Y.; writing original draft preparation, X.Y.; Funding, B.C.; writing-review and editing, B.C. All authors have read and agreed to the published version of the manuscript.

Acknowledgments: The work is supported by Huazhong University of Science and Technology and Jiangsu Huawei Machinery Manufacturing Co. Ltd.

Abbreviations

TPU   traditional power units
PFS   prefill system
FF   fast falling
PS   pressure with slow speed
PM   pressure maintaining
UL   unloading
FR   fast returning
SR   slow returning
WT   waiting stage

References

1. Li, L.; Huang, H.H.; Liu, Z.F.; Li, X.Y.; Triebe, M.J.; Zhao, F. An energy-saving method to solve the mismatch between installed and demanded power in hydraulic press. *J. Clean. Prod.* **2016**, *139*, 636–645. [CrossRef]
2. Duflou, J.R.; Sutherland, J.W.; Dornfeld, D.; Herrmann, C.; Jeswiet, J.; Kara, S.; Hauschild, M.; Kellens, K. Towards energy and resource efficient manufacturing: A processes and systems approach. *CIRP Ann. Manuf. Technol.* **2012**, *61*, 587–609. [CrossRef]
3. Lin, T.; Chen, Q.; Ren, H.; Huang, W.; Chen, Q.; Fu, S. Review of boom potential energy regeneration technology for hydraulic construction machinery. *Renew. Sustain. Energy Rev.* **2017**, *79*, 358–371. [CrossRef]
4. Xu, Z.; Liu, Y.; Hua, L.; Zhao, X.; Guo, W. Energy analysis and optimization of main hydraulic system in 10,000 kN fine blanking press with simulation and experimental methods. *Energy Convers. Manag.* **2019**, *181*, 143–158. [CrossRef]
5. Mousa, E.; Kazemi, M.; Larsson, M.; Karlsson, G.; Persson, E. Potential for Developing Biocarbon Briquettes for Foundry Industry. *Appl. Sci.* **2019**, *9*, 5288. [CrossRef]
6. Zhao, K.; Liu, Z.F.; Yu, S.R.; Li, X.Y.; Huang, H.H.; Li, B.T. Analytical energy dissipation in large and medium-sized hydraulic press. *J. Clean. Prod.* **2015**, *103*, 908–915. [CrossRef]
7. Huang, H.H.; Zou, X.; Li, L.; Li, X.Y.; Liu, Z.F. Energy-Saving Design Method for Hydraulic Press Drive System with Multi Motor-Pumps. *Int. J. Precis. Eng. Manuf. Green Technol.* **2019**, *6*, 223–234. [CrossRef]
8. Quan, L.; Helduser, S. Energy Saving and High Dynamic Hydraulic Power Unit Based on Speed Variable Motor and Constant Hydraulic Pump. *China Mech. Eng.* **2003**, *14*, 606–609.
9. Su, C.L.; Chung, W.L.; Yu, K.T. An Energy-Savings Evaluation Method for Variable-Frequency-Drive Applications on Ship Central Cooling Systems. *IEEE Trans. Ind. Appl.* **2014**, *50*, 1286–1294. [CrossRef]
10. Camoirano, R.; Dellepiane, G. Variable frequency drives for MSF desalination plant and associated pumping stations. *Desalination* **2005**, *182*, 53–65. [CrossRef]
11. Wang, H.; Wang, H.Y.; Zhu, T. A new hydraulic regulation method on district heating system with distributed variable-speed pumps. *Energy Convers. Manag.* **2017**, *147*, 174–189. [CrossRef]
12. Ge, L.; Quan, L.; Zhang, X.G.; Zhao, B.; Yang, J. Efficiency improvement and evaluation of electric hydraulic excavator with speed and displacement variable pump. *Energy Convers. Manag.* **2017**, *150*, 62–71. [CrossRef]
13. Lovrec, D.; Kastrevc, M.; Ulaga, S. Electro-hydraulic load sensing with a speed-controlled hydraulic supply system on forming-machines. *Int. J. Adv. Manuf. Technol.* **2009**, *41*, 1066–1075. [CrossRef]

14. Feng, L.; Yan, H. Nonlinear Adaptive Robust Control of the Electro-Hydraulic Servo System. *Appl. Sci.* **2020**, *10*, 4494. [CrossRef]

15. Chen, B.J.; Huang, S.H.; Jin, L.; Gao, J.F. Control Strategy for Free Forging Hydraulic Press. *Chin. J. Mech. Eng.* **2008**, *44*, 304–307. [CrossRef]

16. Ferreira, J.A.; Sun, P.; Grácio, J.J. Close loop control of a hydraulic press for springback analysis. *J. Mater. Process. Technol.* **2006**, *177*, 377–381. [CrossRef]

17. Gao, F.; Pan, S.X. Experimental study on model of negative control of hydraulic system. *China Mech. Eng.* **2005**, *41*, 49–54. [CrossRef]

18. Yan, X.P.; Chen, B.J.; Zhang, D.W.; Wu, C.X.; Luo, W.X. An energy-saving method to reduce the installed power of hydraulic press machines. *J. Clean. Prod.* **2019**, *233*, 538–545. [CrossRef]

19. Dai, M.Q.; Zhao, S.D.; Yuan, X.M. The Application Study of Accumulator Used in Hydraulic System of 20MN Fast Forging Machine. *J. Appl. Mech.* **2011**, *80*, 870–874. [CrossRef]

20. Triet, H.H.; Ahn, K.K. Comparison and assessment of a hydraulic energy- saving system for hydrostatic drives. *Proc. Inst. Mech. Eng. Part I J. Syst. Control Eng.* **2011**, *225*, 21–34. [CrossRef]

21. Van de Ven, J.D. Constant pressure hydraulic energy storage through a variable area piston hydraulic accumulator. *Appl. Energy* **2013**, *105*, 262–270. [CrossRef]

22. Xia, L.P.; Quan, L.; Ge, L.; Hao, Y.X. Energy efficiency analysis of integrated drive and energy recuperation system for hydraulic excavator boom. *Energy Convers. Manag.* **2018**, *156*, 680–687. [CrossRef]

23. Lin, T.l.; Huang, W.P.; Ren, H.L.; Fu, S.J.; Liu, Q. New compound energy regeneration system and control strategy for hybrid hydraulic excavators. *Autom. Constr.* **2016**, *68*, 11–20. [CrossRef]

24. Fu, S.; Chen, H.; Ren, H.; Lin, T.; Miao, C.; Chen, Q. Potential Energy Recovery System for Electric Heavy Forklift Based on Double Hydraulic Motor-Generators. *Appl. Sci.* **2020**, *10*, 3996. [CrossRef]

25. Fan, Y.J.; Mu, A.L.; Ma, T. Design and control of a point absorber wave energy converter with an open loop hydraulic transmission. *Energy Convers. Manag.* **2016**, *121*, 13–21. [CrossRef]

26. Huang, N.; Yang, S.; Chen, Y. Development of machine tools and large tonnage hydraulic automatic hydraulic fine blanking press. *Mach. Tool Hydraul.* **2011**, *2011*, 4.

8

Motion Planning and Coordinated Control of Underwater Vehicle-Manipulator Systems with Inertial Delay Control and Fuzzy Compensator

Han Han [1], Yanhui Wei [1,*] , Xiufen Ye [1] and Wenzhi Liu [2]

[1] College of Automation, Harbin Engineering University, Harbin 150001, China; hanhanheu@163.com (H.H.); yexiufen@hrbeu.edu.cn (X.Y.)

[2] College of Information and Communication Engineering, Harbin Engineering University, Harbin 150001, China; liuwenzhi@hrbeu.edu.cn

* Correspondence: wyheauini@hrbeu.edu.cn

Abstract: This paper presents new motion planning and robust coordinated control schemes for trajectory tracking of the underwater vehicle-manipulator system (UVMS) subjected to model uncertainties, time-varying external disturbances, payload and sensory noises. A redundancy resolution technique with a new secondary task and nonlinear function is proposed to generate trajectories for the vehicle and manipulator. In this way, the vehicle attitude and manipulator position are aligned in such a way that the interactive forces are reduced. To resist sensory measurement noises, an extended Kalman filter (EKF) is utilized to estimate the UVMS states. Using these estimates, a tracking controller based on feedback Linearization with both the joint-space and task-space tracking errors is proposed. Moreover, the inertial delay control (IDC) is incorporated in the proposed control scheme to estimate the lumped uncertainties and disturbances. In addition, a fuzzy compensator based on these estimates via IDC is introduced for reducing the undesired effects of perturbations. Trajectory tracking tasks on a five-degrees-of-freedom (5-DOF) underwater vehicle equipped with a 3-DOF manipulator are numerically simulated. The comparative results demonstrate the performance of the proposed controller in terms of tracking errors, energy consumption and robustness against uncertainties and disturbances.

Keywords: underwater vehicle-manipulator system; motion planning; coordinated motion control; inertial delay control; fuzzy compensator; extended Kalman filter; feedback linearization

1. Introduction

With increasing interest in the field of marine research, autonomous underwater vehicle manipulator systems (UVMSs) [1] have rapidly developed into important devices for exploring the ocean, completing underwater tasks, underwater sampling and so on. It is a challenging problem to accurately control the UVMS in an energy-efficient manner due to the kinematic redundancy and underwater environment with hydrodynamic uncertainties, unknown external disturbances (such as ocean currents) and inaccurate sensor information. For solving these problems, inverse kinematics and robust coordinated control techniques have been developed for the UVMS.

For the inverse kinematics of the UVMS, the solution can be obtained through mapping the end-effector's velocities to the velocities of the vehicle and manipulator. As the UVMS has redundant degrees of freedom (DOFs), there are various combinations of vehicle and manipulator velocities without affecting the end-effector velocities. A common solution is to adopt the pseudo-inverse Jacobian matrix of the UVMS or its weighted form [2]. However, this method is not desirable for redundant exploration to avoid joint limits, improve system manipulability or save energy. Therefore, the task-priority redundancy resolution technique [3] was proposed in such a way that the fulfillment of the primary task has a higher priority than that of a secondary task. Generally, the secondary task is to optimize the performance index through assigning additional motion in the null space of the primary task. Sarkar and Podder [4] solved the inverse kinematics of the UVMS on the acceleration level to minimize the total hydrodynamic drag; however, the performance index of this method requires dynamic equations which can not be modeled exactly. Han et al. [5] proposed a new performance index designed to minimize restoring moments without using dynamic equations. However, this method was implemented for a specific configuration of the UVMS.

The task-priority strategy can be extensible to chain multiple tasks which have a lower order of priority (Siciliano and Slotine [6]). Antonelli et al. [7] used a fuzzy inference system (FIS) to handle multiple secondary tasks, such as reduction of fuel consumption and improvement of system manipulability. In such a way, a secondary task can be activated by FIS when the corresponding variable is without the safe range. Wang et al. [8] used a fuzzy logic algorithm to decide the priorities of secondary objectives, such as manipulator singularity avoidance and attitude optimization of the UVMS. The experimental validation of three difference kinematic control schemes was presented in [9]. In [10], a multitask kinematic control of the underwater biomimetic vehicle-manipulator system (UBVMS) was designed. A unifying framework for the kinematic control of UVMSs was proposed in [11]. A very recent work dedicated to motion planning for the UVMS was presented in [12].

To achieve trajectory tracking, it is very important to design a coordinated motion controller for the UVMS. The simple control methods (e.g., proportional-integral-derivative (PID) control) are not suitable for the UVMS due to the inherent nonlinear and coupled dynamics of the system [13–15]. Schjlberg and Fossen [16] proposed a control strategy in terms of feedback linearization. Sarkar and Podder [4] utilized a computed torque controller (CTC) for trajectory tracking of the UVMS. Taira et al. [17] proposed a model-based motion control for the UVMS, which can be applicable to three types of servo systems; i.e., a voltage-controlled, a torque-controlled and a velocity-controlled servo system. Korkmaz et al. [18] presented a trajectory tracking control for an underactuated underwater vehicle manipulator system (U-UVMS) based on the inverse dynamics. However, these model-based controllers are poor in terms of robustness against model uncertainties. In [19], a fuzzy logic control method was designed for a hybrid-driven UVMS to grasp marine products on the seabed. A model reference adaptive control approach for an UVMS was proposed in [20]. Antonelli et al. [21] proposed an adaptive controller based on virtual decomposition; however, a regressor matrix corresponding to parameter vector is required in this method. An indirect adaptive controller based on the extended Kalman filter (EKF) was proposed in [22]; meanwhile the performance would be degraded due to the estimated error via EKF. To eliminate the bias from the EKF estimation, Dai et al. [23] introduced a H∞ control in the indirect adaptive controller to achieve robust performance. However, this method results in a residual error when the bounds of the disturbance cannot be known prior. To reduce or omit the estimation error of the uncertainties and disturbances, a fuzzy compensator based on estimations was utilized [24]. To handle state and input constraints of the UVMS, a robust predictive control (RMPC) [24], a nonlinear MPC (NMPC) [25], a tube-based robust MPC [26], a fast MPC (FMPC) [27] and a fast tube MPC (FTMPC) [28] were used for the UVMS trajectory tracking, but these MPC approaches do not permit the self-motion utilized to perform energy efficient trajectory tracking.

For achieving precise and robust performance, the control design should be enhanced by the estimations of the system's uncertainties and disturbances. The popular estimation techniques include time delay control (TDC) [29,30], the extended state observer (ESO) [31], the disturbance observer [32], the nonlinear disturbance observer (NDO) [33,34] the uncertainty and disturbance estimator (UDE) [35,36] (redefined as inertial delay control (IDC) [37]) and so on. Among them, because IDC is simple in design and easy to complete, it is widely used to estimate the effect of the lumped uncertainties and disturbances. Generally, the IDC is applied to the sliding mode control (SMC) for ensuring precise and robust performance. The combined method does not require the bounds of uncertainties and disturbances, and it does not use the discontinuous function in the control law.

However, the above-mentioned methods are based on joint-space variables, which may not be suitable for a variety of underwater tasks with high-precision end-effector position requirements [38]. These task space control schemes can easily adapt to the online modification of the end-effector's motion [39,40]. However, the task-space controllers also have disadvantages. (a) The kinematic redundancy of the UVMS cannot be exploited. (b) The output of the task-space controller should be mapped into the joint space so as to be realized by thrusters and actuators. Li et al. [34] proposed a hybrid strategy-based coordinated controller for the UVMS. The hybrid strategy is to transform the joint-space controller (to exploit the system's redundancy) to the task-space controller (to ensure high-accuracy tracking performance).

Inspired by the above studies, new motion planning and coordinated control schemes of the UVMS are proposed in this paper. The contribution of this work is that the proposed scheme can ensure precise, energy-efficient and robust performance in the presence of model uncertainties, external disturbances, payload and sensory noises. First, a new redundancy resolution technique is proposed, where a new secondary task with a nonlinear function is inserted for generating energy-saving trajectories for the vehicle and manipulator. Second, an EKF estimation system is employed for resisting sensory noises. Third, a coordinated motion control with joint-space errors, end-effector errors, IDC and a fuzzy compensator is proposed as a robust tracking controller against uncertainties and disturbances. Last, the effectiveness of the proposed scheme is verified through numerical simulations.

The rest of the paper is organized as follows. Section 2 is concerned with the kinematic and dynamic modeling of the UVMS. In Section 3, an improved redundancy resolution technique is presented. The proposed control scheme is proposed in Section 4. Numerical simulations and the detailed performance analysis are presented in Section 5. Section 6 holds the conclusions.

2. Modeling

The UVMS investigated in this paper is composed of an underwater vehicle with a 3 DOFs manipulator. The coordinate system of the UVMS is shown in Figure 1. In the body-fixed frame $\{B\}$, we define that the vectors of vehicle's linear and angular velocities are $\boldsymbol{v}_1$ and $\boldsymbol{v}_2$, where $\boldsymbol{v}_1 = [u, v, w]^T$, $\boldsymbol{v}_2 = [p, q, r]^T$ and $\boldsymbol{v} = [\boldsymbol{v}_1^T, \boldsymbol{v}_2^T]^T$. The vector of joint positions is assumed to be $\boldsymbol{q} = \left[q_1, q_2 \cdots q_n\right]^T$, where n is the number of manipulator's joints. The position and orientation vector of the UVMS relative to the body-fixed frame is assumed to be $\boldsymbol{\zeta} = [\boldsymbol{v}_1^T, \boldsymbol{v}_2^T, \dot{\boldsymbol{q}}^T]^T$. In the inertial frame $\{I\}$, the vectors of end-effector's position and orientation are defined as $\boldsymbol{\eta}_{E1}$ and $\boldsymbol{\eta}_{E2}$, and assume $\boldsymbol{x}_E = [\boldsymbol{\eta}_{E1}^T, \boldsymbol{\eta}_{E2}^T]^T$.

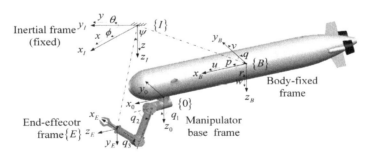

Figure 1. Coordinate systems of the underwater vehicle manipulator system (UVMS).

The kinematic model of the UVMS [21] can be obtained as shown in (1), where the velocities of the UVMS in the body-fixed frame ($\boldsymbol{\zeta}$) are mapped into end-effector velocities ($\dot{\boldsymbol{x}}_E$) via $\boldsymbol{J}(\boldsymbol{R}_B^I, \boldsymbol{q})$.

$$\dot{\boldsymbol{x}}_E = \begin{bmatrix} \boldsymbol{R}_B^I & S(\boldsymbol{R}_B^I \boldsymbol{p}_0^B + \boldsymbol{R}_0^I \boldsymbol{p}_E^0)\boldsymbol{R}_B^I & \boldsymbol{J}_{\text{pos,q}}^I \\ \boldsymbol{0}_{3\times3} & \boldsymbol{R}_B^I & \boldsymbol{J}_{\text{ori,q}}^I \end{bmatrix} \left[\boldsymbol{v}_1^T, \boldsymbol{v}_2^T, \dot{\boldsymbol{q}}^T\right]^T = \boldsymbol{J}(\boldsymbol{R}_B^I, \boldsymbol{q})\boldsymbol{\zeta} \tag{1}$$

where $\boldsymbol{J}(\boldsymbol{R}_B^I, \boldsymbol{q}) \in \mathbb{R}^{6\times(6+n)}$, $\boldsymbol{R}_0^I = \boldsymbol{R}_B^I \boldsymbol{R}_0^B$, $\boldsymbol{J}_{\text{pos,q}}^I = \boldsymbol{R}_0^I \boldsymbol{J}_{\text{pos,q}}^0$ and $\boldsymbol{J}_{\text{ori,q}}^I = \boldsymbol{R}_0^I \boldsymbol{J}_{\text{ori,q}}^0$. $\boldsymbol{R}_0^B$ is the rotational matrix describing the transformation from the manipulator's base frame to the body-fixed frame, and p_E^0 is the position vector from manipulator's base to the center of the body-fixed frame. p_E^0 is the position vector from end-effector to the manipulator's base. $J_{pos,q}^0$ is the manipulator's linear Jacobian matrix, and $J_{ori,q}^0$ is the manipulator's angular Jacobian matrix. $S(\cdot)$ is the cross-product operator.

The vectors of vehicle's position and attitude relative to the inertial frame are defined as η_1 and η_2, where $\eta_1 = [x, y, z]^T$, $\eta_2 = [\phi, \theta, \psi]^T$ and $\eta = [\eta_1^T, \eta_2^T]^T$. The velocity vector of the UVMS defined in the body-fixed frame (ζ) can be obtained by (2).

$$\zeta = \begin{bmatrix} R_I^B & 0_{3\times3} & 0_{3\times3} \\ 0_{3\times3} & J_\nu & 0_{3\times3} \\ 0_{3\times3} & 0_{3\times3} & I_{3\times3} \end{bmatrix} \begin{bmatrix} \dot{\eta}_1 \\ \dot{\eta}_2 \\ \dot{q} \end{bmatrix} = J_\xi(\eta_2)\dot{\xi} \tag{2}$$

where $\nu_1 = R_I^B \dot{\eta}_1$, $\nu_2 = J_\nu \dot{\eta}_2$ and $\xi = [\eta_1^T, \eta_2^T, q^T]^T$. R_I^B is the linear rotational matrix describing the transformation from the inertial frame to the body-fixed frame, and J_ν is the angular rotational matrix. The values of R_I^B and J_ν can be referred to the literature [41]. $J_\xi(\eta_2)$ is the Jacobian matrix which relates the vehicle velocities with respect to the inertial frame and the body-fixed frame.

Dynamic Modeling

The nonlinear dynamic equations of the UVMS expressed in the body-fixed frame $\{B\}$ can be established as [16,21]:

$$M(q,\eta)\dot{\zeta}+C(q,\zeta)\zeta+D(q,\zeta)\zeta+g(q,R_I^B)=\tau+\tau_{dis} \tag{3}$$

where

$$M(q,\eta) = \begin{bmatrix} M_\nu + H_1(q) & H_2(q) \\ H_2^T(q) & M_m(q) \end{bmatrix}, \quad C(q,\zeta) = \begin{bmatrix} C_\nu(\nu) + C_1(q,\dot{q},\nu) & C_2(q,\dot{q}) \\ C_3(q,\dot{q},\nu) & C_m(q,\dot{q}) \end{bmatrix}$$

$$D(q,\zeta)=\begin{bmatrix} D_\nu(\nu)+D_1(q,\dot{q},\nu) & D_2(q,\dot{q},\nu) \\ D_3(q,\dot{q},\nu) & D_m(q)+D_4(q,\dot{q},\nu) \end{bmatrix}, \quad g(q,R_I^B) = \begin{bmatrix} g_\nu(\eta) + g_1(q) \\ g_m(q) \end{bmatrix}, \quad \tau = [\tau_\nu^T, \tau_m^T]^T$$

where $M(q,\eta) \in \mathbb{R}^{(6+n)\times(6+n)}$ is the inertia matrix including added mass terms, and $H_1(q)$ and $H_2(q)$ are matrices of the inertia effects due to the manipulator. $C(q,\zeta) \in \mathbb{R}^{(6+n)}$ is the Coriolis and centripetal matrix, and $C_i(q,\dot{q},\nu)(i = 1,3)/C_2(q,\dot{q})$ is the matrix of Coriolis and centripetal forces due to the coupling effects/due to the manipulator. $D(q,\zeta)\zeta \in \mathbb{R}^{(6+n)}$ is the vector of dissipative effects, and $D_i(q,\dot{q},\nu)$ $(i = 1\cdots4)$ is the matrix of drag effects due to the coupling effects. $g(q,R_B^I) \in \mathbb{R}^{(6+n)}$ is the vector of gravity and buoyancy effects, $g_\nu(\eta)$ is the restoring vector of the vehicle, $g_m(q)$ is the restoring vector of the manipulator and $g_1(q)$ is the restoring vector due to the manipulator. $\tau \in \mathbb{R}^{(6+n)}$ is the vector of generalized forces. τ_{dis} is the vector of disturbances. Generally, in a deep water environment, τ_{dis} comes from ocean currents, payload, etc. In particular, time-varying ocean currents increase the uncertainty of the UVMS hydrodynamic forces, making accurate control of the UVMS difficult.

As for the underwater manipulator, it is assumed that its links are composed of cylindrical elements. The hydrodynamic effects on cylinders can be referred to [16]. For a cylinder, the inertial matrix of added mass and added moment is a diagonal matrix, while the off-diagonal elements are neglected. The drag force can be expressed by a nonlinear function related to the velocity vector of the center of mass of the link. Generally, when calculating the hydrodynamic forces, the linear skin-friction force, quadratic drag force and lift force are considered. The third-order and higher order terms of the drag forces are neglected. In addition, based on the assumption that velocity of the ocean current is constant, the diffraction forces can be neglected.

In the real system, the above model parameters are usually difficult to accurately measure or estimate, especially the hydrodynamic forces acting on the UVMS. Thus, it is advisable to divide the model parameters into two parts: the normal value part and the bias part. The normal value is denoted as $(\cdot)^*$, which can be obtained through using strip theory, pool experiment analysis or CFD computation. The bias term is denoted as $\Delta(\cdot)$, which describes the difference between the real value and the nominal value. Then, (4) can be obtained. For control design, the normal values are available, while the bias parts are considered as model parameter uncertainties.

$$M = M^* + \Delta M, \quad C = C^* + \Delta C, \quad D = D^* + \Delta D \tag{4}$$

Considering that the vehicle is driven by thrusters and the manipulator is driven by motors, the generalized force vector τ is related to the vector of thruster forces and actuator torques F_{td} through (5).

$$\tau = \begin{bmatrix} B_v & 0_{6 \times n} \\ 0_{n \times p_v} & I_n \end{bmatrix} F_{td} = BF_{td} \tag{5}$$

where $F_{td} = [T^T, \tau_m^T]^T \in \mathbb{R}^{p_v+n}$. $T \in \mathbb{R}^{p_v}$ represents the vector of thruster forces, and $\tau_m \in \mathbb{R}^n$ represents the vector of actuator torques. $B_v \in \mathbb{R}^{6 \times p_v}$ is the thruster configuration matrix, and $B \in \mathbb{R}^{(6+n) \times (6+n)}$ is the thruster-actuator configuration matrix. It is known that for an under-actuated underwater vehicle, $p_v < 6$. Generally, for a manipulator, n joint motors are all available.

3. Proposed Redundancy Resolution

This section proposes a new redundancy resolution technique to generate energy-efficient trajectories for the vehicle and the manipulator. It is known that infinite solutions of the UVMS inverse kinematics can be obtained by inverting the mapping (1). The solution using the pseudo inverse of the Jacobian matrix is expressed as [2]

$$\zeta = J^+(R_B^I, q)\dot{x}_E \tag{6}$$

where $\dot{x}_E$ is the end-effector velocity vector. $J^+(R_B^I, q)$ is the pseudo inverse of the Jacobian matrix and $J^+(R_B^I, q) = J^T(R_B^I, q)(J(R_B^I, q)J^T(R_B^I, q))^{-1}$.

However, this solution does not exploit the redundant DOFs of the system, and it is not suitable from the perspective of energy consumption. Therefore, a new task-priority redundancy resolution technique is proposed in this section. In the proposed technique, the primary task is to map the end-effector variables into the joint-space variables, and two secondary tasks are provided to explore the kinematic redundancy for energy savings, joint limit avoidance and small roll and pitch angles kept for the vehicle, as shown in (7).

$$\zeta_d = J_W^+(\dot{x}_{Ed} - K_f e_E) + (I - J_W^+ J_W)[J_s^+(\eta, q)(\dot{\eta}_{sd} - K_s e_s) - \alpha K_\zeta J_\xi(\eta_2)\dot{\xi}] \tag{7}$$

where $J_W^+ = W^{-1}J^T(JW^{-1}J^T)^{-1}$ is the weighted pseudo-inverse Jacobian matrix. J is considered as the primary task Jacobian matrix and $W \in \mathbb{R}^{(6+n) \times (6+n)}$ is the motion distribution matrix with elements belonging to $[0, 1]$. When the diagonal elements of the former three rows of W are close to 1, the diagonal elements of the later n rows of W will be close to 0. This results in greater movement of the vehicle and less movement of the manipulator. Otherwise, when the diagonal elements of the former three rows of W are close to 0, the diagonal elements of the later n rows of W will be close to 1. This results in less movement of the vehicle and greater movement of the manipulator. The diagonal elements of the middle three rows of W correspond to the movement of the vehicle's attitude. The larger they are, the greater the movement of the vehicle's attitude. The off-diagonal elements of W describe the degrees of the coupling effects between the DOFs of the UVMS, which can refer to our previous work [15]. The closer the off-diagonal element of W to 1, the greater the corresponding coupling motion. $J_s(\eta, q)$ and $J_\xi(\eta_2)$ are the secondary task Jacobian matrices. It can be recognized that the secondary tasks are fulfilled in the null space, which will not affect the motion of the primary task. Moreover, the two secondary tasks have the same lower priority relative to the primary task. $\dot{x}_{Ed}$ is the primary-task vector and $e_E = x_E - x_{Ed}$ is the error of the primary task. $\dot{\eta}_{sd}$ is a secondary-task vector to achieve system coordination between its rotational subsystem and translational subsystem, including the vehicle attitudes and joint angles, and $e_s = \dot{\eta}_s - \dot{\eta}_{sd}$ is its error. K_f and K_s are positive definite matrices. The other secondary task vector is the velocity vector of the UVMS $\dot{\xi} = [\dot{\eta}_1^T, \dot{\eta}_2^T, \dot{q}^T]^T$, which contributes to the system's self-motion utilized for reducing energy requirements. K_ζ is a diagonal matrix whose elements belong to $[0, \infty)$. The larger the diagonal element of K_ζ, the greater the corresponding coupling motion. For instance, if the diagonal element of the secondary row of K_ζ is larger, the vehicle will have a larger yaw angle according to the coupling effects. Similarly, if the diagonal element of the third row of K_ζ is larger, the vehicle will have a larger pitch angle. α is a coefficient belonging to $[0, 1]$, which is used to adjust the values of K_ζ.

To effectively utilize the self-motion during the entire UVMS motion, α is defined as a nonlinear function related to time $t \geq 0$, as given in (8).

$$\alpha = \begin{cases} -0.5(1 - e^{-\lambda(t-t_s)}) + 0.5 & t > t_s \\ 0.5(1 - e^{-\lambda(t-t_s)}) + 0.5 & t \leq t_s \end{cases} \tag{8}$$

where t_s relatives to the time at which the system enters deceleration phase. λ is the coefficient and $\lambda > 0$.

The curve of the nonlinear function is shown in Figure 2. It can be recognized that the smaller the value of λ, the smoother the variations of the nonlinear function. Therefore, it is better to choose a small value of λ to ensure smooth movement of the UVMS.

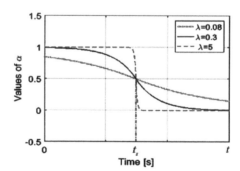

Figure 2. Nonlinear function of α.

4. Control Design

The purpose of control design is to obtain the values of thruster forces and actuator torques in order to drive the UVMS to the desired trajectory. In addition, the robustness of the designed controller is important in the presence of model parameter uncertainties, time-varying external disturbances, payload variations and sensory noises. In this section, for precise and robust control of the UVMS, a new coordinated motion controller including inertial delay control (IDC) and a fuzzy compensator is proposed. Besides, the proposed controller uses the estimated UVMS states via an EKF.

4.1. Design of an EKF

Due to the presence of sensory measurement noises, the vehicle and manipulator positions measured by sensors are not inaccurate. Therefore, it is necessary to utilize a nonlinear filter to estimate the system's states. As the extended Kalman filter (EKF) is simple and easy to complete and has low computational complexity, the EKF is used in this study. It is necessary to obtain a linear model during the KF design process. The dynamic equations of the UVMS can be linearized by ignoring the higher order terms in the expended Taylor series. The state vector of the system, e.g., position/attitude and velocity vectors, is defined as $X = [\xi^T, \zeta^T]^T$. The measurement model can be expressed as $Z = h(X) = \xi$.

Based on (2) and (3), the time derivative of the system state vector X can be obtained as

$$\dot{X} = f(X, t) = \begin{bmatrix} J_{\xi}^{-1}(\eta_2)\dot{\zeta} \\ M^{-1}(\tau - \tau_d - b) \end{bmatrix} \tag{9}$$

where $f(X, t)$ is considered as the estimated model of the system.

With the additive Gaussian white noise, the predicted system state vector $\hat{X}_{k+1}^-$ at $t = t_{k+1}$ is given in (10), and the predicted measurement state vector Z_{k+1}^- at $t = t_{k+1}$ is shown in (11). Then the covariance matrix of the predicted state vector can be obtained as shown in (12).

$$\hat{X}_{k+1}^- = \hat{X}_k + f(\hat{X}_k, t) + Q_k \tag{10}$$

$$Z_{k+1}^- = h(\hat{X}_{k+1}^-) + \Gamma_k \tag{11}$$

$$P_{k+1}^- = \phi_k P_k \phi_k^T + Q_k \tag{12}$$

where $\hat{X}_k$ is the predicted state vector at t_k. Q_k is the vector of system noises. Γ_k is the vector of measurement noises. P_k is the covariance matrix of the estimated system states.

Therefore, the estimated system states can be obtained through the following correction step:

$$K_{k+1} = P_{k+1}^- H_{k+1}^T (H_{k+1} P_{k+1}^- H_{k+1}^T + \Gamma_{k+1})^{-1} \tag{13}$$

$$\hat{X}_{k+1} = \hat{X}_{k+1}^- + K_{k+1}(Z_{k+1} - Z_{k+1}^-) \tag{14}$$

$$P_{k+1} = (I - K_{k+1} H_{k+1}) P_{k+1}^- \tag{15}$$

where $\phi_k = \left.\dfrac{\partial f(X,t)}{\partial X^T}\right|_{X=\hat{X}_k}$ and $H_{k+1} = \left.\dfrac{\partial h(X)}{\partial X^T}\right|_{X=\hat{X}_{k+1}^-}$.

4.2. Design of a Tracking Controller

The control objective is to make sure that the tracking errors quickly converge to zero under the conditions of model parameter uncertainties, time-varying external disturbances and payload. First, the tracking errors of the system are defined. The vector of end-effector tracking errors is shown in (16), and the vectors of tracking errors in the joint space are shown in (17)–(19).

$$e_E = \hat{x}_E - x_{Ed} \tag{16}$$

$$e = \begin{bmatrix} R_I^B(\hat{\eta}_1 - \eta_{1d}) \\ \eta\varepsilon_d - \eta\varepsilon_d + S(\varepsilon)\varepsilon_d \\ \hat{q} - q_d \end{bmatrix} \tag{17}$$

$$\dot{e} = \hat{\zeta} - \zeta_d \tag{18}$$

$$\ddot{e} = \dot{\hat{\zeta}} - \dot{\zeta}_d \tag{19}$$

where the superscript $(\hat{\cdot})$ denotes the corresponding estimated values via EKF, and the subscript $(\cdot)_d$ denotes the corresponding desired values. $[\varepsilon, \eta]$ and $[\varepsilon_d, \eta_d]$ are the quaternions of $\hat{\eta}_2$ and η_{2d}.

Then, the tracking controller based on feedback linearization is given as

$$u = -M^*(k_1 e + k_2 \dot{e} - \dot{\zeta}_d) + C^*\hat{\zeta} + D^*\hat{\zeta} + g - J^T k_E e_E - \hat{\delta} \tag{20}$$

where $J = J(R_B^I, q)$ is the Jacobian matrix of the system, as given in (1). k_1, k_2 and k_E are the positive symmetric matrices. $\hat{\delta}$ is the estimated vector of the lumped uncertainties and disturbances, which is described in the next subsection.

4.3. Inertial Delay Control (IDC)

Due to the underwater circumstances, the dynamic equations of the UVMS include unknown external disturbances and an amount of parameter uncertainties caused by identification errors. These lumped uncertainties and disturbances can be expressed as (21) with reference to (3) and (4).

$$\delta = -(\Delta M \dot{\hat{\zeta}} + \Delta C \hat{\zeta} + \Delta D \hat{\zeta}) + \tau_{dis} \tag{21}$$

where $\dot{\hat{\zeta}}$ and $\hat{\zeta}$ denote the estimates of the system states via EKF.

Then, based on (3), (4) and (21), the acceleration vector of the system can be obtained as

$$\dot{\hat{\zeta}} = -(M^*)^{-1}(C^*\hat{\zeta} + D^*\hat{\zeta} + g) + (M^*)^{-1}(u + \delta) \tag{22}$$

Substituting the proposed control law (20) in (22), dynamical equation of the tracking errors is

$$\ddot{e} + k_2\dot{e} + k_1 e + (M^*)^{-1}J^T k_E e_E = (M^*)^{-1}e_\delta \tag{23}$$

where $e_\delta = \delta - \hat{\delta}$ is the estimated error vector.

It is assumed that a slow-varying signal can be approximated and estimated by a filter with appropriate bandwidth [37]. Based on this assumption, the uncertainty and disturbance estimator (UDE) is proposed for estimating slow-varying uncertainties [35,37]. Then, the estimations of lumped uncertainties and disturbances $\hat{\delta}$ can be given as

$$\hat{\delta} = G_f(s)\delta \tag{24}$$

where $G_f(s)$ is a strictly proper low-pass filter possessing a uniform steady-state gain and a sufficiently large bandwidth. Based on (24), it is found that by passing the lumped uncertainties and disturbances δ through a inertial filter $G_f(s)$, the estimation vector $\hat{\delta}$ can be obtained. The UDE method is redefined as inertial delay control (IDC) [37], because it is analogous to the time delay control (TDC) which delays the plant signals in time to obtain the estimates.

Based on (23) and (24), we can obtain

$$\hat{\delta} = G_f(s)[M^*(\ddot{e} + k_2\dot{e} + k_1 e) + J^T k_E e_E + \hat{\delta}] \tag{25}$$

A choice of $G_f(s)$ with first order is given by

$$G_f(s) = \frac{I}{I + Ts} \tag{26}$$

where T is a diagonal matrix with small positive constant. I is the identity matrix.

Then (25) can be rewritten as

$$T\dot{\hat{\delta}} + \hat{\delta} = M^*(\ddot{e} + k_2\dot{e} + k_1 e) + J^T k_E e_E + \hat{\delta} \tag{27}$$

Therefore, the estimates of the lumped uncertainties and disturbances can be obtained as

$$\hat{\delta} = T^{-1}M^*(\dot{e} + k_2 e + k_1 \int_0^t e \, dt) + T^{-1}J^T k_E \int_0^t e_E \, dt \tag{28}$$

From (25) and (26), the equation of estimated errors can be written as

$$\dot{e}_\delta = -T^{-1} e_\delta + \dot{\delta} \tag{29}$$

If the lumped uncertainties and disturbances δ are slowly varying, then $\dot{\delta}$ is small and $\dot{\delta} \approx 0$. Therefore, the estimated errors (e_δ) go to zero asymptotically. If $\dot{\delta}$ is not small, but $\ddot{\delta}$ is small, e_δ is ultimately bounded and the estimated accuracy can be improved by estimating δ and $\dot{\delta}$.

4.4. Fuzzy Compensator

Based on the estimates via IDC, the fuzzy compensator is given as

$$u_{fuzzy} = \rho \hat{\delta} + \epsilon \tag{30}$$

where $\rho = \text{diag}(\rho_1, \rho_2 \cdots \rho_{6+n})$ is the parameter of the fuzzy compensator, and ϵ is a constant vector.

The fuzzy compensator is a multiple-inputs-single-output fuzzy logic controller (FLC) with the joint-space system errors e_i and e_j as two input variables and ρ_i after defuzzification and denormalization as an output variable. Denote the system error vector e (as given in (17)) as $e = [e_1, e_2 \cdots e_i \cdots e_{6+n}]$. The main advantage of this fuzzy compensator is that the required fuzzy rules take the dynamic coupling between the vehicle and the manipulator [15,16] into account. It is known that the roll, pitch and yaw motions of the vehicle are coupled with its surge, sway and heave motions. As the roll and pitch angles should be kept small for properly working of the bottom sensors, it is assumed that the surge and sway motions are mostly affected by the yaw angle. Note that the pitch and heave motions are interactive, and the manipulator's joints 2 and 3 are interactive. The position of manipulator's joint 1 is mostly affected by the sway motion. Based on these analysis, the fuzzy rules are given in Table 1. Table 2 shows the relationships between an output and two input variables.

Table 1. Rule base for ρ_i.

ρ_i \ $|e_j|$ / $|e_i|$	ZE	PS	PM	PB
ZE	ZE	ZE	PS	PM
PS	PS	PS	PM	PB
PM	PM	PM	PB	PB
PB	PB	PB	PB	PB

The following symbols are used in Table 1: ZE (zeros), PS (positive small), PM (positive medium), PB (positive big). Figure 3a,b shows the member functions of the normalized input and output variables respectively. After the fuzzification stage, the Mamdani inference method is used for fuzzy implication, and then the centroid method is used for defuzzification. Finally, based on denormalization the actual output variables can be obtained.

Table 2. Relationships between two input variables and an output variable.

inputs	$|e_i|$	$|e_1|$	$|e_2|$	$|e_3|$	$|e_5|$	$|e_6|$	$|e_7|$	$|e_8|$	$|e_9|$
	$|e_j|$	$|e_6|$	$|e_6|$	$|e_5|$	$|e_3|$	$|e_2|$	$|e_2|$	$|e_9|$	$|e_8|$
outputs	ρ_i	ρ_1	ρ_2	ρ_3	ρ_5	ρ_6	ρ_7	ρ_8	ρ_9

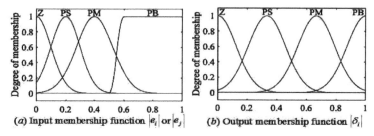

Figure 3. Input and output membership functions.

Incorporating the fuzzy compensator, the proposed coordinated motion controller is given as

$$u_c = u - u_{fuzzy} \tag{31}$$

Then we can obtain the vector of thruster forces and actuator torques F_{td}, as shown in (32).

$$F_{td} = B^+ u_c = \begin{bmatrix} B_v^+ & 0_{p_v \times n} \\ 0_{n \times 6} & I_n \end{bmatrix} u_c \tag{32}$$

where B_v^+ is the pseudo inverse of B_v and B^+ is the pseudo inverse of B.

Therefore, the generalized force vector τ can be obtained based on (5). For an under-actuated UVMS, $\tau = u$ except that the elements of τ corresponding to the underacted motions are zeros.

The proposed control system is schematically represented by a block diagram in Figure 4. The controller block includes five sub blocks to calculate a control vector; i.e., the tracking controller, IDC, fuzzy compensator, B^+ and B blocks. In addition to system dynamics, the tracking controller also requires tracking errors of end-effector positions and joint-space states. The end-effector position tracking errors are calculated according to the desired end-effector positions derived from the trajectory planning block and the estimated end-effector positions obtained from the forward kinematics block using the estimated joint-space states. The estimates of joint-space states are obtained from the EKF block. The proposed redundancy resolution block generates the required joint-space trajectories for the desired tasks. The IDC block estimates the lumped uncertainties and disturbances of the system. The fuzzy compensator reduces the influences of perturbation on the UVMS.

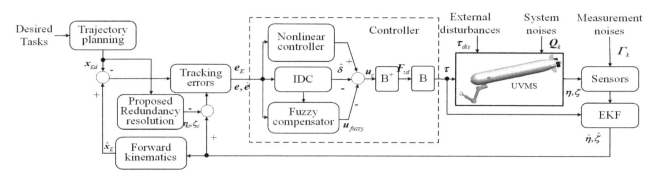

Figure 4. Block diagram of the proposed controller.

4.5. Stability Analysis

We define a Lyapunov function which is positive definite as:

$$V(e, \dot{e}, e_E, e_\delta) = \frac{1}{2} e^T k_1 k_2 e + \frac{1}{2} \dot{e}^T k_2 \dot{e} + \frac{1}{2} e_\delta{}^T e_\delta + \frac{1}{2} e_E^T k_2 (M^*)^{-1} k_E e_E \tag{33}$$

Differentiating $V(x_1, x_2, e_E)$ yields

$$V(e,\dot{e},e_E,e_\delta) = e^T k_1 k_2 \dot{e} + \dot{e}^T k_2 \ddot{e} + e_\delta{}^T \dot{e}_\delta + e_E^T k_2 (M^*)^{-1} k_E \dot{e}_E \tag{34}$$

where M^* is assumed to be constant.

Substituting the proposed control law (31) in (22), and taking into account (29), (34) can be rewritten as

$$
\begin{aligned}
\dot{V} &= e^T k_1 k_2 \dot{e} + e_E^T k_2 (M^*)^{-1} k_E \dot{e}_E + e_\delta{}^T (-T^{-1} e_\delta + \dot{\delta}) \\
&\quad + \dot{e}^T k_2 (-k_2 \dot{e} - k_1 e - (M^*)^{-1} J^T k_E e_E + (M^*)^{-1} (e_\delta - \rho \hat{\delta} - \epsilon)) \\
&= -\dot{e}^T \|k_2\|^2 \dot{e} - e_\delta{}^T T^{-1} e_\delta + e_\delta{}^T \dot{\delta} + \dot{e}^T k_2 (M^*)^{-1} ((1+\rho) e_\delta - \rho \delta - \epsilon) \\
&= -\dot{e}^T k_2 [k_2 \dot{e} - (M^*)^{-1} ((1+\rho) e_\delta - \rho \delta - \epsilon)] - e_\delta (T^{-1} e_\delta - \dot{\delta})
\end{aligned}
\tag{35}
$$

By choosing large enough values of k_2, small values of ρ and small enough values of T such that

$$k_2 \dot{e} \geq (M^*)^{-1}((1+\rho)e_\delta - \rho\delta - \epsilon), \ \|\rho\delta\| \leq \sigma, \ T^{-1}e_\delta \geq \dot{\delta}, \tag{36}$$

$V(e,\dot{e},e_E,e_\delta)$ is negative semi-definite, where $\sigma \to 0$ is a vector with smaller positive values. Consequently, the tracking errors and estimated errors of the system all converge to zero asymptotically; i.e.,

$$\lim_{t\to\infty} e \to 0, \ \lim_{t\to\infty} \dot{e} \to 0, \ \lim_{t\to\infty} e_E \to 0, \ \lim_{t\to\infty} e_\delta \to 0 \tag{37}$$

Therefore, the closed-loop system is asymptotically stable in the entire state space.

5. Simulation Studies

To verify the performance of the proposed technique, numerical simulations were performed on a UVMS with a torpedo-type AUV and a 3-DOF underwater manipulator [15] shown in Figure 1. The AUV is driven by five thrusters in total and its thruster configuration is shown in Figure 5. The thruster configuration matrix B_v and its pseudo inverse B_v^+ are shown in the Appendix A. The parameters for the AUV and manipulator are given in the Appendix A and Tables 3 and 4. From Table 4, it can be seen that the whole system is neutrally buoyant, while the manipulator's links have negative buoyancy. Thus, the UVMS used for numerical simulations in this paper can approximate to the real system.

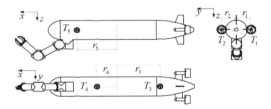

Figure 5. Thruster distribution of the AUV.

Table 3. D-H parameters of the manipulator.

Joint (k)	Offset (α_{k-1})	Length (a_{k-1})	Distance (d_k)	Angle (θ_k)
1	$0°$	0	0	q_1
2	$90°$	l_{q1}	0	q_2
3	$0°$	l_{q2}	0	q_3
4	$0°$	l_{q3}	0	$q_4 = 0°$

Table 4. The list of UVMS parameters.

	Length (m)	Diameter (m)	Mass (kg)	Buoyancy (N)	Weight (N)
AUV	1.78	0.26	78.2	771.59	767.14
Link1	0.1	0.06	1.28	11.43	12.54
Link2	0.3	0.0425	1.92	17.2	18.87
Link3	0.3	0.0425	1.92	17.2	18.87

5.1. Simulation Conditions

In the simulations, the UVMS's end-effector is commanded to follow a spatial circle with diameter 2.24 m and a straight line of length 7.0 m. The simulation time of the circular trajectory is 50 s, where the initial 10 s is used for acceleration, 30 s is used to follow the circle and the final 10 s is used for deceleration. The simulation time of the straight-line trajectory is 30 s, where in the initial 10 s the acceleration is a half-period sine function, and then it maintains zeros, and in the final 10 s it is a half-period negative sine function. The UVMS's end-effector maintains orientation during the two trajectory tracking tasks. The initial desired positions and orientations are the same as the initial actual positions and orientations. The initial desired and actual velocities and accelerations are zeros. The average speeds of the two trajectories both are 0.23 m/s. The sampling time for the simulation is $t = 20$ ms.

In this case, the model uncertainties, external disturbances, payload and sensory noises in position and orientation measurements are introduced for simulating the real working environment. To reflect the uncertainties, it is assumed that the modeling inaccuracy for each parameter is 10%. The vector of time-varying ocean currents in the inertial frame is assumed to be governed by (38). It is supposed that the end-effector of the manipulator is attached with a payload of 1 kg (in water). The following sensory noises are introduced: gaussian noise of 0.01 m mean and 0.01 m standard deviation for the vehicle position measurements; $0.5°$ mean and $0.5°$ standard deviation for the vehicle attitude measurements; and $0.05°$ mean and $0.05°$ standard deviation for the manipulator's joint position measurements. In addition, the thruster dynamic characteristics are inserted into the simulation. Suppose that the thruster response delay time is 50 ms, and its efficiency is 95%.

$$v_c = \begin{bmatrix} 0.15 + 0.1\cos(0.3t), & 0.05\cos(0.1\pi t), & 0.1\cos(0.2t), & 0, & 0, & 0 \end{bmatrix}^T \text{ m/s} \qquad (38)$$

To implement solution (7), the primary task vector is $x_{Ed} = [\eta_{E1d}^T, \eta_{E2d}^T]^T$. $k_f = \text{diag}(4, 4, 4, 6, 6, 6)$ and $k_s = 25I$ for the two trajectories. Other parameters are shown in Table 5. A secondary task is designed to align the vehicle orientation and the joint position with the primary task in terms of reducing the coupling effects, as shown in (39).

$$\dot{\eta}_{sd} = \begin{bmatrix} \dot{\phi} \\ \dot{\theta} \\ \dot{\psi} \\ \dot{q}_1 \\ \dot{q}_2 \end{bmatrix} = \begin{bmatrix} 0 & 0 & 0 \\ 0 & 0 & -\alpha_1 \\ 0 & \alpha_2 & 0 \\ 0 & -\alpha_3 & 0 \\ 0 & 0 & -\alpha_4 \end{bmatrix} \dot{\eta}_{E1d} \qquad (39)$$

where $\dot{\eta}_{E1d}$ is the linear part of x_{Ed}.

Table 5. Parameters for the proposed redundancy resolution technique.

Trajectory	α_1	α_2	α_3	α_4	λ	t_s (s)	K_ζ
Straight line	−0.02	0.13	0	0.05	0.3	15	$\text{diag}(0, 2, 0.2, 0, 1, 1, 1, 1)$
Circle	−0.06	0.2	−0.2	0.15	0.3	45	$\text{diag}(0.1, 0.3, 0.1, 0, 1, 1, 1, 1, 1)$

The distribution matrix is defined as

$$\boldsymbol{W}^{-1} = \begin{bmatrix} 0.02\boldsymbol{I}_{3\times3} & \boldsymbol{0}_{3\times3} & \boldsymbol{0}_{3\times3} \\ \boldsymbol{W}_1 & 0.4\boldsymbol{I}_{3\times3} & \boldsymbol{0}_{3\times3} \\ \boldsymbol{W}_2 & \boldsymbol{0}_{3\times3} & 0.98\boldsymbol{I}_{3\times3} \end{bmatrix}, \ \boldsymbol{W}_1 = \begin{bmatrix} 0 & 0 & 0 \\ 0 & 0 & 0.2 \\ 0 & 0.35 & 0 \end{bmatrix}, \boldsymbol{W}_2 = \begin{bmatrix} 0 & 0.25 & 0 \\ 0 & 0 & 0.5 \\ 0 & 0 & 0 \end{bmatrix} \quad (40)$$

To illustrate the effectiveness of the proposed redundancy resolution technique in terms of energy savings, the comparative redundancy resolution technique is given in (41).

$$\zeta_d = \boldsymbol{J}_W^+(\dot{\boldsymbol{x}}_{Ed} - \boldsymbol{K}_f e_E) + (\boldsymbol{I} - \boldsymbol{J}_W^+ \boldsymbol{J}_W)[\boldsymbol{J}_s^+(\boldsymbol{\eta}, \boldsymbol{q})(\dot{\boldsymbol{\eta}}_{sd} - \boldsymbol{K}_s e_s)] \quad (41)$$

where the difference between (7) and (41) is that the secondary task vector $\dot{\boldsymbol{\xi}}$ is not included in the compared technique.

The proposed control scheme is compared with the H∞-EKF method [23] which is given by

$$\boldsymbol{u}_c = \boldsymbol{M}(\boldsymbol{q})(\ddot{\boldsymbol{q}}_d - \boldsymbol{V}\dot{\boldsymbol{e}} - \boldsymbol{P}\boldsymbol{e}) + \boldsymbol{H}(\boldsymbol{q}, \dot{\boldsymbol{q}}) + \boldsymbol{M}(\boldsymbol{q})(-\boldsymbol{R}^{-1}\boldsymbol{B}^T\boldsymbol{P}\boldsymbol{e}) + \boldsymbol{\tau}_c \quad (42)$$

where $\boldsymbol{V}$ and $\boldsymbol{P}$ are the derivative and the proportional gain matrices. $\boldsymbol{q}$ and $\dot{\boldsymbol{q}}$ are the estimated vectors from EKF. $\boldsymbol{\tau}_c$. is the disturbance estimation from EKF as well. $\boldsymbol{R}$ is the given positive definite matrix.

For simple representation, the proposed redundancy resolution is termed case 1 (c1), and the comparative redundancy resolution is termed case 2 (c2). Hence, the proposed control scheme based on the proposed redundancy resolution is termed proposed control$_{c1}$; the H∞-EKF method based on the proposed redundancy resolution is termed H∞-EKF$_{c1}$; and the proposed control based on the comparative redundancy resolution is termed proposed control$_{c2}$.

5.2. Results and Discussion

The results of numerical simulations are shown in Figures 6–13. Figures 6–8 present the desired and actual spatial trajectories and their tracking errors. From these results it is observed that the proposed controller drives the UVMS to track the desired spatial linear and circular trajectories quite satisfactorily in both the proposed redundancy resolution technique (c1) and the comparative redundancy resolution technique (c2). Moreover, the proposed control scheme outperforms the H∞-EKF method, and has smaller tracking errors in both positions and orientations under the conditions of model uncertainties, time-varying ocean currents, payload and sensory noises. Even though the H∞-EKF method adopted a H∞ robust controller to compensate the estimated bias from the EKF, the residual tracking errors can not be fully eliminated, as shown in Figure 6b,c and Figure 6e,f. The proposed controller performs better than the H∞-EKF method in terms of robustness, which is dedicated to the IDC and fuzzy compensator for reducing the perturbation effects.

Figure 9 plots the norm of the vector $\boldsymbol{F}_{td}$ (i.e., thruster forces and actuator torques) and energy consumption of the UVMS. It can be noted that the comparative redundancy resolution technique (c2) is consuming more energy in generating trajectories for the vehicle and manipulator during both the linear and circular trajectories tracking. However in the proposed redundancy resolution technique (c1), the UVMS states are adjusted by self-motion to minimize interaction effects between the vehicle and the manipulator. This is because of the introduction of the secondary task vector $\dot{\boldsymbol{\xi}}$ and the nonlinear function.

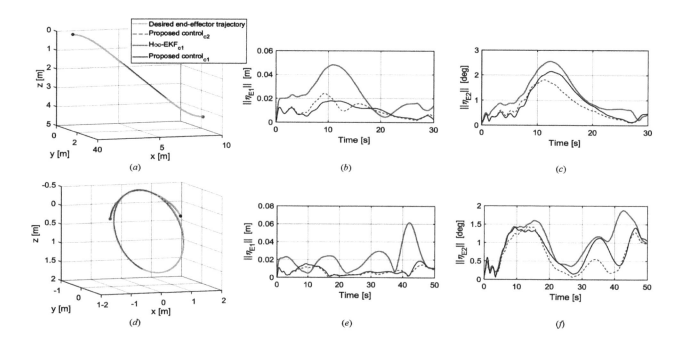

Figure 6. Spatial linear and circular trajectories and their tracking errors. (*a*) Desired linear trajectory and tracking control results; (*b,c*) position tracking errors in positions and orientations; (*d*) desired circular trajectory and tracking control results; (*e' f*) tracking errors in positions and orientations.

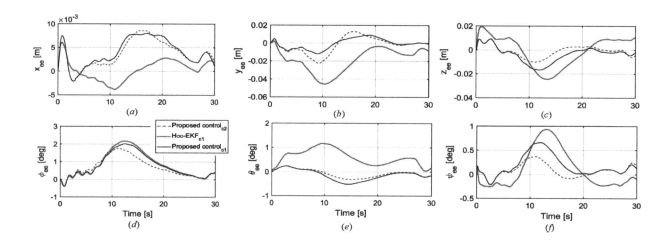

Figure 7. End-effector errors when tracking the linear trajectory. (*a*) x_{ee} error, (*b*) y_{ee} error, (*c*) z_{ee} error, (*d*) tracking errors in the end-effector roll direction, (*e*) tracking errors in the end-effector pitch direction, (*f*) tracking errors in the end-effector yaw direction.

For better understanding, the generated trajectories for the vehicle positions/attitudes and manipulator positions are presented in Figures 10 and 11. It can be seen from the results that the generated trajectories have larger differences on vehicle attitudes and joint angles than vehicle positions. This is because the adjustment of the vehicle position has little effect on reducing the interactive forces between the vehicle and the manipulator without affecting the primary task. Consequently, the energy consumption can be reduced by changing the vehicle attitude and joint angles. In addition, it is observed from Figures 10 and 11 that the small roll and pitch angles of the vehicle are kept in the proposed control scheme, which contributes to properly working of the vehicle's onboard sensors.

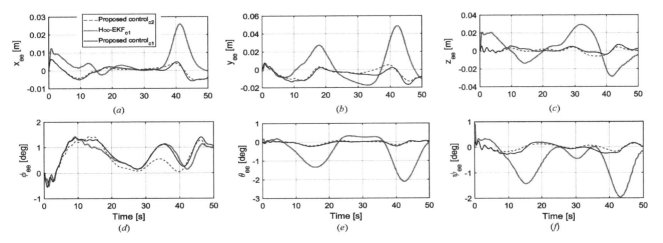

Figure 8. End-effector errors when tracking the circular trajectory. (*a*) x_{ee} error, (*b*) y_{ee} error, (*c*) z_{ee} error, (*d*) tracking errors in the end-effector roll direction, (*e*) tracking errors in the end-effector pitch direction, (*f*) tracking errors in the end-effector yaw direction.

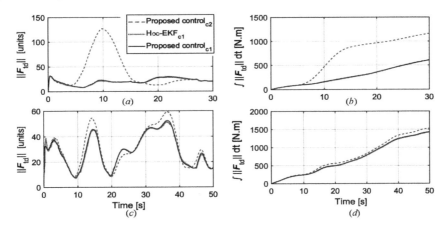

Figure 9. Time histories of the norm of the vector F_{td} and UVMS energy consumption. (*a,b*) Results from the linear trajectory tracking, (*c,d*) results from the circular trajectory tracking, (F_{td}; i.e., the vector of thruster forces and actuator torques).

Figures 12 and 13 show the required thruster forces for the vehicle and actuator torques for the manipulator during the linear and circular trajectory tracking. It is observed that the thruster forces for the two trajectories are less in the proposed redundancy resolution technique (c1), which results in the reduced energy consumption. In addition, the thruster forces and actuator torques for both trajectories in the proposed control$_{c1}$ are within their constraints (± 60 N for the thrusters and ± 3 N·m for the actuators).

The quantitative indexes of the time integral of tracking errors and energy consumption are listed in Table 6. From these indices, it is indicated that the tracking error in the proposed control$_{c1}$ is smaller than that in the H∞-EKF$_{c1}$ method, and the energy consumption in the proposed control$_{c1}$ is less than that in the proposed control$_{c2}$. Overall, the proposed control scheme based on the proposed redundancy resolution technique (c1) ensures the precise and robust performance with a reduced energy requirement under the conditions of model parameter uncertainties, time-varying ocean currents, payload and sensory noises.

In the simulations, we have taken the model parameter uncertainties, time-varying external disturbances, payload and sensory noises into consideration. However, in a practical case, these lumped uncertainties and disturbances may be more complicated, and hence can not be simulated. Even though the results from computer simulations are promising, it is necessary to validate the effectiveness of the proposed control scheme and the proposed redundancy resolution technique through experiments in a water pool or at sea. This is our future work.

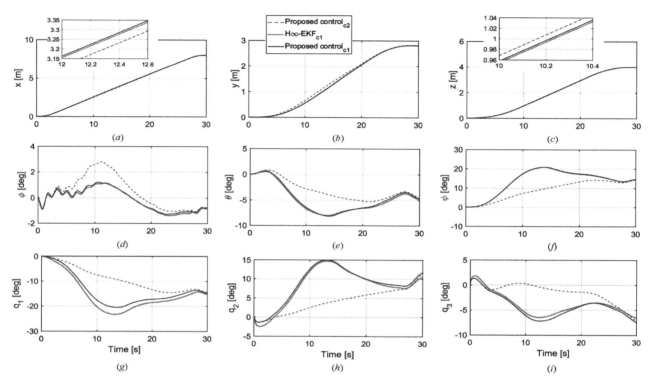

Figure 10. Joint-space positions for the straight line trajectory. (*a*) X position, (*b*) y position, (*c*) z position, (*d*) roll angle, (*e*) pitch angle, (*f*) yaw angle, (*g*) joint **1** angle, (*h*) joint **2** angle, (*i*) joint **3** angle.

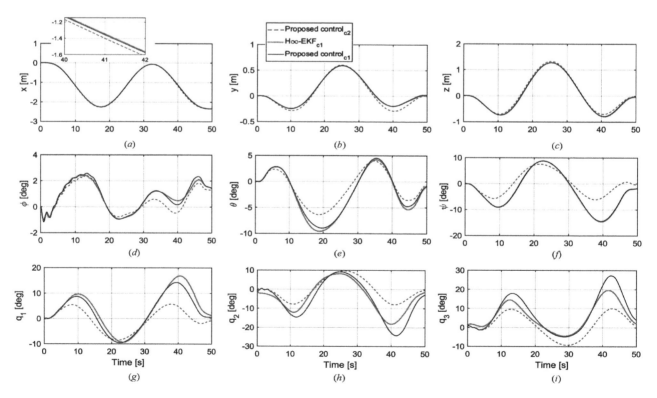

Figure 11. Joint-space positions for the circular trajectory. (*a*) X position, (*b*) y position, (*c*) z position, (*d*) roll angle, (*e*) pitch angle, (*f*) yaw angle, (*g*) joint **1** angle, (*h*) joint **2** angle, (*i*) joint **3** angle.

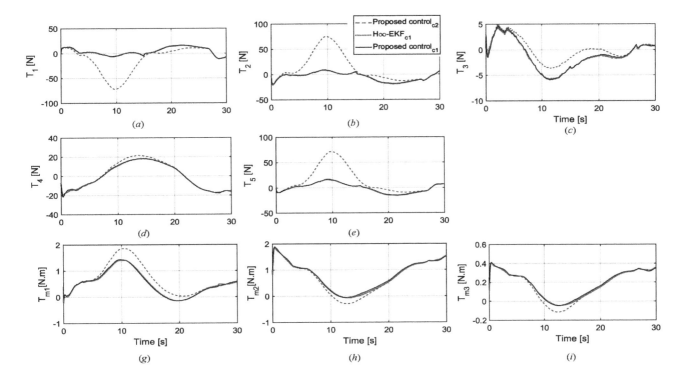

Figure 12. Thruster forces and actuator torques for the line trajectory tracking. (a–e) Thruster forces T_1, T_2, T_3, T_4 and T_5; (g–i) actuator torques T_{m1}, T_{m2} and T_{m3}.

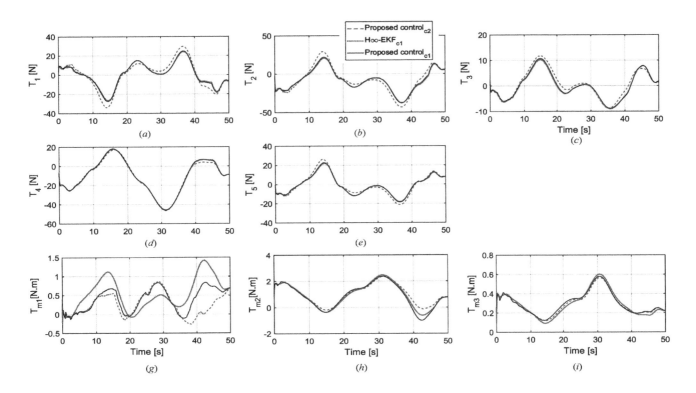

Figure 13. Thruster forces and actuator torques for the circular trajectory tracking. (a–e) Thruster forces T_1, T_2, T_3, T_4 and T_5; (g–i) actuator torques T_{m1}, T_{m2} and T_{m3}.

Table 6. Performance analysis of the UVMS for the linear and circular trajectories tracking.

Control Schemes	Linear Trajectory			Circular Trajectory		
	ISE_p	ISE_o	$\int \|\|F_{td}\|\|dt$	ISE_p	ISE_0	$\int \|\|F_{td}\|\|dt$
Proposed control$_{c2}$	0.3083	22.5597	1161.6	0.3318	32.2834	1532.6
H∞ − EKF$_{c1}$	0.6869	34.5853	607.2053	1.0775	63.1945	1420.4
Proposed control$_{c1}$	0.2905	26.1319	603.8734	0.3708	40.1401	1428.3

Note : $e_E = [(\tilde{\eta}_{E1})^T, (\tilde{\eta}_{E2})^T]^T$, $ISE_p = \int \|\|\tilde{\eta}_{E1}\|\|dt$, $ISE_o = \int \|\|\tilde{\eta}_{E2}\|\|dt$.

6. Conclusions

This paper presents a motion planning and coordinated control scheme for the trajectory tracking of the UVMS. A new secondary task with a nonlinear coefficient for redundancy resolution of the UVMS is proposed. In this way, the interactive effects between the vehicle and the manipulator can be minimized and the energy consumption of the UVMS is reduced. Simulation results show that the energy consumption based on the proposed redundancy resolution technique (proposed control$_{c1}$) is reduced by 48% (in the linear trajectory tracking) and 6% (in the circular trajectory tracking), compared with the comparative redundancy resolution technique (proposed control$_{c2}$). The proposed redundancy resolution technique is simple in design and easy to implement. Furthermore, a control scheme including a fuzzy compensator and a tracking control with joint-space errors, end-effector errors and inertial delay control (IDC) is proposed.

The proposed control scheme ensures precise and robust tracking performance in the presence of model uncertainties, time-varying ocean currents, payload and sensory noises. Simulation results show that the position and orientation tracking precisions based on the proposed control$_{c1}$ are reduced by 57.7% and 24.4% (in the linear trajectory tracking) and 65.6% and 36.5% (in the circular trajectory tracking), compared with the H$_\infty$−EKF$_{c1}$ method [23]. Even though the effectiveness of the proposed redundancy resolution technique and coordinated motion control scheme were validated through numerical simulations, experiments should be carried out on the real UVMS to further enhance the computer simulation results, which will be done in the future.

Author Contributions: Conceptualization, H.H., Y.W. and X.Y.; data curation, H.H.; formal analysis, H.H.; funding acquisition, Y.W. and X.Y.; investigation, H.H.; methodology, H.H.; project administration, Y.W. and X.Y.; resources, Y.W. and W.L.; software, H.H.; supervision, Y.W., X.Y. and W.L.; validation, H.H., Y.W., X.Y. and W.L.; visualization, H.H.; writing—original draft, H.H.; writing—review editing, Y.W. and X.Y. All authors have read and agreed to the published version of the manuscript.

Acknowledgments: The authors would like to thank Shuai Li, Fan Zhang and Shitong Du, who provided insight and expertise that greatly assisted the writing of this manuscript.

Appendix A. Simulation Data of the AUV

For the torpedo-type AUV, its center of buoyancy (CB) is $(0,0,0.02)$ m, its center of gravity (CG) is $(0,0,0)$ m and its moment of inertia is $(0.69, 16.82, 16.82)$ kg · m^2. The AUV is equipped with an underwater manipulator, and the manipulator's base position is $p_0^B = (0.6, 0, 0.2)^T$ m. Some model parameters of the AUV given in (3) are composed of rigid-body terms and hydrodynamic terms [42], as shown in the following equations.

$$M_V = M_{RB} + M_A, \quad C_V(\boldsymbol{v}_r) = C_{RB}(\boldsymbol{v}_r) + C_A(\boldsymbol{v}_r), \quad D_V(\boldsymbol{v}_r) = D_{NL}\mathrm{diag}\,(|\boldsymbol{v}_r|) + D_L\mathrm{diag}(|\boldsymbol{v}_r|)$$

where M_{RB} and C_{RB} are the rigid-body terms which represent the inertia matrix and the Coriolis and centripetal matrix. M_A, C_A, D_{NL} and D_L are the matrices related to the hydrodynamic forces. M_A and C_A are the added mass matrix and the added Coriolis and centripetal matrix. D_{NL} and D_L are the quadratic damping matrix and the lift matrix. M_A, C_A, D_{NL} and D_L are given in the following equations [15].

$$M_A = -\begin{bmatrix} X_{\dot{u}} & 0 & 0 & 0 & 0 & 0 \\ 0 & Y_{\dot{v}} & 0 & 0 & 0 & Y_{\dot{r}} \\ 0 & 0 & Z_{\dot{w}} & 0 & Z_{\dot{q}} & 0 \\ 0 & 0 & 0 & K_{\dot{p}} & 0 & 0 \\ 0 & 0 & M_{\dot{w}} & 0 & M_{\dot{q}} & 0 \\ 0 & N_{\dot{v}} & 0 & 0 & 0 & N_{\dot{r}} \end{bmatrix} = \begin{bmatrix} A_{11} & A_{12} \\ A_{21} & A_{22} \end{bmatrix}, C_A = \begin{bmatrix} \mathbf{0}_{3\times3} & -S(A_{11}\boldsymbol{v}_1 + A_{12}\boldsymbol{v}_2) \\ -S(A_{11}\boldsymbol{v}_1 + A_{12}\boldsymbol{v}_2) & -S(A_{21}\boldsymbol{v}_1 + A_{22}\boldsymbol{v}_2) \end{bmatrix}$$

$$D_{NL} = -\begin{bmatrix} X_{u|u|} & 0 & 0 & 0 & 0 & 0 \\ 0 & Y_{v|v|} & 0 & 0 & 0 & Y_{r|r|} \\ 0 & 0 & Z_{w|w|} & 0 & Z_{q|q|} & 0 \\ 0 & 0 & 0 & K_{p|p|} & 0 & 0 \\ 0 & 0 & M_{w|w|} & 0 & M_{q|q|} & 0 \\ 0 & N_{v|v|} & 0 & 0 & 0 & N_{r|r|} \end{bmatrix}, D_L = -\begin{bmatrix} 0 & 0 & 0 & 0 & 0 & 0 \\ 0 & Y_{uv} & 0 & 0 & 0 & Y_{ur} \\ 0 & 0 & Z_{uw} & 0 & Z_{uq} & 0 \\ 0 & 0 & 0 & 0 & 0 & 0 \\ 0 & 0 & M_{uw} & 0 & M_{uq} & 0 \\ 0 & N_{uv} & 0 & 0 & 0 & N_{ur} \end{bmatrix}$$

To obtain the above hydrodynamic coefficients, the strip theory is utilized for numerical calculation, where the fluid density is assumed to be $1030\ \mathrm{kg}/\mathrm{m}^3$, the linear-skin coefficient is assumed to be 0.4 and the drag coefficient is assumed to be 1. Moreover, some of the obtained coefficients are adjusted based on comparisons with data of the REMUS AUV according to dynamic similarity [15]. Then, the adjusted hydrodynamic coefficients are shown in Table A1.

Table A1. The list of AUV coefficients.

Added Mass Coefficients					
Force	**Value**	**Units**	**Moment**	**Value**	**Units**
$X_{\dot{u}}$	-2.33	Kg	$K_{\dot{p}}$	-0.3	Kg·m²/rad
$Y_{\dot{v}}$	-90.8	Kg	$M_{\dot{q}}$	-20.3	Kg·m²/rad
$Y_{\dot{r}}$	4.53	Kg·m	$M_{\dot{w}}$	-4.53	Kg·m
$Z_{\dot{w}}$	-90.8	Kg	$N_{\dot{r}}$	-20.3	Kg·m²/rad
$Z_{\dot{q}}$	-4.53	Kg·m	$N_{\dot{v}}$	4.53	Kg·m
Drag Coefficients					
Force	**Value**	**Units**	**Moment**	**Value**	**Units**
$X_{u|u|}$	-2.96	Kg/m	$K_{p|p|}$	-0.558	Kg·m²/rad²
$Y_{v|v|}$	-2346	Kg/m	$M_{q|q|}$	-807	Kg·m²/rad²
$Y_{r|r|}$	0.759	Kg·m/rad²	$M_{w|w|}$	8.76	Kg
$Z_{w|w|}$	-242	Kg/m	$N_{r|r|}$	-404	Kg·m²/rad²
$Z_{q|q|}$	-0.759	Kg·m/rad²	$N_{v|v|}$	-8.76	Kg
Lift Coefficients					
Force	**Value**	**Units**	**Moment**	**Value**	**Units**
Y_{uv}	-56.5	Kg/m	M_{uq}	-8.9	Kg·m/rad
Y_{ur}	11.8	Kg/rad	M_{uw}	-24.9	Kg
Z_{uw}	-56.5	Kg/m	N_{ur}	-8.9	Kg·m/rad
Z_{uq}	-11.8	Kg/rad	N_{uv}	24.9	Kg

For the AUV, the thruster configuration matrix and its pseudo inverse are given as

$$
\boldsymbol{B}_v = \begin{bmatrix} 1 & 1 & 0 & 0 & 0 \\ 0 & 0 & 0 & 0 & 1 \\ 0 & 0 & 1 & 1 & 0 \\ 0 & 0 & 0 & 0 & 0 \\ 0 & 0 & r_3 & -r_4 & 0 \\ r_1 & -r_2 & 0 & 0 & r_5 \end{bmatrix}, \boldsymbol{B}_v^+ = \begin{bmatrix} \frac{r_2}{r_1+r_2} & -\frac{r_5}{r_1+r_2} & 0 & 0 & 0 & \frac{1}{r_1+r_2} \\ \frac{r_1}{r_1+r_2} & \frac{r_5}{r_1+r_2} & 0 & 0 & 0 & -\frac{1}{r_1+r_2} \\ 0 & 0 & \frac{r_4}{r_3+r_4} & 0 & \frac{1}{r_3+r_4} & 0 \\ 0 & 0 & \frac{r_3}{r_3+r_4} & 0 & -\frac{1}{r_3+r_4} & 0 \\ 0 & 1 & 0 & 0 & 0 & 0 \end{bmatrix}
$$

where $r_1 = 0.18$ m, $r_2 = 0.18$ m, $r_3 = 0.525$ m, $r_4 = 0.245$ m, $r_5 = 0.485$ m.

References

1. Podder, T.K.; Sarkar, N. Unified Dynamics-based Motion Planning Algorithm for Autonomous Underwater Vehicle-Manipulator Systems (UVMS). *Robotica* **2004**, *22*, 117–128. [CrossRef]
2. Antonelli, G.; Chiaverini, S. Fuzzy redundancy resolution and motion coordination for underwater vehicle-manipulator systems. *IEEE Trans. Fuzzy Syst.* **2003**, *11*, 109–120. [CrossRef]
3. Antonelli, G.; Chiaverini, S. Task-priority redundancy resolution for underwater vehicle-manipulator systems. In Proceedings of the IEEE International Conference on Robotics and Automation, Leuven, Belgium, 20 May 1998; pp. 768–773.
4. Sarkar, N.; Podder, T.K. Coordinated motion planning and control of autonomous underwater vehicle-manipulator systems subject to drag optimization. *IEEE J. Ocean. Eng.* **2001**, *26*, 228–239. [CrossRef]
5. Han, J.; Park, J.; Chung, W.K. Robust coordinated motion control of an underwater vehicle-manipulator system with minimizing restoring moments. *Ocean Eng.* **2011**, *38*, 1197–1206. [CrossRef]
6. Siciliano, B.; Slotine, J.-J.E. A general framework for managing multiple tasks in highly redundant robotic systems. In Proceedings of the Fifth International Conference on Advanced Robotics, Pisa, Italy, 19–22 June 1991; pp. 1211–1216.
7. Antonelli, G.; Chiaverini, S. A fuzzy approach to redundancy resolution for underwater vehicle-manipulator systems. *Control Eng. Pract.* **2003**, *11*, 445–452. [CrossRef]
8. Wang, Y.; Jiang, S.; Yan, F.; Gu, L.; Chen, B. A new redundancy resolution for underwater vehicle—Manipulator system considering payload. *Int. J. Adv. Robot. Syst.* **2017**, *14*. [CrossRef]
9. Haugalokken, B.O.A.; Joergensen, E.K.; Schjolberg, I. Experimental validation of end-effector stabilization for underwater vehicle-manipulator systems in subsea operations. *Robot. Auton. Syst.* **2018**, *109*, 1–12. [CrossRef]
10. Tang, C.; Wang, Y.; Wang, S.; Wang, R.; Tan, M. Floating Autonomous Manipulation of the Underwater Biomimetic Vehicle-Manipulator System: Methodology and Verification. *IEEE Trans. Ind. Electron.* **2018**, *65*, 4861–4870. [CrossRef]
11. Simetti, E.; Casalino, G.; Wanderlingh, F.; Aicardi, M. Task priority control of underwater intervention systems: Theory and applications. *Ocean Eng.* **2018**, *164*, 40–54. [CrossRef]
12. Youakim, D.; Ridao, P. Motion planning survey for autonomous mobile manipulators underwater manipulator case study. *Robot. Auton. Syst.* **2018**, *107*, 20–44. [CrossRef]
13. Mcmillan, S.; Orin, D.E.; Mcghee, R.B. Efficient dynamic simulation of an underwater vehicle with a robotic manipulator. *IEEE Trans. Syst. Man Cybern.* **1995**, *25*, 1194–1206. [CrossRef]
14. Dannigan, M.W.; Russell, G. Evaluation and reduction of the dynamic coupling between a manipulator and an underwater vehicle. *IEEE J. Ocean. Eng.* **1998**, *23*, 260–273. [CrossRef]
15. Han, H.; Wei, Y.; Ye, X.; Liu, W. Modelling and Fuzzy Decoupling Control of an Underwater Vehicle-Manipulator System. *IEEE Access* **2020**, *8*, 18962–18983. [CrossRef]
16. Schjolberg, I.; Fossen, T.I. Modelling and Control of Underwater Vehicle-Manipulator Systems. In Proceedings of the 3rd Conference on Marine Craft Maneuvering and Control, Southampton, UK, 7–9 September 1994; pp. 45–57.
17. Taira, Y.; Sagara, S.; Oya, M. Model-based motion control for underwater vehicle-manipulator systems with one of the three types of servo subsystems. *Artif. Life Robot* **2019**, *6*, 1–16. [CrossRef]

18. Korkmaz, O.; Ider, S.K.; Ozgoren, M.K. Trajectory Tracking Control of an Underactuated Underwater Vehicle Redundant Manipulator System. *Asian J. Control* **2016**, *18*, 1593–1607. [CrossRef]

19. Cai, M.; Wang, Y.; Wang, S.; Wang, R.; Ren, Y.; Tan, M. Grasping Marine Products with Hybrid-Driven Underwater Vehicle-Manipulator System. *IEEE Trans. Autom. Sci. Eng.* **2020**. [CrossRef]

20. Santhakumar, M.; Kim, J. Robust Adaptive Tracking Control of Autonomous Underwater Vehicle-Manipulator Systems. *J. Dyn. Syst. Meas. Control* **2014**, *136*, 054502. [CrossRef]

21. Antonelli, G.; Caccavale, F.; Chiaverini, S. Adaptive Tracking Control of Underwater Vehicle-Manipulator Systems Based on the Virtual Decomposition Approach. *IEEE Trans. Robot. Autom.* **2004**, *20*, 594–602. [CrossRef]

22. Mohan, S.; Kim, J. Indirect adaptive control of an autonomous underwater vehicle-manipulator system for underwater manipulation tasks. *Ocean Eng.* **2012**, *54*, 233–243. [CrossRef]

23. Dai, Y.; Yu, S. Design of an indirect adaptive controller for the trajectory tracking of UVMS. *Ocean Eng.* **2008**, *151*, 234–245. [CrossRef]

24. Esfahani, H.N. Robust Model Predictive Control for Autonomous Underwater Vehicle–Manipulator System with Fuzzy Compensator. *Pol. Marit. Res.* **2019**, *26*, 104–114. [CrossRef]

25. Cai, M.; Wang, Y.; Wang, S.; Wang, R.; Cheng, L.; Tan, M. Prediction-Based Seabed Terrain Following Control for an Underwater Vehicle-Manipulator System. *IEEE Trans. Syst. Man Cybern. Syst.* **2019**. [CrossRef]

26. Nikou, A.; Verginis, C.K.; Dimarogonas, D.V. A Tube-based MPC Scheme for Interaction Control of Underwater Vehicle Manipulator Systems. In Proceedings of the 2018 IEEE/OES Autonomous Underwater Vehicle Workshop (AUV), Porto, Portugal, 6–9 November 2018; pp. 1–8.

27. Dai, Y.; Yu, S.; Yan, Y. An Adaptive EKF-FMPC for the Trajectory Tracking of UVMS. *IEEE J. Ocean. Eng.* **2019**, 1–15. [CrossRef]

28. Dai, Y.; Yu, S.; Yan, Y.; Yu, X. An EKF-Based Fast Tube MPC Scheme for Moving Target Tracking of a Redundant Underwater Vehicle-Manipulator System. *IEEE ASME Trans. Mechatron.* **2019**, *24*, 2803–2814. [CrossRef]

29. Esfahani, H.N.; Azimirad, V.; Danesh, M. A Time Delay Controller included terminal sliding mode and fuzzy gain tuning for Underwater Vehicle-Manipulator Systems. *Ocean Eng.* **2015**, *107*, 97–107. [CrossRef]

30. Hosseinnajad, A.; Loueipour, M. Time Delay Controller Design for Dynamic Positioning of ROVs based on Position and Acceleration Measurements. In Proceedings of the 6th International Conference on Control, Instrumentation and Automation, Sanandaj, Iran, 30–31 October 2019.

31. Cui, R.; Chen, L.; Yang, C.; Chen, M. Extended State Observer-Based Integral Sliding Mode Control for an Underwater Robot With Unknown Disturbances and Uncertain Nonlinearities. *IEEE Trans. Ind. Electron.* **2017**, *64*, 6785–6795. [CrossRef]

32. Chen, W.H. Disturbance Observer Based Control for Nonlinear Systems. *IEEE ASME Trans. Mechatron.* **2005**, *9*, 706–710. [CrossRef]

33. Chen, W.; Wei, Y.; Zeng, J.; Han, H.; Jia, X. Adaptive Terminal Sliding Mode NDO-Based Control of Underactuated AUV in Vertical Plane. *Discret. Dyn. Nat. Soc.* **2016**. [CrossRef]

34. Li, J.; Huang, H.; Wan, L.; Zhou, Z.; Xu, Y. Hybrid Strategy-based Coordinate Controller for an Underwater Vehicle Manipulator System Using Nonlinear Disturbance Observer. *Robotica* **2019**, *37*, 1710–1731. [CrossRef]

35. Londhe, P.S.; Dhadekar, D.D.; Patre, B.M.; Waghmare, L.M. Uncertainty and disturbance estimator based sliding mode control of an autonomous underwater vehicle. *Intern. J. Dyn. Control* **2017**, *5*, 1122–1138. [CrossRef]

36. Han, H.; Wei, Y.; Guan, L.; Ye, X.; Wang, A. Trajectory Tracking Control of Underwater Vehicle-Manipulator Systems Using Uncertainty and Disturbance Estimator. In Proceedings of the OCEANS 2018 MTS/IEEE, Charleston, SC, USA, 22–25 October 2018.

37. Suryawanshi, P.V.; Shendge, P.D.; Phadke, S.B. Robust sliding mode control for a class of nonlinear systems using inertial delay control. *Nonlinear Dyn.* **2014**, *78*, 1921–1932. [CrossRef]

38. Mohan, S.; Kim, J. Coordinated motion control in task space of an autonomous underwater vehicle–manipulator system. *Ocean Eng.* **2015**, *104*, 155–167. [CrossRef]

39. Londhe, P.S.; Santhakumar, M.; Patre, B.M.; Waghmare, L.M. Task Space Control of an Autonomous Underwater Vehicle Manipulator System by Robust Single-Input Fuzzy Logic Control Scheme. *IEEE J. Ocean. Eng.* **2016**, *42*, 13–28. [CrossRef]

40. Londhe, P.S.; Mohan, S.; Patre, B.M.; Waghmare, L.M. Robust task-space control of an autonomous underwater vehicle- manipulator system by PID-like fuzzy control scheme with disturbance estimator. *Ocean Eng.* **2017**, *139*, 1–13. [CrossRef]
41. Antonelli, G. *Underwater Robots: Motion and Force Control of Vehicle-Manipulator Systems*, 3rd ed.; Springer: Berlin/Heidelberg, Germany, 2013; pp. 52–55.
42. Fossen, T.I. *Handbook of Marine Craft Hydrodynamics and Motion Control*; John Wiley and Sons: Hoboken, NJ, USA, 2011; pp. 128–131.

9

Adaptive Feedforward Control of a Pressure Compensated Differential Cylinder

Konrad Johan Jensen *, Morten Kjeld Ebbesen and Michael Rygaard Hansen

Department of Engineering Sciences, University of Agder, 4879 Grimstad, Norway;
morten.k.ebbesen@uia.no (M.K.E.); michael.r.hansen@uia.no (M.R.H.)
* Correspondence: konrad.j.jensen@uia.no

Abstract: This paper presents the design, simulation and experimental verification of adaptive feedforward motion control for a hydraulic differential cylinder. The proposed solution is implemented on a hydraulic loader crane. Based on common adaptation methods, a typical electro-hydraulic motion control system has been extended with a novel adaptive feedforward controller that has two separate feedforward states, i.e, one for each direction of motion. Simulations show convergence of the feedforward states, as well as 23% reduction in root mean square (RMS) cylinder position error compared to a fixed gain feedforward controller. The experiments show an even more pronounced advantage of the proposed controller, with an 80% reduction in RMS cylinder position error, and that the separate feedforward states are able to adapt to model uncertainties in both directions of motion.

Keywords: adaptive control; hydraulics; differential cylinder; feedforward; motion control

1. Introduction

For hydraulically actuated systems such as cranes, the hydraulic cylinder is the most common actuator since it can provide a linear motion with, generally speaking, a large force to volume ratio, a high efficiency and at a modest price. For systems which require a cylinder force in both directions, a double acting cylinder is needed, and the differential cylinder is an obvious choice due to its low cost and simple design. The main disadvantage is the difference in effective hydraulic area which leads to a jump in both velocity and force gain when changing sign of direction, i.e., around zero velocity.

For many hydraulic systems, the pressure compensated directional control valve is a practical choice due to the fact that it provides load independent flow control of the actuators. The pressure compensator senses the load pressure, and adjusts the pressure drop over the directional control valve to give a load independent flow. Since the velocity of the actuator is proportional to the hydraulic flow through the valve, this translates to load independent velocity control. For manually operated systems, the velocity control makes it easy for an operator to control systems that are subjected to large variations in external load.

For closed loop control systems, the load independent velocity control can be utilized in a control system using feedforward [1]. In this case, both a position reference and a velocity reference are generated in the control system. An example of a typical closed loop electro-hydraulic motion control system with feedforward is shown in Figure 1. The feedback controller uses the position reference and the measured cylinder position, whereas the feedforward controller uses the velocity reference. The pressure compensator is connected to a supply line which is shared with other actuators. The red dashed lines show the hydraulic pilot lines for the counterbalance valve and the pressure compensator.

It should be noted that feedforward control cannot be used alone. A feedback controller is also needed to help track the position reference, to eliminate steady state position error, and to counteract any drift. Normally the feedforward gain is based on system components, and is defined as the ratio of valve opening to actuator velocity. With this in mind, it follows that modeling errors and model uncertainties, in addition to external disturbances and system dynamics, may yield sub-optimal performance with a fixed feedforward gain.

This paper focuses on modeling and motion control of a hydraulic loader crane with pressure compensated differential cylinders. An adaptive feedforward controller is investigated to improve performance of the motion control system. Two different approaches to feedforward control have been implemented, the first is based on the MIT-rule [2], and the second is based on the sign-sign algorithm [3].

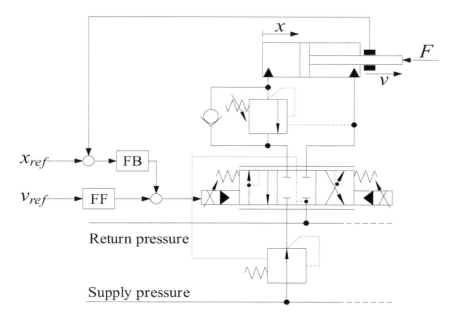

Figure 1. Electro-hydraulic motion control system with feedforward.

2. Background and Method

Adaptive systems have long been used for system identification and parameter estimation. One of the first methods is described in [4]. Another common method is the least mean squares algorithm, which was developed in [5]. An example of this is shown in Equations (1)–(3). Given the linear system:

$$Y = \theta^T \cdot X \tag{1}$$

$$E = Y - \hat{\theta}^T \cdot X \tag{2}$$

$$\dot{\hat{\theta}} = \gamma \cdot X \cdot E^T \tag{3}$$

where

Y = system output;
θ = system parameters;
X = system input;
E = estimation error;
$\hat{\theta}$ = estimated parameters;
γ = adaptation gain, constant.

The estimated parameters will converge towards the system parameters. The idea of using the sign function in the adaptive law comes from the sign-sign least mean squares algorithm, and was first introduced by [3]. Equation (3) then becomes:

$$\dot{\theta} = \gamma \cdot \text{sign}(X) \cdot \text{sign}(E^T) \tag{4}$$

By taking the sign of the estimation error and system input, the adaptation becomes insensitive to the magnitudes of E and X, and as such only the adaptation gain γ sets the adaptation speed.

The MIT rule is also used for adaptive control, and is described in [2]. A typical application is model reference adaptive control, shown in Figure 2. Based on the model output y_m, an additional control output $\hat{u}$ is multiplied with the command signal u_c to shape the plant output y. The equations for the model reference adaptive control is shown in Equations (5) and (6).

$$\hat{u} = -\gamma \cdot y_m \cdot (y - y_m) \tag{5}$$
$$u = u_c \cdot \hat{u} \tag{6}$$

where

u = control output;
$\hat{u}$ = adaptive control output;
u_c = command signal;
y_m = model output;
y = plant output.

Early work in adaptive control can be found in [6–10]. Other work on adaptive control include [11] which investigates adaptive feedback and feedforward control of robot manipulators, Reference [12] which models and implements adaptive control of a flexible arm, and [13] which uses model reference adaptive control on linear time-varying plants. Adaptive fuzzy sliding mode control is investigated and implemented on an inverted pendulum in [14].

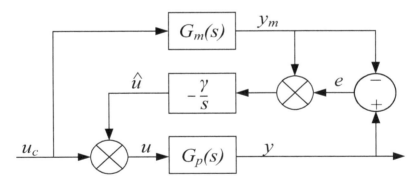

Figure 2. Model reference adaptive control based on MIT-rule.

Newer applications of adaptive control systems include adaptive friction compensation with an adaptive velocity estimator to compensate for the estimated non-linear friction force [15]. In [16], a fuzzy model reference adaptive control of an active magnetic bearing for a milling process is investigated to reduce the milling dynamics. Adaptive integral robust control of an electro-hydraulic servo system is investigated in [17], using parameter estimation and integral control to compensate for disturbances and plant uncertainties. Adaptive control of quadrotors is investigated in [18], which uses an cerebellar model arithmetic computer to adapt to model uncertainties and disturbances. In [19], adaptive control based on least-mean-fourth is implemented for a three-phase grid connected solar system, which is able to provide load balancing and power factor correction.

As for motion control of hydraulic systems, different approaches have previously been investigated, including vector control [20], pressure control [21,22], force control [23,24], and feedforward control [25].

To the knowledge of the authors, adaptive feedforward motion control of hydraulic cylinders has not previously been investigated, and this paper will focus on this novel concept.

In this paper, two adaptive controllers have been tested on a hydraulic differential cylinder and compared to a fixed gain feedforward controller. Based on a typical fixed gain feedforward controller, an adaptive controller can be made by extending it with the MIT rule. An illustration of a control system with feedforward with fixed gain is shown in Figure 3.

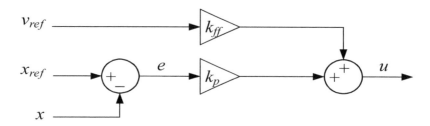

Figure 3. Feedforward with fixed gain.

Defining the position error e as the position reference x_{ref} minus the measured position x, the control output for this control system is given in Equation (7)

$$u = k_p \cdot e + k_{ff} \cdot v_{ref} \tag{7}$$

where

u = controller output;
k_p = proportional gain;
e = position error;
k_{ff} = feedforward gain;
v_{ref} = velocity reference.

Extending the traditional feedforward controller into an adaptive feedforward controller is done by replacing the fixed feedforward gain with the MIT-rule. An illustration of the adaptive feedforward scheme is shown in Figure 4.

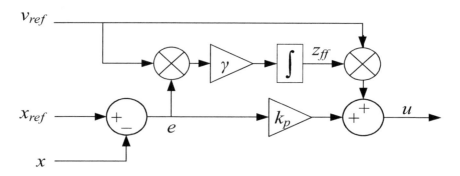

Figure 4. MIT-rule adaptive feedforward.

The MIT-rule adaptive feedforward controller uses the position error, the velocity reference, and the constant γ to update the feedforward gain. The update law and the control output for this adaptive control system is then given in Equations (8) and (9).

$$\dot{z}_{ff} = \gamma \cdot v_{ref} \cdot e \tag{8}$$
$$u = k_p \cdot e + z_{ff} \cdot v_{ref} \tag{9}$$

where

γ = adaptation gain;
z_{ff} = feedforward gain.

Extending this controller to use sign-sign is then straightforward. An illustration of the sign-sign adaptive feedforward scheme is shown in Figure 5.

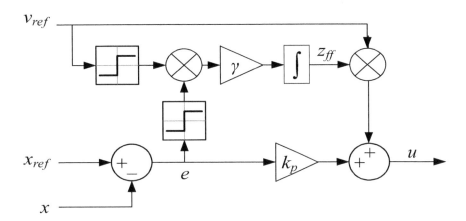

Figure 5. Sign-sign adaptive feedforward.

The update law and the control output for this adaptive control system is shown in Equations (10) and (11).

$$\dot{z}_{ff} = \gamma \cdot \text{sign}(v_{ref}) \cdot \text{sign}(e) \tag{10}$$
$$u = k_p \cdot e + z_{ff} \cdot v_{ref} \tag{11}$$

It should be noted that the sign function can produce unnecessary chattering when the input is oscillating around zero, due to the inherent discontinuity. Therefore the sign function has been replaced with the tanh function, shown in Equation (12).

$$\text{sign}(e) \approx \tanh(k \cdot e) \tag{12}$$

This gives a smooth output when the input is oscillating around zero. Increasing the parameter k gives a sharper rise and a closer approximation to sign(e). Another advantage of using tanh is that the adaptation stops when the position error is zero. The parameter k has been set to $k = 100 \text{ m}^{-1}$ and $k = 100 \text{ s} \cdot \text{m}^{-1}$ for the position error and velocity reference, respectively.

3. Considered System

In this paper an 2020K4 loader crane made by HMF Group A/S, Højbjerg, Denmark has been used for experiments. An illustration of the crane is shown in Figure 6. This crane has two hydraulic differential cylinders: the main cylinder, and the knuckle cylinder. For this paper, the knuckle cylinder has been used for simulation and experiments, since it can experience both resistive and assistive loads in both directions of motion, equivalent to four quadrant operation. The relevant data for the knuckle cylinder is shown in Table 1, and the data for the knuckle boom is given in Figure 7 and Table 2.

Each actuator is controlled via a pressure compensated proportional directional control valve which ensures load independent flow control of the actuators. Counterbalance valves made by Oil Control S.p.A, Modena, Italy are also used for load holding, assisting in lowering of the booms, and pressure relief of pressure surges. An illustration of the hydraulic system for the knuckle cylinder is shown in Figure 8.

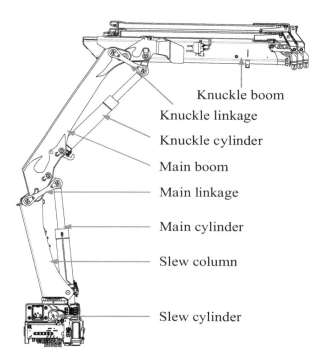

Figure 6. Illustration of the HMF 2020K4 loader crane.

Table 1. Knuckle cylinder data.

Name	Parameter	Value
Piston diameter	D_p	0.15 m
Piston area	A	0.0177 m²
Rod diameter	D_r	0.1 m
Annulus area	A_a	0.0098 m²
Piston area ratio	$\phi = \frac{A_a}{A}$	0.5556
Valve maximum flow	Q_{max}	40 L/min

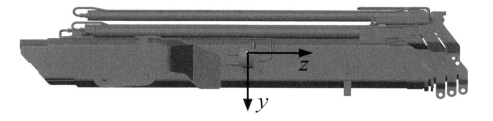

Figure 7. Knuckle boom center of mass.

Table 2. Knuckle boom data.

Name	Parameter	Value		
Mass	m_k	851.972 kg		
Inertia matrix	I_k	$\begin{bmatrix} 579.552 & 8.74629 & 11.5456 \\ 8.74629 & 573.285 & 0.174433 \\ 11.5456 & 0.174433 & 32.2491 \end{bmatrix}$ kg·m²		

The control system is implemented on a CompactRIO 9075 controller made by National Instruments, Austin, TX, USA. The CompactRIO contains the reference generator and feedforward motion controllers. The block diagram of the connections is shown in Figure 9.

The CompactRIO communicates with a PC, sends control signals to the valves, and reads the sensors on the crane.

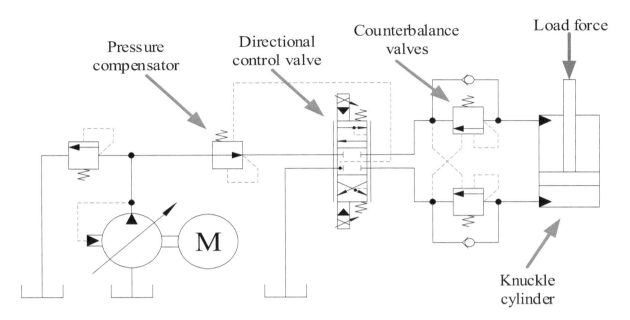

Figure 8. Hydraulic system for the knuckle cylinder.

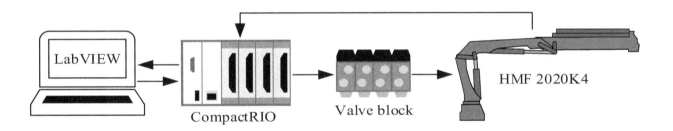

Figure 9. Connection between the crane and CompactRIO controller.

4. Modelling

A dynamic model of the crane has been made in Simscape™ by MathWorks®, Natick, MA, USA. 3D computer-aided design (CAD) models have been imported into the model using the Multibody library. The hydraulic circuit has been made using the hydraulic library of Simscape™. A picture of the CAD model is shown in Figure 10.

In the configuration shown in Figure 10, the knuckle cylinder experiences both resistive and assistive loads in both directions of motion when retracting fully, and extending back out again. The knuckle cylinder is controlled by a pressure compensated directional control valve, shown in Figure 11.

The pressure compensator ensures that there is a constant pressure drop over the directional control valve, which gives a load independent flow. The governing equations of the pressure compensator are given in Equations (13)–(15).

$$u_{pc} = \frac{p_{set} + p_{load} - p_p}{\Delta p} \tag{13}$$

$$p_{load} = \begin{cases} p_a & \text{if } u_{spool} \geq 0 \\ p_b & \text{otherwise} \end{cases} \tag{14}$$

$$Q_{pc} = k_{pc} \cdot u_{pc} \cdot \sqrt{p_i - p_p} \tag{15}$$

where

u_{pc} = opening of compensator, $0 \leq u_{pc} \leq 1$

p_p = compensated pressure at port p;

Δp = pressure difference between fully closed and fully open;

p_a = pressure at port a;

p_b = pressure at port b;

p_t = tank pressure;

p_{set} = spring pressure setting;

p_{load} = load pressure;

u_{spool} = position of the main spool, $-1 \leq u_{spool} \leq 1$;

Q_{pc} = flow in pressure compensator;

k_{pc} = flow gain of compensator;

p_i = compensator inlet pressure.

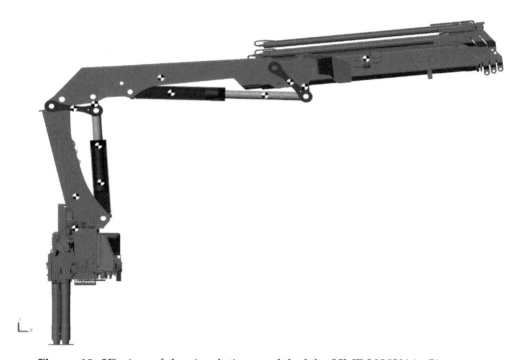

Figure 10. 3D view of the simulation model of the HMF 2020K4 in Simscape.

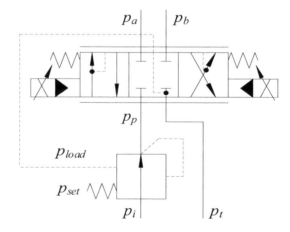

Figure 11. Hydraulic pressure compensated directional control valve for the knuckle cylinder.

The steady state of p_p is then given by Equation (16).

$$p_p = p_{load} + p_{set} \qquad (16)$$

The sensing of the load pressures p_a and p_b ensures that the pressure drop over the directional control valve always equals p_{set}, and that the flow is load independent. This is shown in the orifice equation in Equation (17).

$$
\begin{aligned}
Q &= C_d \cdot A_d \cdot u_{spool} \cdot \sqrt{\frac{2}{\rho} \cdot (p_p - p_{load})} \\
&= C_d \cdot A_d \cdot u_{spool} \cdot \sqrt{\frac{2}{\rho} \cdot p_{set}} \qquad (17) \\
&= Q_{max} \cdot u_{spool}
\end{aligned}
$$

where

Q = flow in the valve;
C_d = discharge coefficient;
A_d = maximum discharge area;
ρ = mass density;
Q_{max} = maximum valve flow;

Double counterbalance valves are used on the knuckle cylinder. An illustration of the counterbalance valves is shown in Figure 12.

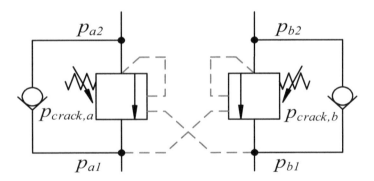

Figure 12. Double counterbalance valve.

The unitless openings of the counterbalance valves are calculated in Equations (18) and (19).

$$u_a = \frac{p_{a2} + \psi \cdot p_{b1} - p_{crack,a}}{\Delta p} \qquad (18)$$

$$u_b = \frac{p_{b2} + \psi \cdot p_{a1} - p_{crack,b}}{\Delta p} \qquad (19)$$

where

u_a = opening of valve a, $0 \le u_a \le 1$;
u_b = opening of valve b, $0 \le u_b \le 1$;
p_{a1} = pressure at valve a input side;
p_{a2} = pressure at valve a actuator side;
p_{b1} = pressure at valve b input side;
p_{b2} = pressure at valve b actuator side;

$p_{crack,a}$ = crack pressure of valve a;

$p_{crack,b}$ = crack pressure of valve b;

ψ = pilot area ratio;

Δp = pressure difference between fully closed and fully open.

When u_a and u_b are 0, the valves are closed. When they are 1, the valves are fully open. During assistive loads the valves tend to be somewhere between 0 and 1, meaning that they are throttling the flow. The dynamics of the valves are included as a time constant, since the valves have a finite bandwidth.

5. Adaptive Control Design

Since the actuator is a hydraulic differential cylinder, two separate states z_{ff}^{+} and z_{ff}^{-} are used for out-stroke and in-stroke motion to handle model uncertainties both directions of motion. Consequently, both the feedforward control output and the update law for the two gains are only active during out-stroke or in-stroke motion respectively. To handle this, some switching logic is introduced based on the sign of the velocity reference. The block diagram for the differential MIT-rule adaptive feedforward is shown in Figure 13.

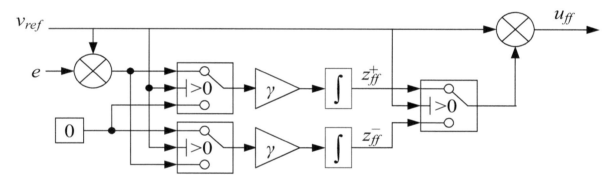

Figure 13. Differential MIT-rule adaptive feedforward.

The governing equations for the differential MIT-rule adaptive feedforward are shown in Equations (20)–(23).

$$\dot{z}_{ff}^{+} = \begin{cases} \gamma \cdot v_{ref} \cdot e, & v_{ref} > 0 \\ 0, & \text{otherwise} \end{cases} \tag{20}$$

$$\dot{z}_{ff}^{-} = \begin{cases} 0, & v_{ref} > 0 \\ \gamma \cdot v_{ref} \cdot e, & \text{otherwise} \end{cases} \tag{21}$$

$$u_{ff} = \begin{cases} z_{ff}^{+} \cdot v_{ref}, & v_{ref} > 0 \\ z_{ff}^{-} \cdot v_{ref}, & \text{otherwise} \end{cases} \tag{22}$$

$$u = k_p \cdot e + u_{ff} \tag{23}$$

where

z_{ff}^{+} = out-stroke feedforward gain;

z_{ff}^{-} = in-stroke feedforward gain;

u_{ff} = feedforward controller output.

Extending the controller to sign-sign is straightforward. The block diagram for the differential sign-sign adaptive feedforward is shown in Figure 14.

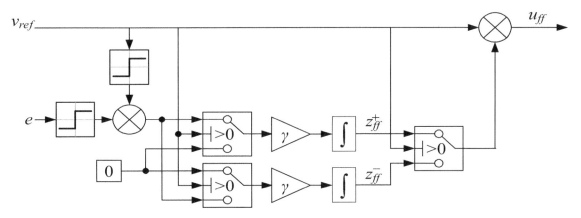

Figure 14. Differential sign-sign adaptive feedforward.

The governing equations for the differential sign-sign adaptive feedforward are shown in Equations (24)–(27).

$$\dot{z}_{ff}^{+} = \begin{cases} \gamma \cdot \text{sign}(v_{ref}) \cdot \text{sign}(e), & v_{ref} > 0 \\ 0, & \text{otherwise} \end{cases} \tag{24}$$

$$\dot{z}_{ff}^{-} = \begin{cases} 0, & v_{ref} > 0 \\ \gamma \cdot \text{sign}(v_{ref}) \cdot \text{sign}(e), & \text{otherwise} \end{cases} \tag{25}$$

$$u_{ff} = \begin{cases} z_{ff}^{+} \cdot v_{ref}, & v_{ref} > 0 \\ z_{ff}^{-} \cdot v_{ref}, & \text{otherwise} \end{cases} \tag{26}$$

$$u = k_p \cdot e + u_{ff} \tag{27}$$

6. Simulation Results

For the simulation, a point-to-point trapezoidal velocity path generator has been used as a reference. The point-to-point path generator has previously been developed in [26]. The path generator operates in actuator space, which eliminates the effects of the non-linearities between the hydraulic cylinder strokes and the joint angles in joint space. A path has been made such that the cylinder experiences both resistive and assistive loads in both directions of motion. The references for position and velocity are shown in Figure 15. The adaptation gain γ is different for the two controllers, due to the use of $\text{sign}(x)$, and has been experimentally set to $\gamma = 200 \, \text{s} \cdot \text{m}^{-3}$ for the MIT-rule feedforward, and $\gamma = 0.1 \, \text{m}^{-1}$ for the sign-sign feedforward. The unit is adapted accordingly to obtain the correct output.

The position error for the MIT-rule feedforward simulation is shown in Figure 16. The position error decreases towards a bounded error of ± 6 mm, which is shown with the dashed lines. The RMS error after convergence is 1.6 mm, showing high performance.

The states z_{ff} for the MIT-rule feedforward simulation are shown in Figure 17. The dashed lines show the theoretical values for a fixed feedforward gain. The states converge to values slightly larger than the theoretical ones. This small discrepancy can be attributed to the constant velocity reference and ramped position reference. When moving with a ramp position reference, there will always be a small constant position error without an integrator in the position controller. Having a slightly larger feedforward gain helps reducing this constant position error by giving the cylinder a small velocity boost. Since the position error is measured, the adaptive controller is able to adapt the feedforward gains to minimize the position error.

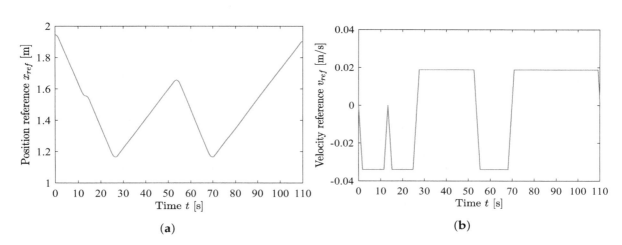

Figure 15. Point-to-point path references for simulation. (**a**) Position reference; (**b**) Velocity reference.

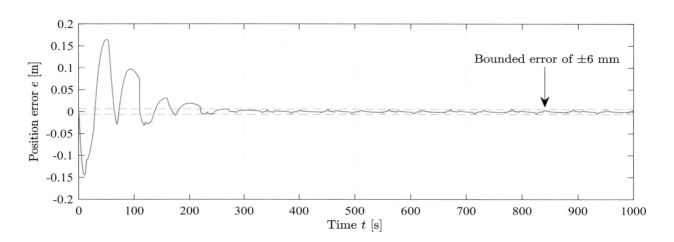

Figure 16. Cylinder position error during MIT-rule feedforward simulation, $\gamma = 200\,\text{s}\cdot\text{m}^{-3}$.

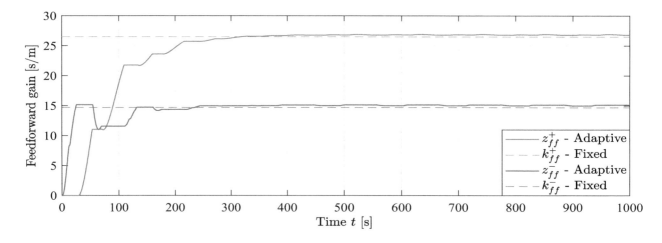

Figure 17. Feedforward states during MIT-rule feedforward simulation, $\gamma = 200\,\text{s}\cdot\text{m}^{-3}$.

Figure 18 shows the control signals u_{ff} and u_{fb} from the feedforward and feedback controller, respectively. Given that the total control signal $u = u_{fb} + u_{ff}$, it can be seen that the contribution from the feedforward controller clearly dominates, providing more than 95% at steady state.

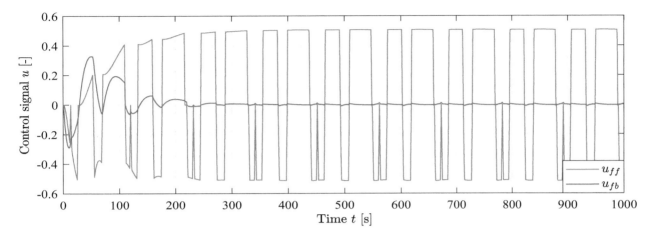

Figure 18. Control signals from feedforward and feedback during simulation, $\gamma = 200 \text{ s} \cdot \text{m}^{-3}$.

The position error for the sign-sign feedforward simulation is shown in Figure 19. The same bounded error of ± 6 mm is shown with the dashed lines. The RMS error after convergence is 2.1 mm.

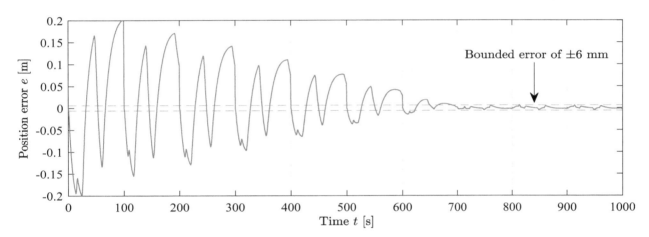

Figure 19. Cylinder position error during sign-sign feedforward simulation, $\gamma = 0.1 \text{ m}^{-1}$.

The states z_{ff} for the sign-sign feedforward simulation are shown in Figure 20. The dashed lines show the theoretical values for a fixed feedforward gain. The same results can be seen here as with the MIT-rule, the states converge to values slightly larger than the theoretical ones, although convergence is slower with 700 s compared to 400 s.

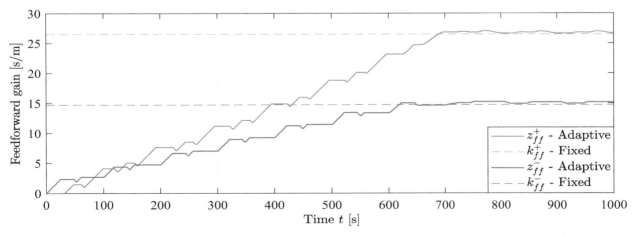

Figure 20. Feedforward states during sign-sign feedforward simulation, $\gamma = 0.1 \text{ m}^{-1}$.

To show the difference in performance between the fixed gain controller and the adaptive controllers, a simulation with fixed gain feedforward has been made and compared with the MIT-rule feedforward at a simulation time where the states z_{ff} have converged, at $t = 800$ s. This is shown in Figure 21. It can be seen that the position error for the MIT-rule feedforward is lower compared to the fixed gain feedforward, showing that the MIT-rule feedforward controller outperforms the fixed gain controller even with an ideal model with correlation between cylinder velocity and feedforward gain.

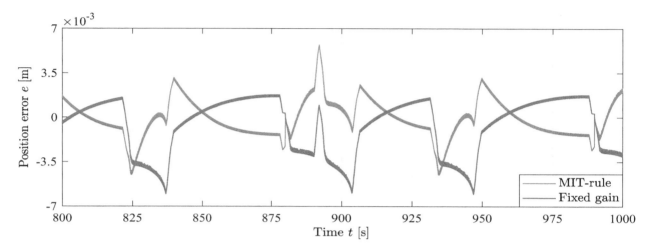

Figure 21. Position error comparison between MIT-rule and fixed gain feedforward in simulation.

The RMS position error for each controller after convergence of the states z_{ff} is shown in Table 3. Even though the fixed gain feedforward is based on an ideal model, the MIT-rule adaptive feedforward controller yields better position tracking with a 23% decrease in RMS position error. This shows the improved performance of the adaptive controller.

Table 3. Comparison of RMS position error after convergence in simulation.

	MIT-Rule	Sign-Sign	Fixed Gain
RMS error	1.6 mm	2.1 mm	2.1 mm

7. Experimental Results

The three controllers have been implemented on the CompactRIO controller in the laboratory. The control laws are implemented in discrete-time based on backward euler integration. A picture of the HMF 2020K4 loader crane in the laboratory is shown in Figure 22. The figure shows the crane in the starting position. During motion the knuckle boom is folded down.

There is some deadband in the valves on the HMF 2020K4 loader crane, and therefore deadband compensation has been implemented for the laboratory experiments. The identified deadbands for the knuckle boom valve are shown in Table 4.

Table 4. Identified deadbands for the knuckle boom valve.

Name	Parameter	Value
Out-stroke deadband	u^+	0.21
In-stroke deadband	u^-	−0.31

The equation for the deadband compensation is shown in Equation (28). By adding a small deadband $\tilde{u}$, it is ensured that the valve will be able to stay closed when no movement is needed.

$$\hat{u} = \begin{cases} u^+ + (1 - u^+) \cdot u, & u > \tilde{u} \\ u^- + (1 + u^-) \cdot u, & u < -\tilde{u} \\ 0, & \text{otherwise} \end{cases} \tag{28}$$

where

$\hat{u}$ = compensated control signal;

u = control signal;

u^+ = Out-stroke deadband;

u^- = In-stroke deadband;

$\tilde{u}$ = desired deadband, 0.001.

Figure 22. HMF 2020K4 loader crane in the laboratory.

The cylinder is running with a point-to-point path in actuator space equal to the simulations. The position error for the MIT-rule feedforward is shown in Figure 23. It is shown that the position error decreases towards a bounded error of ±14 mm. The RMS error after convergence is 5.2 mm. The convergence of the position error is similar to the simulations, showing that the proposed adaptive controller is feasible in a real world scenario, albeit with slightly larger position error.

The states z_{ff} for the MIT-rule feedforward experiment are shown in Figure 24. The dashed lines show the theoretical values for a fixed feedforward gain. The states converge to values that differ from the theoretical ones. The state z_{ff}^+ is higher than the theoretical, while the state z_{ff}^- is lower. This means that there exist some model uncertainties that the controller is able to adapt to. In addition, the ratio of the feedforward gains differs from the cylinder area ratio ϕ, i. e. $\frac{z_{ff}^-}{z_{ff}^+} \neq \frac{A_a}{A}$, showing the importance of using two separate feedforward states. Since the two states are not mathematically linked by the cylinder area ratio ϕ, they are able to converge to values that minimizes position error in both directions of motion regardless of their ratio. This would not be possible if the traditional MIT-rule with a single state was used.

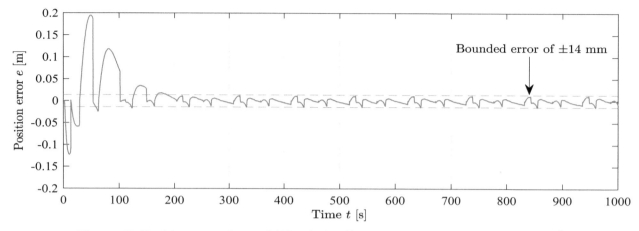

Figure 23. Position error during MIT-rule feedforward experiment, $\gamma = 200 \; \mathrm{s} \cdot \mathrm{m}^{-3}$.

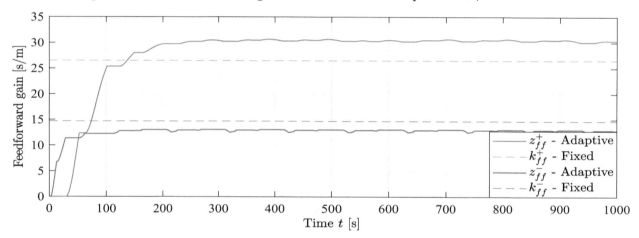

Figure 24. Feedforward states during MIT-rule feedforward experiment, $\gamma = 200 \; \mathrm{s} \cdot \mathrm{m}^{-3}$.

The position error for the sign-sign feedforward is shown in Figure 25. The same bounded error of ± 14 mm is shown. The RMS error after convergence is 5.3 mm.

The states z_{ff} for the sign-sign feedforward experiment are shown in Figure 26. Similar results can be seen here as with the MIT-rule, the states converge to values that differ from the theoretical ones. The dashed lines show the theoretical values for a fixed feedforward gain. The convergence is slower than the MIT-rule feedforward, and even though convergence speed is not critical, it may be a minor disadvantage compared to the MIT-rule feedforward.

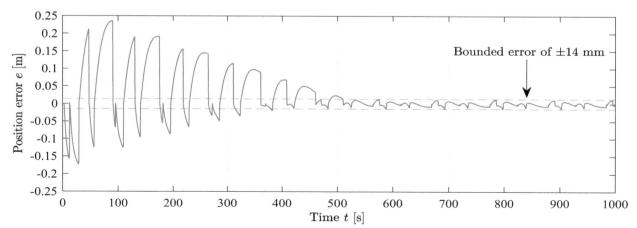

Figure 25. Position error during sign-sign feedforward experiment, $\gamma = 0.1 \; \mathrm{m}^{-1}$.

The same comparison as in the simulations is made in the laboratory. An experiment with fixed gain feedforward has been made and compared with the MIT-rule feedforward at a time where the states z_{ff} have converged, at $t = 800$ s. Figure 27 shows the difference in performance between the fixed gain controller and the adaptive controller, where the position error for the MIT-rule feedforward is significantly lower compared to the fixed gain feedforward.

The RMS position error for each controller after convergence of the states z_{ff} is shown in Table 5. The two adaptive feedforward controllers yield excellent performance with an 80% decrease in RMS position error.

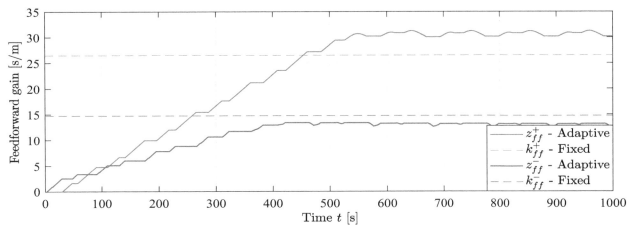

Figure 26. Feedforward states during sign-sign feedforward experiment, $\gamma = 0.1$ m^{-1}.

In general, the RMS position errors are slightly larger than in the simulations, but this is expected and can be attributed to the unmodeled flexibility of the crane, and other unmodeled dynamics. However, the advantage of the adaptive feedforward controller is clear. The independent adaptation of the out-stroke and in-stroke states z_{ff}^+ and z_{ff}^- provides significantly increased performance on a physical system with model uncertainties.

Table 5. Comparison of RMS position error after convergence in experiment.

	MIT-Rule	**Sign-Sign**	**Fixed Gain**
RMS error	5.2 mm	5.3 mm	24.9 mm

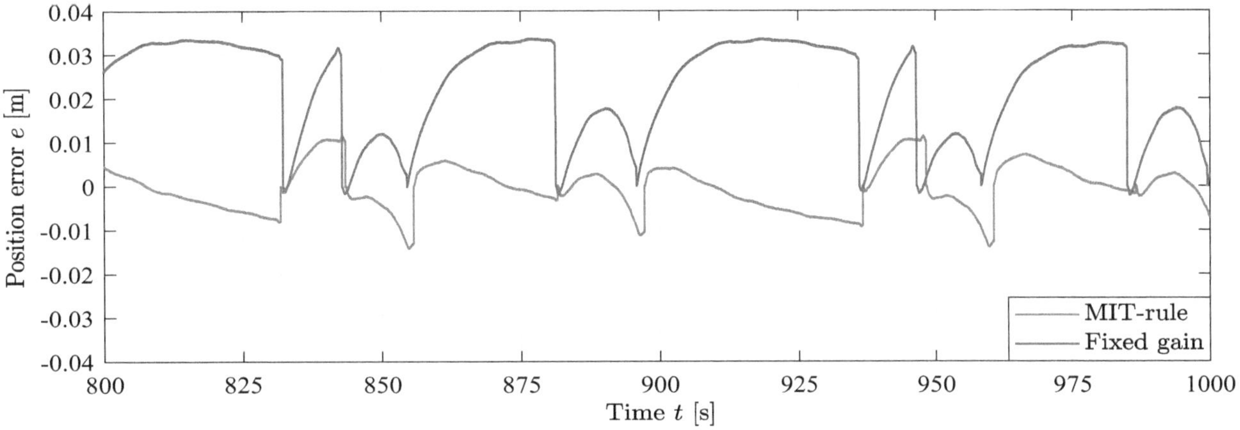

Figure 27. Position error with fixed and adaptive gains.

8. Conclusions

In this paper two adaptive feedforward motion controllers are designed, simulated, evaluated, implemented and experimentally verified on a loader crane with hydraulic differential cylinders. The controllers are based on common and proven adaptation methods to extend a typical electro-hydraulic motion control system into a novel adaptive feedforward motion controller. One of the challenges associated with a differential cylinder, namely the jump in both velocity and force gain when changing sign of direction, is solved by creating two separate feedforward states for out-stroke and in-stroke motion of the hydraulic differential cylinder, respectively. This separation makes the controller able to adapt to model uncertainties where the ratio between the in-stroke and out-stroke feedforward gains is not equal to the cylinder area ratio ϕ. Adaptation of the feedforward states only occurs when the hydraulic cylinder is moving in the direction of motion associated with the feedforward state.

Simulation results show high performance with good position tracking and that the states z_{ff} converge to values slightly higher than the theoretical ones. The cylinder position error is lowest for the MIT-rule controller with an RMS error of 1.6 mm, and shows faster convergence than the sign-sign controller. Compared to a fixed gain feedforward controller, where the gain is equal to the ratio of valve opening to cylinder velocity, the RMS error is reduced by 23%, showing the improved performance of the novel adaptive feedforward controllers.

Experiments in the laboratory show even better results than in the simulations. The adaptive feedforward controllers converge and show good position tracking, while the MIT-rule feedforward converges faster than the sign-sign feedforward. Compared to a fixed gain feedforward, the RMS position error is reduced by 80% to 5.2 mm for the MIT-rule. The results show the feasabillty of the novel adaptive feedforward controllers on a physical system. In addition, the differential structure of the controllers shows its advantage, as the ratio of the feedforward states converges to values different than the cylinder area ratio ϕ, showing the excellent performance of the adaptive feedforward controller and its capability of handling model uncertainties in both directions of motion.

Future work may include stability analysis of the adaptive controllers, since the feedforward gains are dependent on feedback of the cylinder position error e. The effects of the adaptation gain γ may also be investigated to see if there exists an upper boundary where the system becomes unstable.

Author Contributions: Conceptualization, K.J.J., M.K.E. and M.R.H.; methodology, K.J.J.; software, K.J.J.; validation, K.J.J.; formal analysis, K.J.J.; investigation, K.J.J.; data curation, K.J.J.; writing–original draft preparation, K.J.J.; writing–review and editing, K.J.J., M.K.E. and M.R.H.; visualization, K.J.J.; supervision, M.K.E. and M.R.H. All authors have read and agreed to the published version of the manuscript.

References

1.   Bak, M.K.; Hansen, M.R. Analysis of Offshore Knuckle Boom Crane—Part Two: Motion Control. *Model. Identif. Control.* **2013**, *34*, 175–181. [CrossRef]
2.   Mareels, I.M.; Anderson, B.D.; Bitmead, R.R.; Bodson, M.; Sastry, S.S. Revisiting the Mit Rule for Adaptive Control. *IFAC Proc. Vol.* **1987**, *20*, 161–166. [CrossRef]
3.   Lucky, R.W. Techniques for adaptive equalization of digital communication systems. *Bell Syst. Tech. J.* **1966**, *45*, 255–286. [CrossRef]
4.   Widrow, B. Adaptive sampled-data systems. *IFAC Proc. Vol.* **1960**, *1*, 433–439. [CrossRef]
5.   Widrow, B.; Hoff, M.E. Adaptive Switching Circuits. *1960 IRE Wescon Conv. Rec.* **1960**, 96–104.
6.   Unbehauen, H. Theory and Application of Adaptive Control. *IFAC Proc. Vol.* **1985**, *18*, 1–17. [CrossRef]
7.   Truxal, J.G. Adaptive control. *IFAC Proc. Vol.* **1963**, *1*, 386–392. [CrossRef]
8.   Strietzel, R.; Töpfer, H. Feedforward Adaption to Control Processes in Chemical Engineering. *IFAC Proc. Vol.* **1985**, *18*, 115–120. [CrossRef]
9.   M'Saad, M.; Duque, M.; Landau, I. Robust LQ Adaptive Controller for Industrial Processes. *IFAC Proc. Vol.* **1985**, *18*, 91–97. [CrossRef]
10. Unbehauen, H.D. Adaptive Systems for Process Control. *IFAC Proc. Vol.* **1986**, *19*, 15–23. [CrossRef]

11. Oh, B.; Jamshidi, M.; Seraji, H. Two Adaptive Control Structures of Robot Manipulators. *IFAC Proc. Vol.* **1989**, *22*, 371–377. [CrossRef]
12. Van den Bossche, E.; Dugard, L.; Landau, I. Adaptive Control of a Flexible Arm. *IFAC Proc. Vol.* **1987**, *20*, 271–276. [CrossRef]
13. Tsakalis, K.; Ioannou, P. Adaptive control of linear time-varying plants. *Automatica* **1987**, *23*, 459–468. [CrossRef]
14. Hušek, P. Adaptive fuzzy sliding mode control for uncertain nonlinear systems. *IFAC Proc. Vol.* **2014**, *47*, 540–545. [CrossRef]
15. Sato, K.; Tsuruta, K. Adaptive Friction Compensation for Linear Slider with adaptive differentiator. *IFAC Proc. Vol.* **2010**, *43*, 467–472. [CrossRef]
16. Lee, R.M.; Chen, T.C. Adaptive Control of Active Magnetic Bearing against Milling Dynamics. *Appl. Sci* **2016**, *6*, 52. [CrossRef]
17. Yang, G.; Yao, J.; Le, G.; Ma, D. Adaptive integral robust control of hydraulic systems with asymptotic tracking. *Mechatronics* **2016**, *40*, 78–86. [CrossRef]
18. Nicol, C.; Macnab, C.; Ramirez-Serrano, A. Robust adaptive control of a quadrotor helicopter. *Mechatronics* **2011**, *21*, 927–938. [CrossRef]
19. Agarwal, R.K.; Hussain, I.; Singh, B. LMF-Based Control Algorithm for Single Stage Three-Phase Grid Integrated Solar PV System. *IEEE Trans. Sustain. Energy* **2016**, *7*, 1379–1387. [CrossRef]
20. Krus, P.; Palmberg, J.O. Vector Control of a Hydraulic Crane. *Int. Off Highw. Powerpl. Congr. Expo.* **1992**. [CrossRef]
21. Sørensen, J.K.; Hansen, M.R.; Ebbensen, M.K. Boom Motion Control Using Pressure Control Valve. In Proceedings of the 8th FPNI Ph.D Symposium on Fluid Power, Lappeenranta, Finland, 11–13 June 2014. [CrossRef]
22. Sørensen, J.K.; Hansen, M.R.; Ebbesen, M.K. Load Independent Velocity Control on Boom Motion Using Pressure Control Valve. In Proceedings of the Fourteenth Scandinavian International Conference on Fluid Power, Tampere, Finland, 20–22 May 2015.
23. Beiner, L. Identification and Control of a Hydraulic Forestry Crane. *Mechatronics* **1997**, *7*, 537–547. [CrossRef]
24. Mattila, J.; Virvalo, T. Energy-efficient Motion Control of a Hydraulic Manipulator. In Proceedings of the 2000 IEEE International Conference on Robotics and Automation, San Francisco. CA, USA, 24–28 April 2000. [CrossRef]
25. Zhang, Q. Hydraulic Linear Actuator Velocity Control Using a Feedforward-plus-PID Control. *Int. J. Flex. Autom. Integr. Manuf.* **1999**, *7*, 277–292.
26. Jensen, K.J.; Kjeld Ebbesen, M.; Rygaard Hansen, M. Development of Point-to-Point Path Control in Actuator Space for Hydraulic Knuckle Boom Crane. *Actuators* **2020**, *9*, 27. [CrossRef]

10

Design of a Two-DOFs Driving Mechanism for a Motion-Assisted Finger Exoskeleton

Giuseppe Carbone [1,2,*], Eike Christian Gerding [3], Burkard Corves [3], Daniele Cafolla [4], Matteo Russo [5] and Marco Ceccarelli [6]

[1] DIMEG, University of Calabria, 87036 Rende, Italy
[2] CESTER, Technical University of Cluj-Napoca, 400114 Cluj-Napoca, Romania
[3] IGMR, RWTH Aachen University, 52062 Aachen, Germany; eike.gerding@rwth-aachen.de (E.C.G.); corves@igmr.rwth-aachen.de (B.C.)
[4] IRCCS Istituto Neurologico Mediterraneo Neuromed, 86077 Pozzilli, Italy; contact@danielecafolla.eu
[5] Faculty of Engineering, University of Nottingham, Nottingham NG7 2RD, UK; matteo.russo@nottingham.ac.uk
[6] LARM2, Laboratory of Robot Mechanics, University of Rome "Tor Vergata", 00133 Rome, Italy; marco.ceccarelli@uniroma2.it
* Correspondence: giuseppe.carbone@unical.it

Abstract: This paper presents a novel exoskeleton mechanism for finger motion assistance. The exoskeleton is designed as a serial 2-degrees-of-freedom wearable mechanism that is able to guide human finger motion. The design process starts by analyzing the motion of healthy human fingers by video motion tracking. The experimental data are used to obtain the kinematics of a human finger. Then, a graphic/geometric synthesis procedure is implemented for achieving the dimensional synthesis of the proposed novel 2 degrees of freedom linkage mechanism for the finger exoskeleton. The proposed linkage mechanism can drive the three finger phalanxes by using two independent actuators that are both installed on the back of the hand palm. A prototype is designed based on the proposed design by using additive manufacturing. Results of numerical simulations and experimental tests are reported and discussed to prove the feasibility and the operational effectiveness of the proposed design solution that can assist a wide range of finger motions with proper adaptability to a variety of human fingers.

Keywords: bionic mechanism design; synthesis; exoskeleton; finger motion rehabilitation

1. Introduction

Aging of population and stroke incidence are expected to significantly increase in the coming decades and become the second leading cause of disability in Europe as forecast, for example, in [1,2]. Usually, a stroke produces neuro-motory disabilities, including finger impairments. Since the movement of fingers is fundamental in activities of daily life, there is a strong motivation in focusing on finger rehabilitation as a high priority following an injury or a stroke.

Several studies have shown that the rehabilitation after a stroke is faster and more cost effective when using a robotic system as compared to conventional rehabilitation methods, as reported, for example, in [3]. Accordingly, researchers have widely addressed the topic for developing finger exoskeletons and/or similar wearable devices for finger rehabilitation and exercising, as reported, for example, in [4–21].

The "index finger exoskeleton" reported by Agarwal et al. [4] consists of eight linkages that are actuated by two cable drives. The exoskeleton has three DOFs (degrees of freedom) in total, as each linkage between the exoskeleton and finger has one DOF. Each linkage consists of four links and four joints with one DOF each. The exoskeleton allows flexion and extension of all phalanxes of the finger as well as passive abduction and adduction of the first phalanx. This exoskeleton is quickly adjustable and has a low resistance against finger movement when the motors are not activated. The device applies 80 g to the finger, and it has five angle-sensors to monitor link orientations. The index-finger exoskeleton uses a closed-loop torque-control with maximum torque at the first phalanx equal to 250 Nmm. The maximum torque at the second phalanx is 50 Nmm [4].

The exoskeleton reported by Bataller et al. [5] is optimized by an evolutionary synthesis algorithm. The synthesis determines the link length of the mechanism design to fulfill a given coupler curve with high accuracy. The kinematic structure of the device is set before the synthesis. It consists of seven links that are fixed on the back of the hand and on each phalanx, resulting in one DOF in total. A video analysis of the healthy finger motion acquires the coupler curve. The phalanx lengths of a patient's finger are acquired by an X-ray. The proposed exoskeleton is manufactured for each patient individually by 3D-printing and it has one servo motor.

Amadeo is a commercial hand and finger rehabilitation device that has been developed by Tyromotion GmbH [6]. The fingertips can be connected to the caps of the exoskeleton while being adjustable to different finger sizes. Each cap is attached to an automated linear slide. The fingertips can be pushed with a predefined force, and the device can move the finger while they apply no force. In this way, a grasping activity is simulated. The arm and wrist are fixed on the device frame, and the finger caps are connected to a slider. All fingers and the thumb can be treated. The device also allows for several measurements, such as force and range of motion (ROM). In [7], it is reported that 70% of the ROM of a healthy finger is sufficient for rehabilitation motions.

The Script SPO-F exoskeleton [8] is a passive device with no motor but a spring as the actuator. It consists of six links and seven joints with only one DOF. Earlier versions of this exoskeleton have been bulky, complex, and reached a weight of 1.5 kg. The Script SPO-F is actually an exoskeleton for both the wrist and the hand. It is designed for home-usage during rehabilitation treatment. The finger is flexed by a user, and a spring exerts a force on the finger due to the deflection. In contrast to other exoskeletons, it has no predetermined trajectory. The finger can be moved due to a cable connection between the fingertip and the exoskeleton. As the fingertip is fixed, only first and second phalanxes can move. A torque of about 125 Nmm is needed for a 90° flexion of the second phalanx.

The hand exoskeleton version HX [11] is suitable for both the index finger and the thumb. The exoskeleton has five DOFs and is driven by cables. The weight of the index finger module is 118 g, and the total weight lying on the hand is 438 g. The exoskeleton is made of a 3D-printed titanium alloy.

At LARM (Laboratory of Robotics and Mechatronics), a specific research line has been addressing the development of exoskeletons for motion assistance, as reported, for example, in [22–24]. Moreover, [25–27] focus on the fundamentals of the mechanics of grasping as well as the design and validation of anthropomorphic robotic hands. The LARM robotic hands are based on a driving mechanism with linkages that remain within the finger body duringthe finger operation, as reported in [28–30]. The design of such a driving mechanism is the conceptual reference for the exoskeleton solution that is reported in [24] and in preliminary exoskeleton designs, as reported in [31–34].

The main problem with existing exoskeletons is that they are often not wearable by different patients, as in 4,5], are bulky, and the overall equipment is not easily transportable, such as in [6] or is heavy, such as in [11]. Commercial robots, as in [6], are considered too expensive for home rehabilitation use. Further, the Amadeo device is not able to fulfill a complete grasping movement. The solution in [8] has no defined trajectory to move all joints of the finger in a defined way. Accordingly, the authors believe there is still a need for a design procedure that can lead to novel design solutions as based on kinematic analysis and a proper mechanism synthesis referring to the specific task of finger motion assistance.

This paper aims at a systematic design approach towards a novel two-DOFs driving linkage mechanism for a motion assistance finger exoskeleton by presenting a novel design solution for a finger exoskeleton with adaptability to the finger size, as well as cost-oriented design and user-friendly features. The design process is carried out within a specific design procedure. As a first step, the movement of a human finger was characterized by video motion tracking to identify the desired reference finger motions. Then, the relative kinematics of a human finger were obtained based on the acquired data. As a next step, a type synthesis was carried out to identify a mechanism consisting of linkages with two active DOFs as the most convenient solution to assist the motion of a finger along the desired trajectory, as also preliminarily discussed in [33]. This paper also provides FEM analyses that are integrated in the proposed design approach. A graphic/geometric synthesis procedure has been implemented for achieving the dimensional synthesis of the proposed linkage mechanism. Numerical simulations and experimental tests have been carried out and discussed to prove the feasibility and effectiveness of the proposed design solution. The main contribution of this work can be recognized in the design of a proposed novel linkage mechanism that, unlike other existing designs, can drive the three finger phalanxes by using two independent active DOFs that are both driven by rotary servomotors placed on the back of the palm. This configuration allows for a wide range of motion assistance with proper adaptability to a variety of human fingers. The paper content can be outlined as follows: the first section addresses the design requirements for achieving a device for finger exercising/rehabilitation of multiple users; next, the paper deals with the kinematic design of the proposed new device based on a two-DOFs driving linkage mechanism, whose synthesis is described in Section 4; the following section focuses on the mechanical design and construction of a prototype; Section 6 describes an experimental validation with comparisons of numerical and experimental results to assess the feasibility and performance of the proposed device.

2. Design Requirements for a Finger Exoskeleton

To expand the range of suitable patients, a novel proposed exoskeleton should be easy to attach to a finger, adaptable to a wide range of users, and easily portable for home use.

Exoskeletons driven by cables are a common solution in the literature. However, they show a range of drawbacks such as high losses due to friction as well as a high risk of cable failures. Further, cables need to be kept under tension during motion. The cable management system has a negative effect on the portability, as it is often bulky as in [12]. Because of that, servo motors with linkage transmissions are preferred in this work as they can be robust, lightweight, compact, and easy to control. The linkage parts can be easy and cheap to manufacture even with commercial 3D printers. However, the design of such a linkage mechanism requires full understanding of the desired human finger motion assistance.

A human hand is composed of fingers, metacarpus, and carpus. The fingers consist of three phalanxes, except for the thumb, which consists of two phalanxes. The metacarpus is connected to the proximal phalanx. On the fingers, the second link is the medial phalanx. The third link is the distal phalanx. A detailed description of the joints and functionalities of a human hand can be found, for example, in [35,36]. The joints between the intermediate and distal phalanxes are called distal interphalangeal joints (DIP), and the joints between the proximal and intermediate phalanxes are called proximal interphalangeal joints (PIP). The metacarpophalangeal joints (MCP) connect the proximal and metacarpal phalanxes, and the carpometacarpal joints (CMC) connect metacarpal phalanxes and the carpal bones, as shown in the scheme of Figure 1, [36]. In general, flexion reduces the angle between bones or parts of the body, whereas extension increases the angle between the bones of a limb. Abduction is an outward movement of a limb, and adduction is an inward movement of a limb [9]. The MCP joints have two DOFs and they allow flexion and extension as well as abduction and adduction of a finger. The interphalangeal joints PIP and DIP have one DOF each. However, the axis of rotation of these joints is not constant during flexion and extension [10], and the ligaments restrict the movements of the joints. Moreover, DIP and PIP joints cannot be moved independently of each other [35].

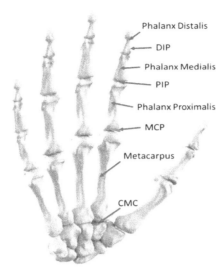

Figure 1. A scheme of bones and joints in a human hand.

A finger exoskeleton can be conveniently designed for motion exercising of the muscles after an injury or a stroke. For this application, the motion of a finger will be assisted by the exoskeleton with a motion trajectory similar to a healthy finger of the same size as proposed in Figure 2. The motion trajectory does not require high accuracy. For example, in [11], a misalignment of a phalanx up to ±8° is deemed acceptable. However, unnatural movements of the finger must be avoided by the mechanism to prevent potential injuries. As an additional design requirement for a finger exoskeleton, its installation space needs to be small enough to be attached onto the finger and not interfere with its motion. For the same reason, the total weight of the system should not exceed 500 g as reported, for example, in [7,13,17]. This value should be considered an upper bound, since a heavy exoskeleton can limit the effectiveness of motion exercising. Indeed, a high weight makes a patient easily tired and not willing to continue the treatment. Thus, it is advisable to limit the number of motors as they are the main source of weight. Motors need to provide a minimum torque of around 200 Nmm as reported previously [3,7].

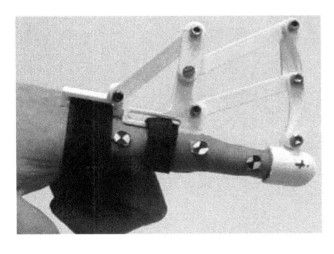 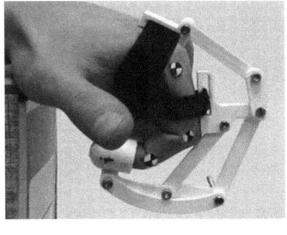

(a)   (b)

Figure 2. Desired motion assistance for a finger that moves from fully open (**a**) to fully closed (**b**).

One of the main peculiarities of the proposed novel linkage mechanism design solution is that it can move all three phalanxes of a human finger with two active DOFs. This is mainly achieved by coupling the motion of the last two phalanxes. Accordingly, only two DOFs are needed to replicate whole finger motions. The proposed mechanism is optimized to perform a specific desired motion but it is actually able to perform a wide range of other motion combinations, given its two active degrees of freedom. The proposed design is limited to flexion and extension movements of a finger. However, flexion and extension are identified in literature as the most important for activities of daily life and the first recovery priority in case of finger injury, as also mentioned in [35,36]. Therefore, a motion assistance mechanism at first instance does not need to focus on abduction and adduction movements, and these movements can be safely neglected to achieve a device with a light and simple design.

3. Kinematic Design

The index finger of a healthy subject has been used as a reference to develop and validate the proposed novel exoskeleton. First, grasping motions have been analyzed by using the Computer Vision Toolbox of MATLAB (Mathworks, Natick, MA, USA). Markers have been attached to the joints of the index finger and its fingertip (FT). Distances between the markers have been also measured with a digital caliper for calibration purposes. The camera has been aligned perpendicular to the side of the finger under study, and a fixing frame has been used both for the hand and camera in order to achieve a planar motion of the finger and a fully orthogonal video recording. The acquired videos have been post-processed to collect positions of the joints versus time. Figure 3 shows the experimental setup for the finger motion tracking as well as the definition of the angles and reference frame with the x-axis horizontal oriented from left to right, and the y-axis vertical oriented from top to bottom. The z-axis is defined according to the right-hand rule. The origin of the reference frame is attached to the center of the MCP joint. The angle of the MCP joint is called φ, the angle of the PIP joint is called ε, and the angle of the DIP joint is called τ. The angles have a clockwise positive direction.

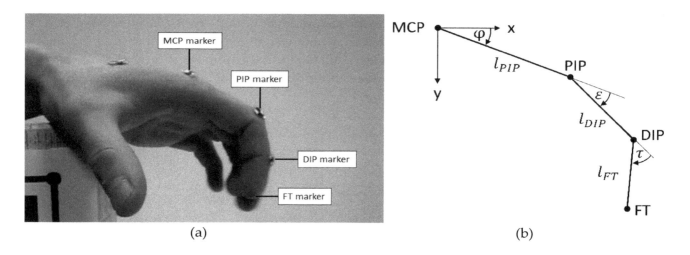

(a)   (b)

Figure 3. Experimental setup for finger motion video tracking (**a**) and definition of angles of the finger joints (**b**).

The lengths of the phalanxes of a test subject were identified from set-up measurements as 25 mm for the distal phalanx, 28 mm for the medial phalanx, and 43 mm for the proximal phalanx. Twenty-two frames have been evaluated from the motion tests for a suitable finger motion characterization. Figure 2 shows a detail of a video-captured motion for a healthy human finger movement referring to a typical human finger motion. In particular, the plot in Figure 4 shows trajectories of PIP, DIP, and FT markers during the motion of an index finger that moves from fully open to fully close. These trajectories will be used as the reference motion trajectories for the linkage mechanism of the proposed novel exoskeleton. However, the desired exercising motion usually does not require the full joint feasible rotation range of the motion range of a healthy human finger but only a subset of such a motion.

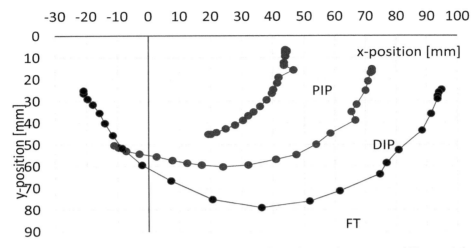

Figure 4. Acquired trajectories of PIP, DIP, and FT markers (as in the set-up of Figure 1 for the finger movement of a healthy human (dots indicate the measured data from the video capture tracking).

Moreover, it is advisable to avoid full finger closing as this would result in an undesired interference with the palm. A specific linkage mechanism has been identified as convenient for mimicking the desired reference finger motion. This proposed mechanism consists of eight links that are arranged as two interconnected four-bar mechanisms. This specific linkage arrangement has been proposed as an Italian patent [34]. The proposed mechanism requires two motors that each drive one of the two four-bar mechanisms. A detailed kinematic scheme is shown in Figure 5. The motors are located at joints D0 and B0. The angle of driving link D0-D is called δ. The angle of the driving link B0-B is called β. Both angles are positive in a clockwise direction. The first four-bar linkage (D0-D-E-MCP) has four links and four revolute joints with one active DOF. The second four-bar linkage (B0-B-A-A0-C) has one active DOF, four links, and four revolute joints.

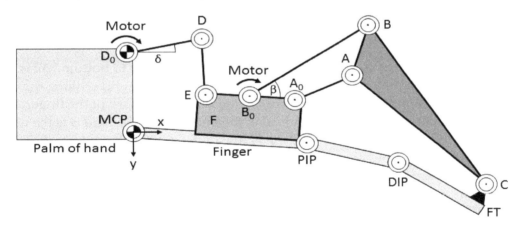

Figure 5. A kinematic scheme of the proposed design with its linkage structure.

The second four-bar linkage drives the point C that is attached to the fingertip (FT). The exoskeleton is also attached to the palm at the fixed frame link D0-MCP and to the first phalanx with the link E-B0-A D, as shown in Figure 5.

For the first linkage, the link lengths need to be chosen by considering the dimensional constraints of the motor and the palm of a hand. Accordingly, the link length D0-MCP is close to the vertical dimension of the selected motor. The link length E-MCP is mostly determined by the geometrical constraints for the attachment of E-B0-A to the first phalanx. The link lengths D0-D and D-E need to be calculated to fulfil the desired motion trajectory of the PIP joint, according to the motion reference in Figure 4. Considering the finger in fully close configuration, one can identify that in this configuration, the longest distance is expected to be D0-E. This distance can be calculated as 63.3 mm from the reference motion data. For achieving this value, the link length $MCP - D_0$ has to be equal to 27.8 mm, the link length D_0–D has to be equal to 32.0 mm, and the link length D-E has to be equal to 58.1 mm. Joints E and A0 on body F have coaxial axes.

To size the link lengths of the second linkage mechanism, a graphic/geometric synthesis dimensional synthesis has been carried out to obtain the desired motion of point C. The procedure is reported as a general formulation, for example, in [35]. The dimensional synthesis starts by calculating the desired relative motion of point C relative to A0 based on the experimental data in Figures 3 and 4 and on the relative position of the joints in the proposed linkage mechanism shown in the scheme of Figure 6. In this scheme, γ is the angle between the horizontal axis and line through A0, α is the angle between the horizontal axis and line through C, and η is the difference between the angles. The relative coordinate system is located in A0 and has its p-axis along the line MCP − A_0, as shown in Figure 6.

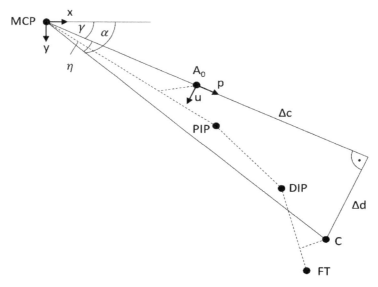

Figure 6. A scheme for determining C coordinates in the second linkage mechanism by considering the known PIP (proximal interphalangeal joint), DIP (distal interphalangeal joint), and FT (fingertip) coordinates.

The u-axis is perpendicular to the p-axis with direction from top to bottom. Δc is the distance between A0 and C along the p-axis, and Δd is the distance between A0 and C along the u-axis.

The coordinates of point PIP are known from the measured trajectory of the finger movement in Figure 4. The displacement of the PIP point and A0 is shown in Figure 7 where φ is the angle between the horizontal axis and line through PIP, while ψ is the angle of MCP from A0 to PIP. The distance between MCP and A0 is l_{A0}, and the distance between MCP and PIP is l_{PIP}. Parameters a and b represent the displacement between A0 and PIP. The coordinates of point A0 can be calculated by using the parameters a and b and the angle γ. Since the angle φ and the length l_{PIP} are known from the calculations of the reference trajectory in Figure 4, ψ can be calculated as

$$\psi = \arctan\left(\frac{b}{l_{PIP} - a}\right) \tag{1}$$

Therefore, referring to Figure 7, the angle γ is given by

$$\gamma = \varphi - \psi \tag{2}$$

The coordinates of A0 can be calculated as

$$x_{A0} = \sqrt{(l_{PIP} - a)^2 + b^2}\, \cos\gamma \tag{3}$$

$$y_{A0} = \sqrt{(l_{PIP} - a)^2 + b^2}\, \sin\gamma \tag{4}$$

By squaring and summing Equations (3) and (4), it is possible to obtain

$$l_{A0} = \sqrt{\left(l_{PIP} - a\right)^2 + b^2} \tag{5}$$

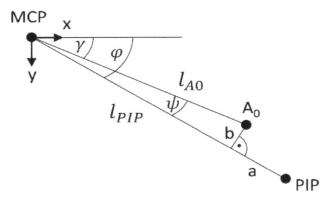

Figure 7. A scheme showing the displacement between A0 and PIP (proximal interphalangeal joint).

A similar procedure is used to calculate the global coordinates of C. Figure 6 shows a sketch for the displacement between FT and C, where λ is the angle between the horizontal axis and the line from DIP to C; θ is the angle between the horizontal axis and the line from DIP to FT. The distance between DIP and FT is l_{FT}, and the distance between DIP and C is l_C. Parameter e is the displacement between FT and C on the line from FT to DIP, and f is the displacement between C and FT perpendicular to that line.

The global x and y coordinates of DIP and FT are known from the captured trajectory in Figure 2. As given in Figure 8, the link orientation, given by angle θ, can be calculated as

$$\theta = \arctan\left(\frac{y_{FT} - y_{DIP}}{x_{FT} - x_{DIP}}\right) \tag{6}$$

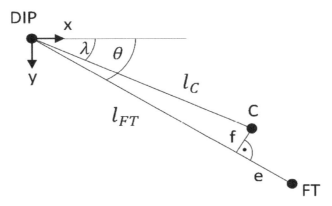

Figure 8. A scheme showing the displacement between FT (fingertip) and C.

Therefore, the angle λ is calculated as

$$\lambda = \theta - \arctan\left(\frac{e}{\overline{DIP - FT} - f}\right) \tag{7}$$

The global coordinates of C can be calculated as

$$x_C = \sqrt{\left(\overline{DIP - FT} - f\right)^2 + e^2} \cos\lambda + x_{DIP} \tag{8}$$

$$y_C = \sqrt{\left(\overline{DIP - FT} - f\right)^2 + e^2} \sin\lambda + y_{DIP} \tag{9}$$

Consequently, η is calculated as

$$\eta = \alpha - \gamma = \arctan\left(\frac{x_C}{y_C}\right) - \gamma \tag{10}$$

Finally, distances Δc and Δd can be computed as

$$\Delta c = \sqrt{x_c^2 + y_c^2} \cos\eta - l_{A0} \tag{11}$$

$$\Delta d = \sqrt{x_c^2 + y_c^2} \sin\eta \tag{12}$$

The resulting parameters from the above-mentioned calculations for the offset of A0 and C are summarized in Table 1. The parameters are obtained from measurements on the finger of the subject and also referring to an early prototype. The diameter of a reference finger has been measured to calculate the distance from the finger joint to the top of the finger, giving parameters b and f. Parameters a and e were identified from a preliminary prototype, which showed that A0 could be placed directly above the PIP joint, whereas C requires some distance from FT to avoid slipping off the finger during motion. With the calculated positions of joint C, the dimensional synthesis of the second linkage mechanism can be conducted.

Table 1. Design parameters from kinematic calculations in Figures 5 and 6.

Parameter	a	b	e	f
Length [mm]	4	17	15	19

No singular configuration is reachable within the used workspace of this mechanism. This has been verified at the design stage by setting up limits at the feasible transmission angles, and it has also been verified experimentally.

4. Mechanism Synthesis of the Second Linkage

The dimensional synthesis of the second linkage can be outlined as the procedure of determining the remaining lengths of a 4-bar mechanism that guides a point on the coupler curve. One possibility for such a synthesis is the graphical method based on the Burmester theory as described in [37]. This synthesis approach has been selected since it can quickly address the desired features in terms of replicating a desired motion trajectory as given by the reference motion trajectory in Figure 4. The expected accuracy for joints (±4 degrees) does not justify the use of more complex and time-consuming synthesis methods such as numerical optimization.

The proposed procedure allows an approximation of the desired movement from the measured trajectory in Figure 4 by defining three points on the coupler curve that are reached precisely by the mechanism. The synthesis can be conducted when two lengths of a mechanism and two positions of joints are known. Figure 9a shows the initial problem with all given parameters. The position of joints A0 and B0, lengths $\overline{A_0B_0}$, $\overline{B_0B}$, $\overline{BC}$, and three positions of joint C are given in the initial problem. From the input, the position of the missing joint can be determined. When the positions of all joints are established, the lengths can be determined. Curve k in Figure 9a shows the original movement of joint C, and curve kc in Figure 9b is the movement after the synthesis. With the synthesis in Figure 9b, joint A can be determined, giving the missing lengths $\overline{A_0A}$, $\overline{AC}$, and $\overline{AB}$.

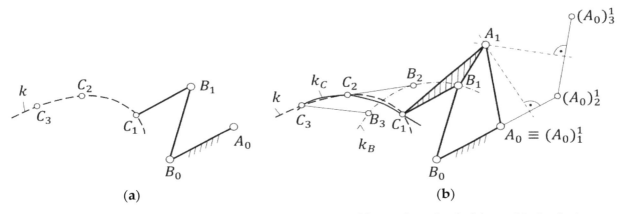

Figure 9. Schemes for a dimensional synthesis: (**a**) the problem to be solved; (**b**) graphical solution as based on the approach that is described in [33].

A pose of a mechanism consists of the positions and the orientations of all links. For example, for pose 1 of the mechanism in Figure 4, joint A is in position A1, joint B is in position B1, and joint C is in Position C1. The graphical procedure for the dimensional synthesis is given in the flow-chart of Figure 10 according to the following steps:

1. The given parameters in Figure 9a are the positions of C1, C2, C3, and B0 as well as the lengths of $\overline{B_0B}$ and $\overline{BC}$;
2. With the information from step 1, joint B is known in pose 1, 2, and 3, as joint B moves on the curve k_B. Therefore, the poses of $\overline{B_1C_1}$, $\overline{B_2C_2}$, and $\overline{B_3C_3}$ are known. At this point, joint A is missing to complete the mechanism sizing:
3. Next, one link in one pose is chosen as a reference (in this example, link BC in pose 1);
4. Now, joint A_0 is virtually moved around $\overline{B_1C_1}$ to create A0 that corresponds to position 2 in pose 1 (by moving A0 with respect to the coordinate system of pose 2 into pose 1). The transformation can be done by creating a triangle between C2, A0, and B2 and moving the triangle into the pose of C1 and B1. The resulting position is A_0 from pose 2 in pose 1, written as $(A_0)_2^1$;
5. The same as step 4 is done with A_0 from pose 3 in pose 1, resulting in $(A_0)_3^1$;
6. Finally, A_1 is found by the intersection of the perpendicular bisectors between three positions of A_0. The resulting mechanism guides joint C along curve k_C. Curve k_C matches curve k in the points C1, C2, and C3 [33];
7. With all known positions of the joints, the link lengths can be determined.
8. Other link lengths/positions can be chosen to improve the matching of curves k and kc.
9. The dimensional synthesis is applied to the second linkage mechanism by using the relative movement of C with respect to A0. The aim is to design a mechanism that matches the desired motion trajectory of point C.

The positions of point C with respect to the u-p coordinate system are calculated by Equations (11) and (12). For the dimensional synthesis, the positions of A0, B0 and the lengths $\overline{A_0A}$ and $\overline{AC}$ are required as well as three reference positions of C within its desired motion trajectory, according to step 1 in Figure 10. A0 and B0 are set as frame joints. A0 has the coordinates $(0; 0)$ with respect to the u-p-coordinate system. B0 is designed with the coordinates $(-8; -17.8)$. The input parameters have been iterated to find a mechanism that fulfills the desired path of C well. After few iterations, sufficient input parameters were found. Positions 1 to 15 of the calculation of point C have been considered for the synthesis, as it gets more difficult to approximate a longer motion path. Furthermore, the identified range of motion is sufficient for rehabilitation purposes.

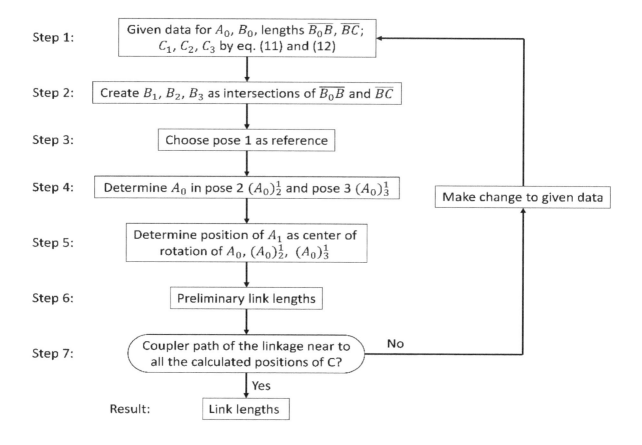

Figure 10. A flowchart for the mechanism synthesis according to the scheme in Figure 9.

The initially chosen link lengths and point coordinates are shown in Figure 11, similar to Figure 4. With this information, positions B2 and B3 can be determined as intersecting points with the known segments of $\overline{B_0B}$ and $\overline{BC}$, as mentioned in step 2. As a result of iterations of the synthesis, $\overline{B_0B}$ has been measured to have a length of 46 mm, and $\overline{BC}$. has been measured to have a length of 53 mm. As input from the positions of C, the first position is chosen as C1, position C2 is in the middle, and position C3 refers to the last used configuration during the finger closing motion. According to step 3, pose 1 is the reference for the synthesis. A0 with respect to position $\overline{B_3C_3}$ is transferred into position $\overline{B_1C_1}$, resulting in $(A_0)_3^1$. $(A_0)_3^1$ is A0 in pose 3 transferred to pose 1. This means that the triangles C3-B3-A0 and C1-B1-$(A_0)_3^1$ are identical. In the same manner, the point $(A_0)_2^1$ can be identified. This corresponds to step 4. The triangles to find the position $(A_0)_3^1$ are given in Figure 12a. The synthesized mechanism and the calculated positions of C are shown in Figure 12b. The previously calculated coordinates for C are marked as crosses, and the computed trajectory of the mechanism is marked as dots. The figure also shows that the trajectory matches well with the desired motion for point C.

The position of A1 is the center point of a circle through the positions A0, $(A_0)_2^1$ and $(A_0)_3^1$, as mentioned in step 5. The resulting position is A1, as pose 1 has been used as a reference. The link lengths can be determined with the obtained position of A1. Since the given data in step 1 have already been identified by a prior iteration, step 7 can be skipped. The complete kinematic design is obtained by combining both linkages. The resulting kinematic design of the whole exoskeleton mechanism is shown in Figure 13. The numerical values are summarized in Table 2, defined according to the scheme in Figure 5. The calculated positions of A0 and C are indicated with plus (+) and cross (x) marks, respectively.

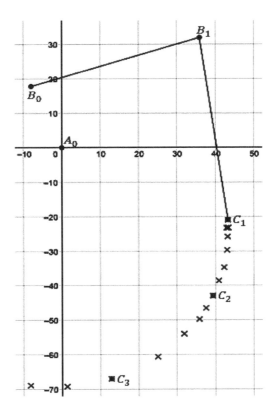

Figure 11. Initial problem of the synthesis for the second linkage with the selected three points C1, C2, C3 along the desired motion trajectory of point C.

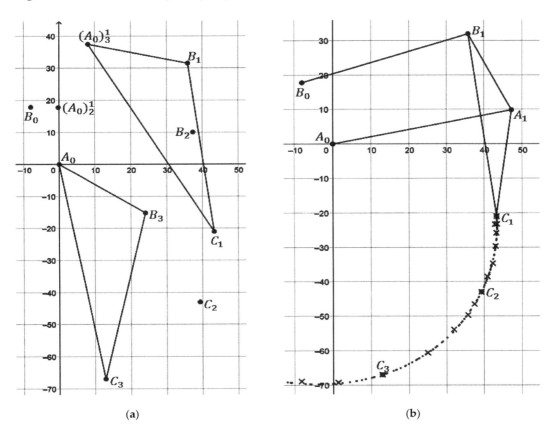

(a)   (b)

Figure 12. A scheme for determining the second linkage: (**a**) Triangles of the relative positions C0; (**b**) synthesized mechanism and resulting trajectory passing through the selected points C1, C2, C3.

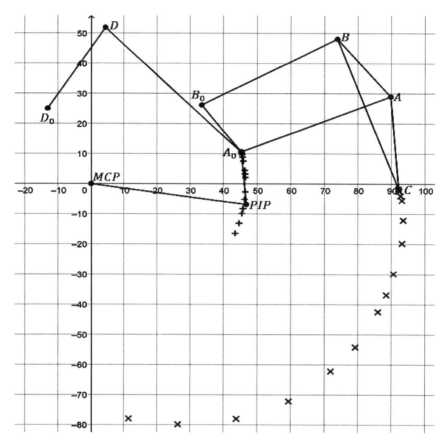

Figure 13. The synthesized kinematic design of the finger exoskeleton in Figure 3 with the trajectories of A0 and C indicated with plus (+) and cross (x) marks.

Table 2. Design parameters and link lengths of the finger exoskeleton for the scheme in Figure 5.

Parameter	Length [mm]	Parameter	Length [mm]	Parameter	Length [mm]
a	4.0	A0-B0	19.5	B-C	53.0
b	17.0	B0-B	46.0	D0-D	32.0
e	15.0	A-B	24.9	D-E	58.1
f	19.0	A-C	30.7	A0-A	48.2
MCP-PIP	43.0	PIP-DIP	28.0	DIP-FT	25.0

This paper reports the proposed graphical procedure for a specific case with nominal biometric measurements. However, the proposed graphical procedure is general in its approach. Accordingly, the same procedure can be performed again for different finger sizes when the phalanx lengths are expected to exceed the adaptability allowed by the proposed design. A chart can be generated with link dimensions for different patient biometrics to adapt the proposed finger exoskeleton to the wearer. Moreover, the proposed synthesis procedure can be also automated by implementing it in a numerical solving algorithm.

5. Mechanical Design and Prototype

Based on the obtained kinematic design, a prototype of the proposed finger exoskeleton has been developed. Since the finger exoskeleton is manufactured by 3D printing, it has been necessary to define the secondary geometric parameters of each linkage, such as link thickness. For this purpose, given the slow speeds, accelerations, and inertias of the application, a specific static analysis has been carried out according to the schemes that are shown in Figure 14. The computation has been carried out by considering as load the maximum motor torque of 216 Nmm, and a maximum force at the connections between finger and exoskeleton equal to about 6 N on the fingertip. This value is calculated by using the principle of virtual powers from the given input torque and also matches previous experiences of similar prototypes.

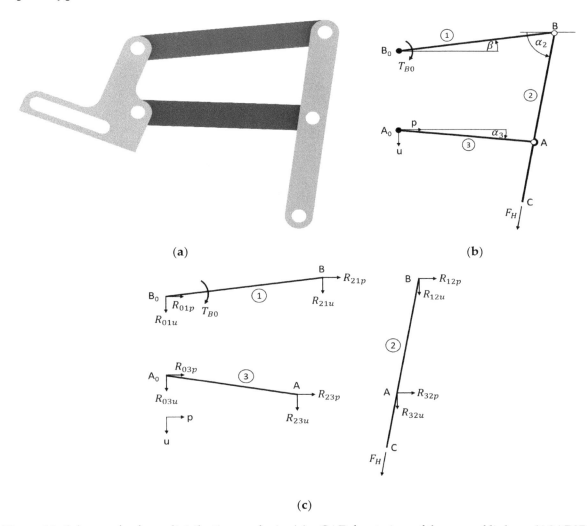

(a)   (b)

(c)

Figure 14. Schemes for force distribution analysis: (**a**) a CAD front view of the second linkage (A0AB0B); (**b**) an overall free body diagram; (**c**) free body diagrams of single elements.

FEM analyses have been carried out iteratively to find a proper linkage cross section and thickness. In particular, final FEM simulations have been carried out by considering the 3D CAD model that is shown in Figure 15. The main link thickness has been set as equal to 2 mm, also based on previous experiences. Similarly, the holes for the joints have been set at a diameter of 4 mm. Therefore, the links need to have a total width of 8 mm and a thickness of 2 mm. Link $\overline{AC}$ is crooked as per Figure 15 in order to avoid collisions with the finger, and link $\overline{BC}$ is crooked to allow for fixation of the joint A. Even though $\overline{ABC}$ behaves kinematically as a single body, it is realized with three different links that can be easily changed to fit the different finger sizes of different users. The link $\overline{BC}$ is manufactured with two beams to increase its stiffness. A CAD design has been elaborated with the above-mentioned design considerations. A functional solution of the mechanism with all its joints is shown in Figure 14. Screws and nuts of M3 size connect the links.

The main merit of the proposed design can be identified in the adaptability to multiple users. This is achieved by slotted holes (Figure 15) in the exoskeleton that allow easy adjustability to users. The slotted holes are used to adapt the finger exoskeleton to the user's biometrics (phalanx lengths) by moving each link to the optimal configuration evaluated through the proposed dimensional synthesis. Cable straps and loop fasteners are used to fix the exoskeleton on the finger, giving some additional adaptability. Accordingly, the proposed exoskeleton is expected to fit users having specific phalanx sizes exceeding ± 10% of the nominal sizes. A design limitation can be identified in the need to replace the links of the device when users are expected to exceed ± 10% of the nominal finger size. This limitation can be partially overcome by preparing sets of replacement links whose sizes are designed to fit with different nominal finger sizes (e.g., for children, male/female adults).

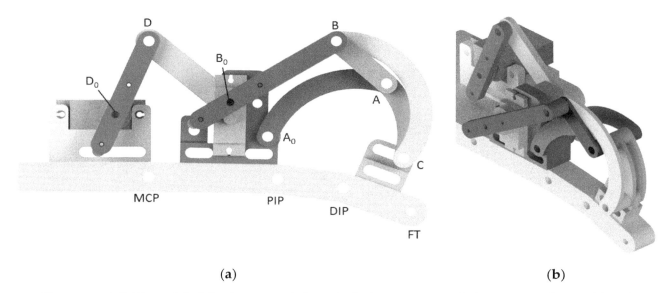

(a)   (b)

Figure 15. A CAD model of the finger exoskeleton including motors with description of joints (**a**) and angular view (**b**).

Final FEM tests have been performed in SolidWorks 2019 to verify the correctness of the selected cross-sections and minimum required thickness to avoid any failure or plastic deformation, as reported in Figures 16 and 17. A minimum safety factor equal to 1 has been considered in static nodal stress analysis for keeping the overall weight as low as possible. A minimum factor of safety equal 2.8 has been found on shear stress. The chosen material is a commercial poly-lactic acid (PLA) filament that is suitable for additive manufacturing with commercial 3D printers. Its main properties are tensile strength equal to $3 \cdot 10^7$ N/m^2, elastic modulus equal to $2 \cdot 10^9$ N/m^2; Poisson's ration equal to 0.394, mass density equal to 1020 Kg/m^3, and shear modulus equal to $3.189 \cdot 10^8$ N/m^2.

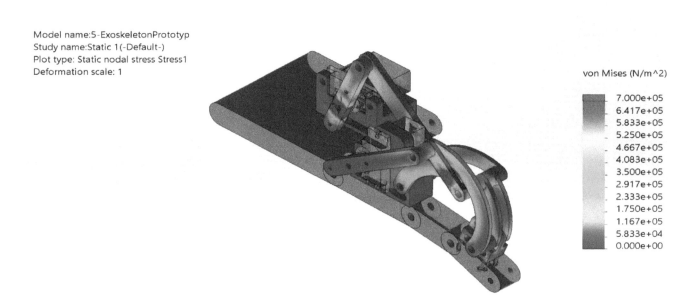

Figure 16. Static nodal FEM analysis of the whole prototype.

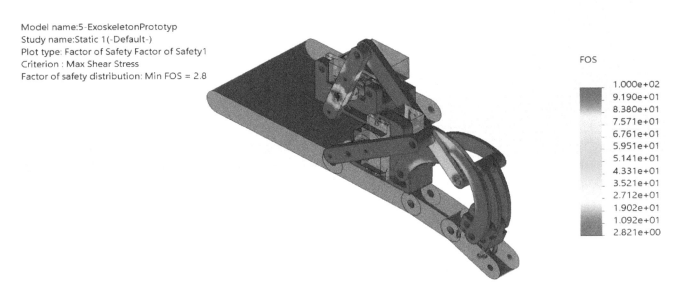

Model name:5-ExoskeletonPrototyp
Study name:Static 1(-Default-)
Plot type: Factor of Safety Factor of Safety1
Criterion : Max Shear Stress
Factor of safety distribution: Min FOS = 2.8

FOS

1.000e+02
9.190e+01
8.380e+01
7.571e+01
6.761e+01
5.951e+01
5.141e+01
4.331e+01
3.521e+01
2.712e+01
1.902e+01
1.092e+01
2.821e+00

Figure 17. FEM factor of safety calculations based on maximum shear stress.

The size of motors has been chosen to match the results of simulations for the designed mechanism as well as by comparison with similar devices in the literature. Both servo motors have been selected with a nominal torque of 216 Nmm while the desired torque was about 200 Nmm for the first joint and about 150 Nmm for the second joint. Servo motors with a torque of 216 Nmm are integrated into body F and lay on the back of the palm of the human hand. An Arduino microcontroller has been chosen to drive the motors.

The total cost of the system is around 50€, including the servomotors and the microcontroller. The motors are connected to an external power supply, which can be a LiPo battery for easy portability. The exoskeleton has a total weight of 64 g for the parts that are mounted on the finger. This includes the linkage, cable straps, hook and loop fasteners, and servo motors. The whole system has a weight of 175 g, including an Arduino board and all cablings. A full prototype has been built as shown in Figure 18.

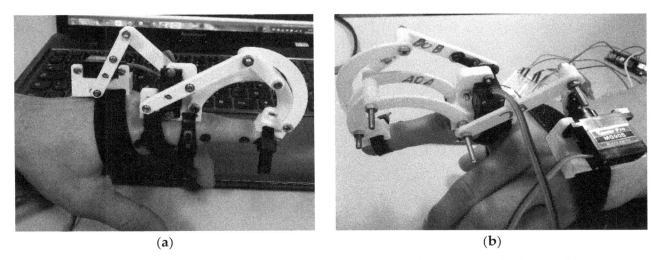

(a)   (b)

Figure 18. The built prototype of the finger exoskeleton, front view (**a**), back view (**b**).

6. Test Results

The built finger exoskeleton prototype has been tested experimentally to prove its feasibility as a finger motion exercising device in terms of it kinematic and operation behaviors. Given the expected slow speed operation, dynamic simulations and tests are not required at this proof-of-concept stage. The finger exoskeleton can be easily worn with Velcro fasteners. The connection between the finger and exoskeleton can be as tight as the subject wishes. Even after long use, it is still comfortable to wear. Also, the calibration procedure is very straightforward. It consists of the following steps: Attaching the exoskeleton to the finger, manually placing the finger in its desired straight configuration; registering this position as the initial configuration; manually placing the finger in its desired fully closed configuration; registering this position as the fully closed configuration. After the above steps, the device is ready to operate within the desired operation range.

The exoskeleton movement has been compared with the reference motion of a healthy human finger, and the angular error has been calculated as also partially reported in [33]. The driving angle of joint D0 is called δ, and the driving angle of joint B0 is called β. The angles of the joints of the finger from the grasping test have been used as an input for a multibody motion simulation of the CAD model on SolidWorks 2019. The simulation acquires the angles of the motor for each position during the finger motion. A comparison of the simulation and the motion of the prototype was used to determine if the exoskeleton prototype moves as planned.

The desired path planning and angular joint coordinates are managed by using an Arduino control board, which is connected to the two servomotors that drive the exoskeleton. The angles of the driving links have been calculated and interpolated for each position that has been experimentally measured during a grasping test, as reported in Figure 4. Then, the obtained motor angles are sent to an Arduino controller, which drives the exoskeleton. The desired motion has been obtained by an offline video post-processing that is carried out with MATLAB, allowing us to measure the angular joint positions in each acquired video frame. This information is converted into a desired set of joint angles versus time. The controller is programmed to move both motors in each desired position with an interpolated step motion. Accordingly, the proposed motion planning consists of passing through a prescribed number of path points with a smooth interpolated motion to reach precision positions. A pause of half a second is set before proceeding with the next step desired precision position to keep the motion safe for the user. The user has the time to easily stop if he/she feels discomfort in any reached configuration. The maximum angular reaches of motors are limited via software limits that are well within the range of the servo motors being equal to 180°. After completing the motion planning, videos of human finger exoskeleton-assisted motions were collected and analyzed. Tests were performed on the same subject as for the initial grasping test in Figures 1 and 3.

Video captures were acquired by tracking both finger joints and exoskeleton joints during a grasping motion. Then, the driving angles and the angles of the finger were determined. The resulting movement of the prototype and the finger is shown in Figure 19. The movement takes approximately 16 s. The proposed tests are performed with a human sitting and his/her forearm fixed on a reference table. The first snapshot Figure 19a refers to the beginning of the movement. Snapshot Figure 19b is taken after 3.2 s, snapshot Figure 19c after 6.5 s, snapshot Figure 19e after 9.7 s, and snapshot Figure 19e after 13.0 s. Snapshot Figure 19f shows the final position of the motion. In this test, the MCP joint moves within a range of 3.1° to 37.6° for a finger flexion. Similarly, the PIP joint goes from 13.0° to 78.7°, and the range of the DIP joint is from 0° to 58.9° for a finger flexion.

The time history of the measured finger joint angles is given in Figure 20. for a finger flexion assisted by the exoskeleton. Namely, Figure 20a–c show the acquired values of angles φ, ε, and τ versus time, respectively. Moreover, (x) markers are reported in Figure 20 to show the simulated angular motions angles of φ, ε, and τ versus time. In the plots of Figure 20, an initial offset can be observed between the measured and simulated values. The maximum velocity and swing frequency of the finger exoskeleton mechanism are defined by means of a reference finger-assisted motion. This motion needs to be very slow to allow safe finger motion assistance, so a fast speed is not desirable. In particular, for the prototype reported in the paper, the average velocity is 5 deg/s, with a swing frequency of 0.07 swings/s. From a practical point of view, this initial offset is mostly due to manufacturing/assembly errors as well as joint clearances. These effects can be seen also as positive in terms of adaptability to a more natural finger trajectory.

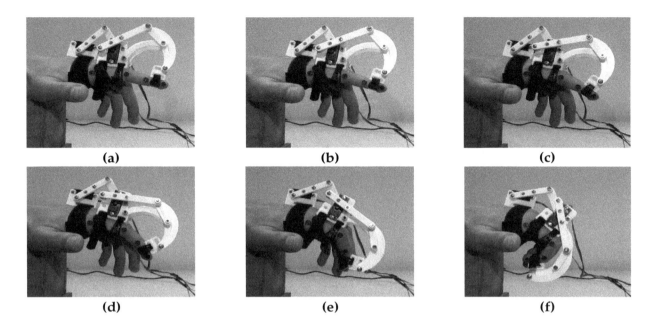

Figure 19. Snapshots of the exoskeleton motion during the test: (**a**) Starting position; (**b**) After 3.2 s; (**c**) After 6.5 s; (**d**) After 9.7 s; (**e**) After 13.0 s; (**f**) Final position.

The absolute error of the angular positions of the finger joints can be calculated by comparing the two curves in Figure 20. To compensate for an initial misalignment, a systematic error can be calculated. Moreover, an angular deviation error can be calculated as the difference between the experimentally measured and simulated values at a given time. For example, the time history of the angle δ is given in Figure 21 including both simulated and experimentally measured values. The experimentally measured angle δ versus time goes from $-47.2°$ to $-12.5°$ in this test. The comparison between the experimentally measured and simulated angle δ versus time shows an initial systematic error of $17.5°$. If this systematic error is compensated, the angular deviation error of angle δ versus time is given in Figure 22. The systematic error is mainly due to the clearance between the exoskeleton and finger, which are connected through velcro straps, whose clearance cannot be eliminated.

The absolute error for the angle δ in Figure 20 is always below ± 4 deg. This absolute error can be considered suitable as it is exactly within the acceptable error given in the design requirements for the proposed motion assistance application. Similarly, the time history of the β angle is given in Figure 21 including both simulated and experimentally measured values. The experimentally measured angle β versus time goes from $-6.5°$ to $64.7°$ in this test. The comparison between the experimentally measured and simulated angle β versus time shows an initial systematic error of $10.8°$. If one removes this systematic error, the angular deviation error of angle β versus time is given in Figure 22. Similarly, a comparison of β from the experimental test and simulation is reported in Figure 23, and the computed deviation error of β is reported in Figure 24. One can note that the absolute error for the angle δ in Figure 22 is below $\pm 4°$ until reaching a critical pose, corresponding to the pose reached after 13 s in the given experiment. Accordingly, operation of the prototype can be considered acceptable within the given design requirements until the closing configuration reaches the critical pose. Further motion needs to be avoided, since it would lead to interference of the human finger with the palm. This limit physically consists of allowing the finger flexion only until the DIP joint is vertically aligned with the MCP joint, referring to the scheme in Figure 5.

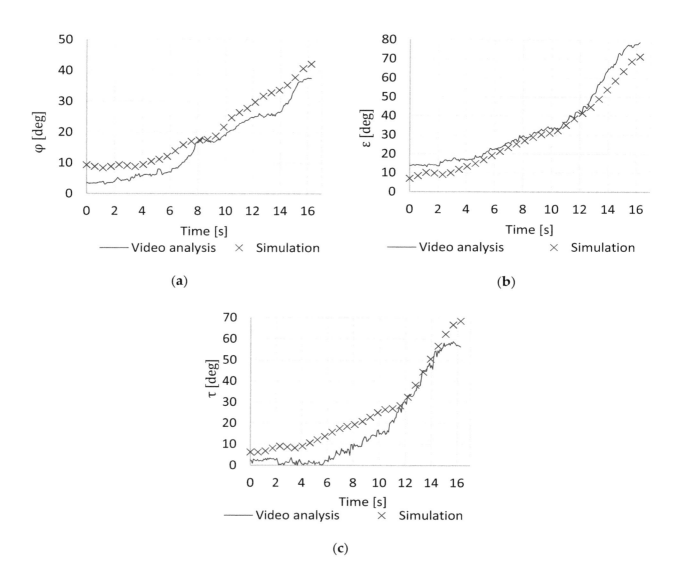

Figure 20. Comparison of finger joint angles in the prototype test (continuous line) and simulation (x marks): (**a**) angle φ versus time; (**b**) angleε versus time; (**c**) angle τ versus time.

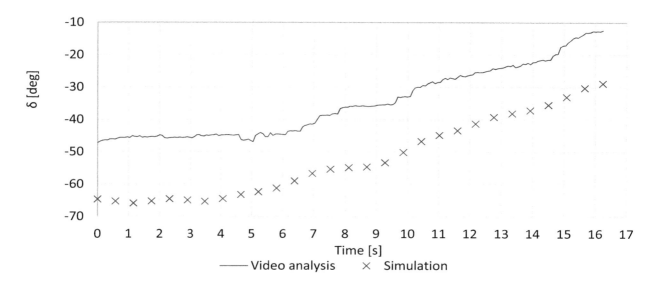

Figure 21. Comparison of δ for the test and simulation versus time.

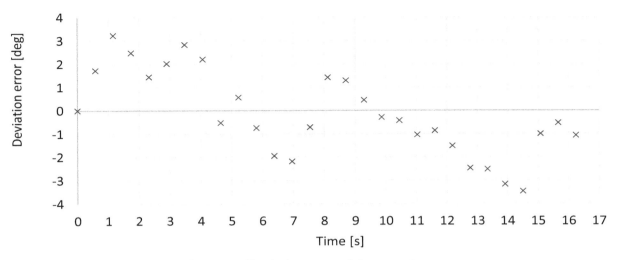

Figure 22. Deviation error of δ versus time.

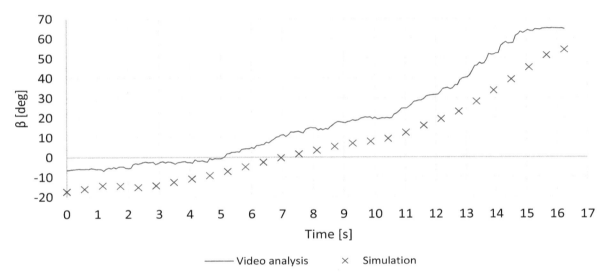

Figure 23. Comparison of β from the experimental test (continuous line) and simulation (x marks) versus time.

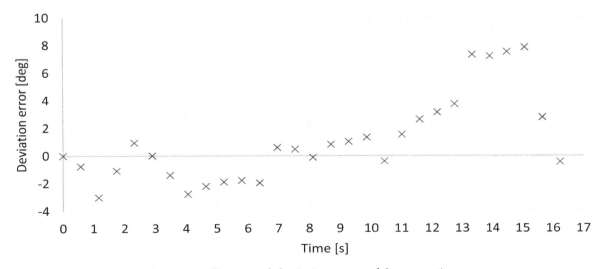

Figure 24. Computed deviation error of β versus time.

Average angular errors can be computed for each joint deviation error, in terms of as a square root average, as reported in Table 3. For all angles, the average error remains within the acceptable range of ± 4 degrees as established in the design requirements. This confirms the proposed finger exoskeleton can provide suitable motion assistance to replicate the desired human finger motion trajectory. The error calculation can be affected by the accuracy of the used method for experimentally measured angles via video capture tracking. This can generate non-negligible errors, in particular, if the tracking camera is not properly aligned or calibrated. For this purpose, the camera has been attached to a fixed frame, and the wrist is fixed, so that the exoskeleton movement is kept fixed in a proper plane. Accordingly, main sources of angular error s can be identified in the used 2D video tracking method as well as in joint clearances and backlashes. However, the clearance and backlash aspects can also provide positive features in terms of flexibility and adaptability of the finger exoskeleton motion to a natural finger motion.

Table 3. Angular errors of the finger joints and driving links.

Angle [deg]	φ	ε	τ	δ	β
	1.8	3.6	5.2	1.8	3.2

7. Conclusions

This paper reports the design of a novel exoskeleton mechanism for finger motion guidance. A human finger motion is analyzed through video motion tracking as a design reference. Then, a novel 2-DOF linkage mechanism is synthesized to mimic the desired finger motion. This linkage mechanism is driven by two independent actuators that can be conveniently placed on the back of the palm. The obtained mechanism with two linkage mechanisms is implemented in a 3D CAD model, and it is rapidly prototyped and assembled into a prototype. The resulting design is achieved with compact and lightweight features, and it can be manufactured in an easy and low-cost manner. The finger exoskeleton has been experimentally validated showing an acceptable error of guiding the desired human finger motion while performing a proper range of motions for assistance and rehabilitation of a large variety of human fingers.

Author Contributions: Conceptualization, G.C. and M.C.; software, E.C.G.; validation, E.C.G., D.C., and M.R.; data curation, E.C.G.; writing—original draft preparation, E.C.G.; writing—review and editing, G.C. and M.R.; supervision, G.C., M.C., and B.C. All authors have read and agreed to the published version of the manuscript.

Acknowledgments: The second author gratefully acknowledges the Erasmus+ program for a period of study he spent at LARM under the supervision of Prof. Marco Ceccarelli and Prof. Giuseppe Carbone.

References

1. United Nations Department of Economic and Social Affairs, Population Division: World Population Prospects: The 2019 Revision, Volume II: Demographic Profiles. 2019. Available online: https://population.un.org/wpp/Publications/Files/WPP2019_PressRelease_EN.pdf (accessed on 9 April 2020).
2. Kaplan, W.; Wirtz, V.; Mantel, A.; Béatrice, P.S.U. Priority medicines for Europe and the world update 2013 report. *Methodology* **2013**, *2*, 99–102.
3. Sale, P.; Lombardi, V.; Franceschini, M. Hand robotics rehabilitation: Feasibility and preliminary results of a robotic treatment in patients with hemiparesis. *Stroke Res. Treat.* **2012**, *2012*, 820931. [CrossRef] [PubMed]
4. Agarwal, P.; Fox, J.; Yun, Y.; O'Malley, M.K.; Deshpande, A.D. An index finger exoskeleton with series elastic actuation for rehabilitation: Design, control and performance characterization. *Int. J. Robot. Res.* **2015**, *34*, 1747–1772. [CrossRef]
5. Bataller, A.; Cabrera, J.A.; Clavijo, M.; Castillo, J.J. Evolutionary synthesis of mechanisms applied to the design of an exoskeleton for finger rehabilitation. *Mech. Mach. Theory* **2016**, *105*, 31–43. [CrossRef]
6. Stein, J.; Bishop, L.; Gillen, G.; Helbok, R. Robot-assisted exercise for hand weakness after stroke: A pilot study. *Am. J. Phys. Med. Rehabil.* **2011**, *90*, 887–894. [CrossRef] [PubMed]

7. Kun, L.I.U.; Hasegawa, Y.; Saotome, K.; Sainkai, Y. Design of a Wearable MRI-Compatible Hand Exoskeleton Robot. In *International Conference on Intelligent Robotics and Applications*; Springer: Cham, Switzerland, 2017; pp. 242–250.

8. Ates, S.; Haarman, C.J.; Stienen, A.H. SCRIPT passive orthosis: Design of interactive hand and wrist exoskeleton for rehabilitation at home after stroke. *Auton. Robot.* **2017**, *41*, 711–723. [CrossRef]

9. Pons, J.L. (Ed.) *Wearable Robots: Biomechatronic Exoskeletons*; John Wiley & Sons: Chichester, UK, 2008.

10. Heo, P.; Gu, G.M.; Lee, S.J.; Rhee, K.; Kim, J. Current hand exoskeleton technologies for rehabilitation and assistive engineering. *Int. J. Precis. Eng. Manuf.* **2012**, *13*, 807–824. [CrossRef]

11. Cempini, M.; Cortese, M.; Vitiello, N. A Powered Finger–Thumb Wearable Hand Exoskeleton with Self-Aligning Joint Axes. *IEEE/ASME Trans. Mechatron.* **2015**, *20*, 705–716. [CrossRef]

12. Li, J.; Zheng, R.; Zhan, Y.; Yao, J. iHandRehab: An Interactive Hand Exoskeleton for Active and Passive Rehabilitation. In Proceedings of the IEEE International Conference on Rehabilitation Robotics, Zurich, Switzerland, 29 June–1 July 2011; pp. 597–602.

13. Aubin, P.M.; Sallum, H.; Walsh, C.; Stirling, L.; Correia, A. A Pediatric Robotic Thumb Exoskeleton for at-Home Rehabilitation: The Isolated Orthosis for Thumb Actuation (IOTA). In Proceedings of the IEEE International Conference on Rehabilitation Robotics, Seattle, WA, USA, 24–26 June 2013; pp. 665–671.

14. Mertz, L. The next generation of exoskeletons: Lighter, cheaper devices are in the works. *IEEE Pulse J.* **2012**, *3*, 56–61. [CrossRef]

15. Jones, C.; Wang, F.; Morrison, R.; Sarkar, N.; Kamper, D. Design and development of the cable actuated finger exoskeleton for hand rehabilitation following stroke. *IEEE/ASME Trans. Mechatron.* **2014**, *19*, 131–140. [CrossRef]

16. Iqbal, J.; Tsagarakis, N.; Caldwell, D. Human hand compatible underactuated exoskeleton robotic system. *Electron. Lett.* **2014**, *50*, 494–496. [CrossRef]

17. Brokaw, E.; Black, I.; Holley, R.; Lum, P. Hand spring operated movement enhancer (handsome): A portable, passive hand exoskeleton for stroke rehabilitation. *IEEE Trans. Neural Syst. Rehabil. Eng.* **2011**, *19*, 391–399. [CrossRef]

18. Polygerinos, P.; Wang, Z.; Galloway, K.C.; Wood, R.J.; Walsh, C.J. Soft robotic glove for combined assistance and at-home rehabilitation. *Robot. Auton. Syst.* **2015**, *73*, 135–143. [CrossRef]

19. Amirabdollahian, F.; Ates, S.; Basteris, A.; Cesario, A.; Buurke, J.; Hermens, H.; Hofs, D.; Johansson, E.; Mountain, G.; Nasr, N.; et al. Design, development and deployment of a hand/wrist exoskeleton for home-based rehabilitation after stroke-script project. *Robotica* **2014**, *32*, 1331–1346. [CrossRef]

20. Nycz, C.J.; Meier, T.B.; Carvalho, P.; Meier, G.; Fischer, G.S. Design Criteria for Hand Exoskeletons: Measurement of Forces Needed to Assist Finger Extension in Traumatic Brain Injury Patients. *IEEE Robot. Autom. Lett.* **2018**, *3*, 3285–3292. [CrossRef]

21. Shen, Z.; Allison, G.; Cui, L. An Integrated Type and Dimensional Synthesis Method to Design One Degree-of-Freedom Planar Linkages with Only Revolute Joints for Exoskeletons. *J. Mech. Des.* **2018**, *140*, 092302. [CrossRef]

22. Copilusi, C.; Ceccarelli, M.; Carbone, G. Design and numerical characterization of a new leg exoskeleton for motion assistance. *Robotica* **2015**, *33*, 1147–1162. [CrossRef]

23. Carbone, G.; Aróstegui Cavero, C.; Ceccarelli, M.; Altuzarra, O. A Study of Feasibility for a Limb Exercising Device. In *Advances in Italian Mechanism Science*; Springer: Cham, Switzerland, 2017; pp. 11–21.

24. Cafolla, D.; Carbone, G. A study of feasibility of a human finger exoskeleton. In *Service Orientation in Holonic and Multi-Agent Manufacturing and Robotics*; Springer: Cham, Switzerland, 2014; pp. 355–364.

25. Ceccarelli, M. *Fundamentals of Mechanics of Robotic Manipulation*; Kluwer/Springer: Dordrecht, The Netherlands, 2004.

26. Carbone, G. (Ed.) *Grasping in Robotics*; Springer: London, UK, 2013.

27. Russo, M.; Ceccarelli, M.; Corves, B.; Hüsing, M.; Lorenz, M.; Cafolla, D.; Carbone, G. Design and Test of a Gripper Prototype for Horticulture Products. *Robot. Comput. Integr. Manuf.* **2017**, *44*, 266–275. [CrossRef]

28. Carbone, G.; Ceccarelli, M. Experimental Tests on Feasible Operation of a Finger Mechanism in the LARM Hand. *Int. J. Mech. Based Des. Struct. Mach.* **2008**, *36*, 1–13. [CrossRef]

29. Carbone, G.; Iannone, S.; Ceccarelli, M. Regulation and Control of LARM Hand III. *Robot. Comput. Integr. Manuf.* **2010**, *26*, 202–211. [CrossRef]

30. Yao, S.; Ceccarelli, M.; Carbone, G.; Zhan, Q.; Lu, Z. Analysis and Optimal Design of an Underactuated Finger Mechanism for LARM Hand. *Front. Mech. Eng.* **2011**, *6*, 332–343. [CrossRef]

31. Carbone, G.; Ceccarelli, M. Design of LARM Hand: Problems and Solutions. In Proceedings of the IEEE

International Conference on Automation, Quality and Testing, Robotics AQTR, Cluj-Napoca, Romania, 22–25 May 2008.

32. Gerding, E.-C.; Carbone, G.; Cafolla, D.; Russo, M.; Ceccarelli, M.; Rink, S.; Corves, B. Design of a Finger Exoskeleton for Motion Guidance. In *7th European Conference on Mechanism Science EuCoMeS 2018*; Springer: Cham, Switzerland, 2018; Volume 59, pp. 11–18.

33. Gerding, E.-C.; Carbone, G.; Cafolla, D.; Russo, M.; Ceccarelli, M.; Rink, S.; Corves, B. Design and Testing of a Finger Exoskeleton Prototype. In *Advances in Italian Mechanism Science, Mechanisms and Machine Science*; Springer: Cham, Switzerland, 2018; Volume 68, pp. 342–349.

34. Gerding, E.; Marco, C.; Carbone, G.; Daniele, C.; Matteo, R. Mechanism for Finger Exoskeleton. Italian Patent No. IT. 102018000003847, 6 April 2020.

35. Levangie, P.K.; Norkin, C.C. *Joint Structure and Function: A Comprehensive Analysis*; FA Davis: Philadelphia, PA, USA, 2005.

36. ASSH Homepage. American Society for Surgery of the Hand. Available online: https://www.assh.org/ (accessed on 11 November 2019).

37. Kerle, H.; Corves, B.; Hüsing, M. *Mechanism Design: Fundamentals, Development and Application of Non-Constantly Transmitting Gears*; Springer: Wiesbaden, Germany, 2015. (In German)

11

A Novel Kinematic Directional Index for Industrial Serial Manipulators

Giovanni Boschetti

Department of Management and Engineering, University of Padova, 36100 Vicenza, Italy;
giovanni.boschetti@unipd.it

Abstract: In the last forty years, performance evaluations have been conducted to evaluate the behavior of industrial manipulators throughout the workspace. The information gathered from these evaluations describes the performances of robots from different points of view. In this paper, a novel method is proposed for evaluating the maximum speed that a serial robot can reach with respect to both the position of the robot and its direction of motion. This approach, called Kinematic Directional Index (KDI), was applied to a Selective Compliance Assembly Robot Arm (SCARA) robot and an articulated robot with six degrees of freedom to outline their performances. The results of the experimental tests performed on these manipulators prove the effectiveness of the proposed index.

Keywords: directional index; serial robot; performance evaluation; kinematics

1. Introduction

Performance evaluation can provide useful information in the design of robots. For this reason, the first performance indexes were introduced as early as 1982. Manipulability [1], condition number [2], minimum singular value, kinematic isotropy [3], conditioning indexes [4] and dexterity analyses [5] provide a variety of information about a robot's performance throughout its workspace. Manipulability allows the robot to be kept far from the kinematic singularity while the condition number (and usually its reverse) is used to achieve a fast measurement of the workspace isotropy, i.e., the ratio between the maximum and the minimum performance of the robot.

These indexes, as with many others, are based on the Jacobian matrix of the velocity kinematic problem. The use of these approaches brought up some problems [6] that have been resolved with the evolution of these methods.

A modern performance index must provide information in a summarized form, such as a value, on the behavior that the manipulator can have in each position that it can reach within its workspace. The key features for a performance index are:

- To be homogeneous/independent of the unit of measurement [7–12].
- To be independent of the reference system [13,14].
- Providing increasingly accurate information [15–18].

A first sample of homogeneous indexes has been proposed in [7], where the main concept consists of directly relating the Cartesian and the actuator velocities. Another approach suggested introducing a novel dimensionally homogeneous Jacobian matrix [8]. In these methods, the velocities of three points of the flange have been considered and related with the motor velocities. A different approach was proposed in [10] where a proper performance index was developed to gather interesting information on a robot's behavior using a non-homogeneous Jacobian matrix.

An interesting approach was proposed in [13,14], where a motion/force transmission index was introduced. In these works, the use of proper virtual coefficients instead of the Jacobian matrix allowed introducing a frame-free performance index. This means that the evaluation is completely independent from the choice of the absolute reference frame in which it is computed.

Recent performance indexes [15–18] give more information compared to previous ones and are usually not affected by the adopted units of measurement and the choice of the reference frame.

Once a performance index can guarantee the above-mentioned features, it can be adopted for several purposes. In recent years, performance evaluation has been mainly conducted for the following uses:

- Optimization of manipulator design [19–21].
- Comparison between different robot architectures [22,23].
- Optimization of robot trajectory planning [24–27].
- Optimization of task locating (or robot positioning) [28,29].

The proposed performance index has been conceived to give useful information for the latter issue, i.e., finding the best location for a generic task, given the direction of the robot's main movements. The first performance index that took the direction of motion into account with respect to the actuator's movements was introduced in [30] and extended in [28]. Such an index, called a directional selective index (DSI), gives accurate information about the performance of parallel robots and allows finding the regions of the workspace where a parallel robot achieves its maximum and minimum performance. However, the DSI formulation cannot be easily extended to serial robots. For this reason, a novel approach, called the kinematic directional index (KDI), has been introduced in this paper. This method allows analyzing the behavior of a serial robot, taking into account the position of the robot and the direction of motion. As such, the index can be adopted for several purposes. In this paper, a performance analysis will be performed, by means of KDI, with two main goals:

- Finding the direction of maximum velocity with respect to a point.
- Finding the area of maximum velocity with respect to a direction of interest.

These two main analyses can provide very useful information for robot programming and the definition of the trajectory planning of industrial tasks. Performing the movements along the suggested directions can drastically reduce the time taken to complete a task.

The main motivations for this work are:

- Defining the novel KDI.
- Presenting two applications of the KDI.
- Proving the effectiveness of the KDI by exploiting two industrial robots.

The paper is organized as follows: in Section 2, the KDI is defined and formulated. In Section 3, two robots are presented for KDI computation and for the experimental tests. In the first part of Section 4, the KDI is computed at a point in the workspace in any direction of the horizontal plane, while in the second part, the index is computed along a direction of interest in the whole horizontal plane. In Section 5, the robot performances have been verified experimentally and are compared with the KDI. The strong correlation between the KDI values and the experimental results proves the effectiveness of the proposed index. Finally, in Section 6, the conclusions are addressed, and further research directions are given.

2. The KDI Performance Index

The KDI performance index allows evaluating the behavior of a serial manipulator in terms of linear velocity. For this purpose, the serial robot's kinematics are considered. As mentioned above, this index is based on the analysis of the Jacobian matrix J. As such, the analyzed problem can be identified by the forward velocity kinematic equation defined in Equation (1):

$$\dot{x} = J\dot{q} \tag{1}$$

where $\dot{x}$ and $\dot{q}$ are the velocity vectors in the Cartesian space and the joint space, respectively.

Since the KDI aims to determine the region in which a robot reaches its maximum translational velocity, only the translational part of the Jacobian matrix has been considered; hence, the motion of the wrist and its motors are not taken into account.

Once the direction of interest (i.e., the direction of motion) is defined, the velocities along the axes that are normal for the direction of interest are set as null values. Without loss of generality, it is possible to consider the direction of interest along the x-axis, and the velocity in this direction can be defined as follows in Equation (2):

$$\left\{\begin{array}{c} \dot{x} \\ \vdots \\ 0 \end{array}\right\} = \left[\begin{array}{cccc} j_{1,1} & j_{1,2} & \cdots & j_{1,m} \\ \vdots & \vdots & \ddots & \vdots \\ j_{n,1} & j_{n,2} & \cdots & j_{n,m} \end{array}\right] \left\{\begin{array}{c} \dot{q}_1 \\ \vdots \\ \dot{q}_m \end{array}\right\} \tag{2}$$

where n is the number of degrees of freedom in the Cartesian space and m is the number of actuators that defines the joint space.

Let us define R as the rotation matrix that aligns the x-axis to the direction of interest (d), J_R as the Jacobian matrix rotated by the matrix R, and $\dot{d}$ as the speed along d. In this way, in Equation (3), the velocity along the direction of interest is identified.

$$\left\{\begin{array}{c} \dot{d} \\ \vdots \\ 0 \end{array}\right\} = \left[\begin{array}{ccc} r_{1,1} & \cdots & r_{1,n} \\ \vdots & \ddots & \vdots \\ r_{n,1} & \cdots & r_{n,n} \end{array}\right] \left[\begin{array}{cccc} j_{1,1} & j_{1,2} & \cdots & j_{1,m} \\ \vdots & \vdots & \ddots & \vdots \\ j_{n,1} & j_{n,2} & \cdots & j_{n,m} \end{array}\right] \left\{\begin{array}{c} \dot{q}_1 \\ \vdots \\ \dot{q}_m \end{array}\right\} = \left[\begin{array}{cccc} j_{R\,1,1} & j_{R\,1,2} & \cdots & j_{R\,1,m} \\ \vdots & \vdots & \ddots & \vdots \\ j_{R\,n,1} & j_{R\,n,2} & \cdots & j_{R\,n,m} \end{array}\right] \left\{\begin{array}{c} \dot{q}_1 \\ \vdots \\ \dot{q}_m \end{array}\right\} \tag{3}$$

The value of KDI is identified by the letter K and is defined as the maximum value taken by $\dot{d}$ as highlighted in Equation (4):

$$K = \max\left(\dot{d}\right) \tag{4}$$

When a robot moves at its maximum speed, some joint motors are also working at their maximum speed. Since Equation (3) is a linear system, it is possible to state that the minimum number of motors in this condition is equal to $m - n + 1$. For example, a solution can be found by means of a linear programming problem. However, most industrial robots are non-redundant, so a proper solution for these robots is proposed in detail. In this case, the number of degrees of freedom (n) is equal to the number of motors (m) $(n = m)$. Hence, the Jacobian matrix is a square matrix as defined in Equation (5).

$$\left\{\begin{array}{c} \dot{d} \\ \vdots \\ 0 \end{array}\right\} = \left[\begin{array}{ccc} j_{R\,1,1} & \cdots & j_{R\,1,n} \\ \vdots & \ddots & \vdots \\ j_{R\,n,1} & \cdots & j_{R\,n,n} \end{array}\right] \left\{\begin{array}{c} \dot{q}_1 \\ \vdots \\ \dot{q}_n \end{array}\right\} \tag{5}$$

It is important to highlight that the maximum velocity in the direction of interest is achieved when at least one motor reaches its maximum velocity. Therefore, in order to find the solution to this problem, the velocity in the direction of interest $(\dot{d})$ can be set to a fictitious value of "1" and the fictitious motor velocities can be easily computed since the following problem is defined by a linear system (Equation (6)) where the number of unknowns is equal to the number of equations.

$$\left\{\begin{array}{c} 1 \\ \vdots \\ 0 \end{array}\right\} = J_R \left\{\begin{array}{c} \dot{q}_1 \\ \vdots \\ \dot{q}_n \end{array}\right\} \tag{6}$$

The motor that limits the robot's performance is that which is nearest to its maximum velocity. In order to easily find this motor (p) and its ratio with respect to its maximum velocity, let us define, by means of Equation (7), the maximum value achieved by the ratios between each joint speed and its maximum.

$$\frac{1}{K} = \frac{\dot{q}_p}{\dot{q}_{p,max}} = \max\left(\frac{\dot{q}_1}{\dot{q}_{1,max}}, \ldots, \frac{\dot{q}_n}{\dot{q}_{n,max}}\right) \tag{7}$$

$1/K$ is the maximum ratio between a joint speed and its maximum velocity, and this value is given by the joint p whose speed is the nearest to its maximum. This value can be also greater than 1, which can only happen because the solution to the problem is proposed using fictitious velocities. Nevertheless, this eventuality will not affect the validity of the solution.

Here, K identifies the KDI and also the maximum velocity that the robot can reach in the chosen direction of interest. This can be easily understood multiplying the velocities of Equation (5) by K and achieving Equation (8).

$$\begin{Bmatrix} K \\ \vdots \\ 0 \end{Bmatrix} = J_R \begin{Bmatrix} K\dot{q}_1 \\ \vdots \\ K\dot{q}_n \end{Bmatrix} \tag{8}$$

The value of K represents the maximum velocity of the robot, since $K\dot{q}_p$ is equal to the maximum value ($\dot{q}_{p,max}$), while the velocities of the other joints are below their maximum values.

From the definition of the KDI, it is possible to observe that it is not affected by the choice of the unit of measurement since only the value of K is used to compare the performances at different points in the workspace. Moreover, the index is also "frame free" because it does not depend on the use of an absolute reference frame but only on the robot configuration and on the direction of interest.

3. System Layout for KDI Validation

Two industrial manipulators were chosen to verify the accuracy of the KDI: an Adept Cobra 600 and an Adept Viper 650 robots (manufactured by Adept Technology, Pleasanton, CA, USA). This choice was made by looking for the most common kinematic architectures for serial robots. The first robot is a Selective Compliance Assembly Robot Arm (SCARA), while the second one is an articulated robot with six degrees of freedom (see Figure 1). In the SCARA robot, the index is used only in the horizontal plane since the vertical movements depend only on a prismatic joint whose performance does not change throughout the workspace; in the second case, the index can be useful along any direction of the workspace. However, in order to achieve comparable results, the performance of the articulated robot will also be investigated in a horizontal plane.

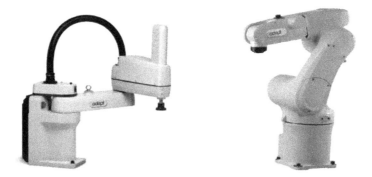

Figure 1. The Selective Compliance Assembly Robot Arm (SCARA) robot (**left**) and the articulated robot (**right**) exploited for the experimental validation of the Kinematic Directional Index (KDI) index.

3.1. SCARA Robot

The first test was performed on a horizontal plane by means of a SCARA robot. The horizontal movements depended on the motion of the first two links as represented in Figure 2, where the first two links are depicted in blue, and their reference frames are represented in violet together with the absolute frame.

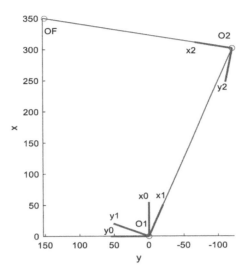

Figure 2. The kinematic scheme of the SCARA robot.

In Table 1, the complete Denavit Hartenberg table is defined in order to highlight the terms that are used for the computation of the Jacobian matrix of this robot.

Table 1. Denavit Hartenberg table of the Selective Compliance Assembly Robot Arm (SCARA) robot.

i	$T_{i,i-1}$	α_{i-1}	a_{i-1}	θ_i	d_i
1	$T_{1,0}$	0	0	θ_1	0
2	$T_{2,1}$	0	L_1	θ_2	0
3	$T_{3,2}$	π	L_2	0	d_3
4	$T_{4,3}$	0	0	θ_4	0

The projection in the horizontal plane of a SCARA manipulator allows highlighting the lengths of the two main links: $L_1 = 325$ mm and $L_2 = 275$ mm. The end-effector horizontal speed is given in function of the actuator speeds by means of the Jacobian matrix defined in Equation (9).

$$\mathbf{J}_S = \begin{bmatrix} z_1 \times (OF - O_1) & z_2 \times (OF - O_2) \end{bmatrix} \tag{9}$$

In this case, it is possible to easily calculate the Jacobian Matrix as follows in Equation (10):

$$\mathbf{J}_S = \begin{bmatrix} -L_1 \sin(\theta_1) - L_2 \sin(\theta_{1,2}) & -L_2 \sin(\theta_2) \\ L_1 \cos(\theta_1) + L_2 \cos(\theta_{1,2}) & L_2 \cos(\theta_2) \end{bmatrix} \tag{10}$$

where the expression $\theta_{i,j} = \theta_i + \theta_j$ has been used.

3.2. Articulated Robot

Figure 3 depicts the kinematic scheme of the articulated robot, without taking its wrist into account. The links that are involved in the translation of the wrist center are depicted in blue and their reference frames are represented in violet together with the absolute frame. The origin of the wrist is also highlighted (OF). In Table 2, all the kinematic parameters are highlighted by means of the Denavit Hartenberg table.

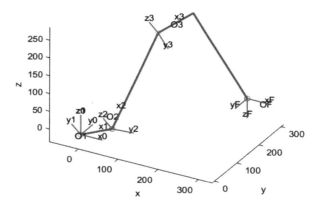

Figure 3. The kinematic scheme of the articulated robot.

Table 2. Denavit Hartenberg table of the articulated robot.

i	$T_{i,i-1}$	α_{i-1}	a_{i-1}	θ_i	d_i
1	$T_{1,0}$	0	0	θ_1	0
2	$T_{2,1}$	$-\pi/2$	a_1	θ_2	0
3	$T_{3,2}$	0	L_1	θ_3	0
4	$T_{4,3}$	$-\pi/2$	a_3	θ_4	L_2
5	$T_{5,4}$	$\pi/2$	0	θ_5	0
6	$T_{6,5}$	$\pi/2$	0	θ_6	0

Where the lengths of the links are $L_1 = 270\ mm$ and $L_2 = 295$ mm, while the non-null offset between the kinematic axes are $a_1 = 75$ mm and $a_3 = 90$ mm.

Using these parameters, the Jacobian matrix of Equation (11) can be computed at each point of the workspace.

$$\mathbf{J}_A = \begin{bmatrix} z_1 \times (OF - O_1) & z_2 \times (OF - O_2) & z_3 \times (OF - O_3) \end{bmatrix} \tag{11}$$

This Jacobian matrix relates the velocity of the wrist center with the ones of the joints.

4. Performance Investigation

Given the proper Jacobian matrix, it is possible to compute the KDI in any point of the workspace and in any direction. To highlight the usefulness of the KDI, two different performance investigations were proposed. The first one consisted of finding the maximum velocity that the robot can reach in a point for each direction of motion. The second one pointed out the areas where the robot reached its maximum velocity with respect to a direction of interest.

4.1. Performance Investigation in a Point

Without loss of generality, the KDI for a SCARA robot is plotted at the point P (350, 150) along any direction of the horizontal plane (see Figure 4). The performance along the x-axis is highlighted with a red line.

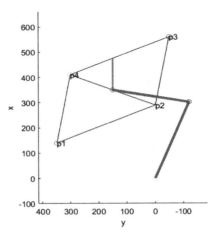

Figure 4. KDI computed for the SCARA robot in point P (350, 150) along any horizontal direction.

As mentioned above, when the maximum speed along any direction of a robot is achieved, at least one motor reaches its maximum speed. Therefore, given a generic point of the workspace, if this type of information is needed, the computation of the KDI can be greatly simplified. In fact, the maximum speed (absolute or relative) is reached when both motors reach their maximum (or minimum) speed. As such, the directions of the maximum Cartesian speed can be calculated by using a proper matrix that contains the four possible combinations of maximum and minimum speed. By multiplying such a matrix with the Jacobian matrix, the maximum speed direction can be identified by Equation (12).

$$P = J_S \begin{bmatrix} \dot{q}_{1,max} & -\dot{q}_{1,max} & -\dot{q}_{1,max} & \dot{q}_{1,max} \\ \dot{q}_{2,max} & \dot{q}_{2,max} & -\dot{q}_{2,max} & -\dot{q}_{2,max} \end{bmatrix} \quad (12)$$

where P is a matrix made up of the four virtual points that represent the vertexes of the parallelogram that delimits the robot speed in the horizontal plane as depicted in Figure 4.

By following the same approach with the articulated robot, the maximum and minimum values of the three joint speeds allowed defining the eight vertexes of a cuboid as depicted in Figure 5. Given these vertexes, the maximum speed of the robot could be gathered in any direction of the workspace. In Figure 5, for example, the performance of the robot along the x-axis is highlighted with a red line.

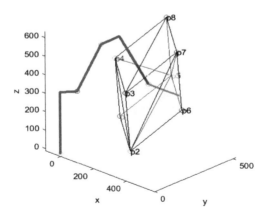

Figure 5. KDI computed for the articulated robot in point P (250, 300, 350) along any direction.

4.2. Performance Investigation in the Workspace

As mentioned above, the performance was evaluated in a horizontal plane. For each robot, a proper set of points was defined, and a direction of interest was chosen. The SCARA set of points was given by nine points along any radial direction of the workspace with an angular step of ten degrees.

Since the articulated robot was also investigated in a horizontal plane, the plane with the highest reach distance was chosen (i.e., where the wrist was at the same height as the shoulder). For this test, the set was given by seven points along any radial direction with an angular step of ten degrees. The points that are outside the workspace were considered. In Figure 6, the two sets of points are depicted together with the workspace limits.

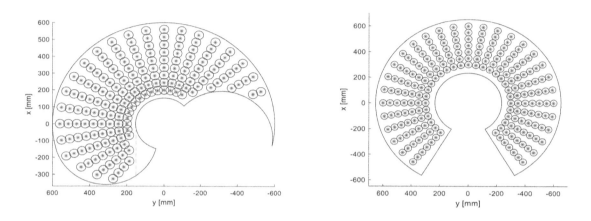

Figure 6. Point sets in which the experimental analyses were performed for the SCARA robot (**left**) and the articulated robot (**right**).

For the SCARA robot, the KDI index was computed along the direction of the y-axis (i.e., with a rotation of 90 degrees about the z-axis by means of the matrix R). The KDI was computed for the articulated robot with a rotation of 45 degrees about the z-axis. The KDI was normalized and plotted in Figure 7 for both robots. The yellow areas indicate the regions where the robots achieved their best performance, while the blue areas show where the robots were slower.

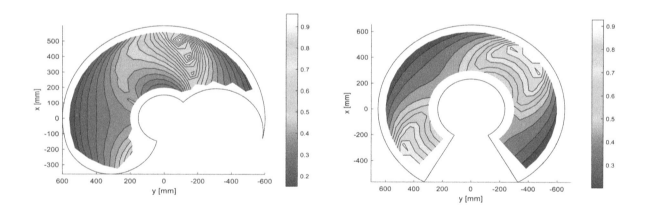

Figure 7. KDI computed for SCARA (**left**) and articulated (**right**) robots.

5. Experimental Setup and Performance Analysis in the Workspace

In order to experimentally verify where the robots can achieve the maximum velocity, the end effector of each robot was equipped with an accelerometer, as shown in Figure 8. The chosen device was a piezometric accelerometer, DeltaTron Type 4508001 (manufactured by Hottinger Brüel & Kjær, Nærum, Denmark), which did not require additional hardware for signal conditioning. The data were acquired by means of an LMS Pimento analyzer (manufactured by LMS International, Leuven, Belgium) at an acquisition frequency of 1000 Hz. The software supplied with the analyzer allowed directly computing the speed of the investigated movements.

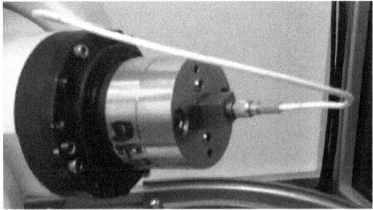

Figure 8. The end effectors of the robots equipped with accelerometers.

The experimental analysis was performed with the same set of points defined above and illustrated in Figure 6, in order to compare the results of the investigation with the values taken by the KDI. At each point, a back-and-forth movement with a length of 60 mm in the direction of interest was repeated five times so that the center of the displacement (where the maximum speed was reached) coincided with the investigated point. Using a data acquisition system, the signals of the accelerometers were gatherd, and the velocity of the robots was computed. In order to reduce the effect of measuring errors, the maximum experimentally measured speed was computed by calculating the average between the peaks of each repeated movement. In this way, the maximum speed reached in each point of the workspace could be collected. Such results have been normalized since the target of the work is to highlight the region of maximum speed without making performance comparisons between different robots. The experimentally gathered normalized performances are plotted in Figure 9 for both robots.

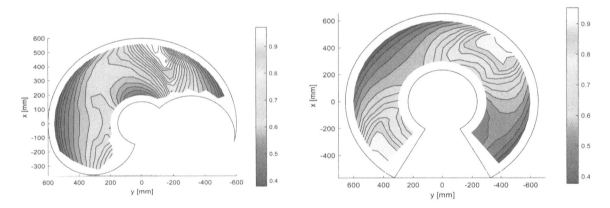

Figure 9. Normalized maximum speeds of the SCARA (**left**) and the articulated (**right**) robots.

It is now possible to appreciate that the values of the KDI perfectly matched the behavior of the robots; the best performance region identified by the KDI has a good correspondence with the measured region. Moreover, the trend of performance given by the KDI met the speed variation throughout the workspace. In fact, the comparison between Figures 7 and 9 highlights that the KDI can be adopted as a useful tool to foresee the regions where a robot can achieve its maximum speed with respect to a direction of interest. In order to better identify the correlation between the index and the real performance, the maximum value of the measured speed has been plotted in Figure 10 for each investigated point with respect to the value given by the KDI at the same point. Moreover, the linear regression between the KDI and the speed of each robot has been computed and plotted by means of a red line. In Figure 10, the proximity of the points on the scatter diagram to the red lines clearly shows a strong correlation with the KDI values of the robots' velocities; therefore, it proves the effectiveness of the novel proposed index.

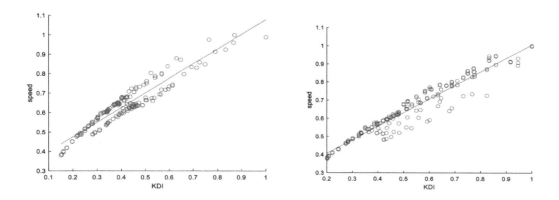

Figure 10. Scatter diagram of the KDI versus speeds for the SCARA (**left**) and articulated (**right**) robots.

6. Conclusions

A novel performance index called the KDI has been proposed. This index is not affected by the non-homogeneous Jacobian matrix and does not depend on the choice of the units of measurement and on the position of the absolute reference frame. The index was computed for a SCARA robot and an articulated robot to analyze the behavior of the manipulators for two main purposes: finding the direction of maximum velocity with respect to a point, and finding the area of maximum velocity with respect to a direction of interest in the workspace. Being a fully kinematic index (i.e., it does not consider the dynamics and the control of the robots), it gives a high-quality description of the robot's behavior, and the experimental tests demonstrate the effectiveness of the proposed index.

Future works can be addressed by following some steps: first, the use of the index for redundant manipulators will be investigated; second, the same reasoning will be followed to synthesize a dynamic performance index that is able to consider the direction of motion; afterwards, the use of these indexes for other issues, such as the optimal design of manipulators and/or robot trajectory planning, will be investigated.

References

1. Yoshikawa, T. Manipulability of Robotic Mechanisms. *Int. J. Robot. Res.* **1985**, *4*, 3–9. [CrossRef]
2. Salisbury, J.K.; Craig, J.J. Articulated hands: Force control and kinematic issues. *Int. J. Robot. Res.* **1982**, *1*, 4–17. [CrossRef]
3. Angeles, J.; López-Cajún, C.S. Kinematic Isotropy and the Conditioning Index of Serial Robotic Manipulators. *Int. J. Robot. Res.* **1992**, *11*, 560–571. [CrossRef]
4. Gosselin, C.; Angeles, J. A Global Performance Index for the Kinematic Optimization of Robotic Manipulators. *J. Mech. Des.* **1991**, *113*, 220–226. [CrossRef]
5. Gao, F.; Liu, X.; Gruver, W.A. Performance evaluation of two-degree-of-freedom planar parallel robots. *Mech. Mach. Theory* **1998**, *33*, 661–668. [CrossRef]
6. Merlet, J.P. Jacobian, Manipulability, Condition Number, and Accuracy of Parallel Robots. *J. Mech. Des.* **2005**, *128*, 199–206. [CrossRef]
7. Gosselin, C. The optimum design of robotic manipulators using dexterity indices. *Robot. Auton. Syst.* **1992**, *9*, 213–226. [CrossRef]
8. Kim, S.-G.; Ryu, J. New dimensionally homogeneous jacobian matrix formulation by three end-effector points for optimal design of parallel manipulators. *IEEE Trans. Robot. Autom.* **2003**, *19*, 731–737. [CrossRef]

9. Kim, S.-G.; Ryu, J.-H. Force transmission analyses with dimensionally homogeneous Jacobian matrices for parallel manipulators. *KSME Int. J.* **2004**, *18*, 780–788. [CrossRef]

10. Mansouri, I.; Ouali, M. A new homogeneous manipulability measure of robot manipulators, based on power concept. *Mechatronics* **2009**, *19*, 927–944. [CrossRef]

11. Mansouri, I.; Ouali, M. The power manipulability – A new homogeneous performance index of robot manipulators. *Robot. Comput. Manuf.* **2011**, *27*, 434–449. [CrossRef]

12. Cardou, P.; Bouchard, S.; Gosselin, C. Kinematic-Sensitivity Indicies for Dimensionally Nonhomogeneous Jacobian Matrices. *IEEE Trans. Robot.* **2010**, *26*, 166–173. [CrossRef]

13. Wang, J.; Liu, X.; Wu, C. Optimal design of a new spatial 3-DOF parallel robot with respect to a frame-free index. *Sci. China Ser. E: Technol. Sci.* **2009**, *52*, 986–999. [CrossRef]

14. Wang, J.; Wu, C.; Liu, X. Performance evaluation of parallel manipulators: Motion/force transmissibility and its index. *Mech. Mach. Theory* **2010**, *45*, 1462–1476. [CrossRef]

15. Brinker, J.; Corves, B.; Takeda, Y. Kinematic performance evaluation of high-speed Delta parallel robots based on motion/force transmission indices. *Mech. Mach. Theory* **2018**, *125*, 111–125. [CrossRef]

16. Wang, C.; Zhao, Y.; Dong, C.; Liu, Q.; Niu, W.; Liu, H.; Liu, H. Kinematic Performance Comparison of Two Parallel Kinematics Machines. In *Advances in Mechanism and Machine Science. IFToMM WC 2019. Mechanisms and Machine Science*; Uhl, T., Ed.; Springer: Cham, Switzerland, 2019; Volume 73.

17. Boschetti, G.; Trevisani, A. Cable Robot Performance Evaluation by Wrench Exertion Capability. *Robotics* **2018**, *7*, 15. [CrossRef]

18. La Mura, F.; Romanó, P.; Fiore, E.; Giberti, H. Workspace Limiting Strategy for 6 DOF Force Controlled PKMs Manipulating High Inertia Objects. *Robotics* **2018**, *7*, 10. [CrossRef]

19. Wu, X. Performance Analysis and Optimum Design of a Redundant Planar Parallel Manipulator. *Symmetry* **2019**, *11*, 908. [CrossRef]

20. Alvarado, R.R.; Castañeda, E.C. Optimum design of the reconfiguration system for a 6-degree-of-freedom parallel manipulator via motion/force transmission analysis. *J. Mech. Sci. Technol.* **2020**, *34*, 1339–1349. [CrossRef]

21. Yu, G.; Wu, J.; Wang, L.; Gao, Y. Optimal Design of the Three-Degree-of-Freedom Parallel Manipulator in a Spray-Painting Equipment. *Robotics* **2019**, *38*, 1064–1081. [CrossRef]

22. Wu, X.; Yan, R.; Xiang, Z.; Zheng, F.; Tan, R. Performance Analysis and Comparison of Three Planar Parallel Manipulators. *Power Transm.* **2019**, *79*, 270–279. [CrossRef]

23. Baena, A.H.; Valdez, S.I.; Romero, F.D.J.T.; Montes, M.M. Comparison of Parallel Versions of SA and GA for Optimizing the Performance of a Robotic Manipulator. *Power Transm.* **2020**, *86*, 290–303. [CrossRef]

24. Boschetti, G. A Picking Strategy for Circular Conveyor Tracking. *J. Intell. Robot. Syst.* **2016**, *81*, 241–255. [CrossRef]

25. Bottin, M.; Rosati, G. Trajectory Optimization of a Redundant Serial Robot Using Cartesian via Points and Kinematic Decoupling. *Robot.* **2019**, *8*, 101. [CrossRef]

26. Bottin, M.; Ceccarelli, M.; Morales-Cruz, C.; Rosati, G. Design and Operation Improvements for CADEL Cable-Driven Elbow Assisting Device. *Mech. Mach. Sci.* **2021**, *91*, 503–511.

27. Bottin, M.; Rosati, G.; Boschetti, G. Working Cycle Sequence Optimization for Industrial Robots. *Mech. Mach. Sci.* **2021**, *91*, 228–236.

28. Boschetti, G.; Rosa, R.; Trevisani, A. Optimal robot positioning using task-dependent and direction-selective performance indexes: General definitions and application to a parallel robot. *Robot. Comput. Manuf.* **2013**, *29*, 431–443. [CrossRef]

29. Sharkawy, A.-N.; Papakonstantinou, C.; Papakostopoulos, V.; Moulianitis, V.; Aspragathos, N. Task Location for High Performance Human-Robot Collaboration. *J. Intell. Robot. Syst.* **2020**, 1–20. [CrossRef]

30. Boschetti, G.; Trevisani, A. Direction Selective Performance Index for Parallel Manipulators. In Proceedings of the First International Conference on Multibody System Dynamics IMSD 2010, Lappeenranta, Finland, 25–27 May 2010.

e the supporting

parameters of the bearing pedestal, the vibration of the rotor was reduced [6,7]. The advantages and disadvantages of active damping feedback control and active stiffness feedback control were analyzed. The experimental results showed that the speed feedback control method can effectively suppress the rotor shaft system [8]. The Smart Spring [9] is an active vibration damping device based on a piezoelectric actuator, which is composed of a main support and an auxiliary support. The main support is the elastic support of shafting itself. By controlling the actuating force of the piezoelectric ceramic actuator on the auxiliary support, positive pressure is generated between friction elements on the main support and the auxiliary support, and vibration energy is consumed through dry friction between friction plates.

According to the research on the vibration reduction technology of Smart Spring, Nitzsche of Carlton University of Canada has done a lot of work on the application of the Smart Spring to helicopter blade vibration reduction. In 2012, Nitzsche et al. applied a set of semi-active Smart Spring systems to the vibration reduction of a helicopter rotor system. The control scheme selected the state switching, and the vibration reduction performance of the Smart Spring was verified by experiments [10]. In 2013, Nitzsche et al. improved the active variable pitch pull rod APL (Active pitch link) of the damping device and verified the damping performance of the improved APL in the rotating tower test. The results showed that the APL effectively attenuated the vibration response of the blade and reduced the transmission force [11]. Other scholars on Nitzsche's team have also done a lot of work on the application of Smart Springs to helicopter blade damping. Afagh et al. placed the Smart Spring mechanism on the load transfer path of the blade to achieve the purpose of vibration reduction and studied the stability of the blade in the Elastohydrodynamic state [12]. Coppotelli et al. studied the dynamic characteristics of the Smart Spring installed on the blade of a non-rotating helicopter, analyzed the influence of the Smart Spring on the modal characteristics, and finally established the finite element model of the blade [13]. Some scholars from the Aeronautical Research Institute of Canada also studied the Smart Spring damping technology. Chen Yong et al. established a mathematical model under harmonic excitation and designed an adaptive notch algorithm to reduce vibration on the DSP (Digital Signal Processor) platform. Wind tunnel tests were carried out to verify the damping effect of the Smart Spring [14]. Wickramasinghe et al. proved the ability of the Smart Spring to control the dynamic impedance characteristics of the structure through experimental research, and the vibration reduction performance was proved by the wind tunnel test [15]. Grewal et al. designed a control scheme to make the stiffness of the Smart Spring device change continuously and compared it with the control effect of the state switching control algorithm [16]. To sum up, the research into Smart Spring has been mainly on the aspect of torsional vibration control of the helicopter blade, and the research into Smart Spring vibration reduction applied to multi-support shafting has been mainly dynamic modeling and characteristic analysis. Ni De [17] carried out dynamic analysis on the shaft system with Smart Spring, studied the design method of intelligent spring parameters, and analyzed the vibration control effect of the intelligent spring in multi-support shafting. Peng Bo [18] made a preliminary exploration of the multi-span shafting vibration reduction of the intelligent spring, established the multi-span shafting dynamic model of the intelligent spring, analyzed the influence of intelligent spring control force and its parameters on vibration reduction performance, and optimized the parameters of the intelligent spring with a genetic algorithm.

At present, there is a lack of theoretical guidance and related experimental verification for the research on using Smart Spring to reduce the vibration of tail drive shafting. The application of Smart Spring to the vibration reduction of the tail drive shaft system needs further research. In this paper, the control strategy of exerting fixed control force and function control force is put forward. By using the ADAMS and MATLAB joint simulation method, the research on over-critical vibration control of three-support shafting with Smart Spring support is carried out, and the accuracy of the joint simulation method is verified by experiments.

2. Research on Vibration Reduction of the Shaft System with Fixed Control Force Applied by Smart Spring

2.1. Establishment of Joint Simulation Model of Three-Support Shafting Based on ADAMS and MATLAB

The double-disk three-support shafting with Smart Spring was taken as the research object, as shown in Figure 1. The three-support shafting consisted of motor, diaphragm coupling, transmission shaft, rigid disk, elastic support, and Smart Spring pair support. The Smart Spring consisted of basic support, auxiliary support, and a piezoelectric ceramic actuator (PZTA). The simplified principle of Smart Spring support is shown in Figure 2a. The radial stiffness and damping of the basic support are k and c, and the radial stiffness and damping of the active support are k' and c', m' is the equivalent mass of the PZTA and other components connected with the active support. F_d is the sliding friction force caused by the control force N_t between the friction pair of the basic support and the auxiliary support. The lower end of the Smart Spring support was fixed on the foundation, and the upper end bore the support reaction force Ft from the shaft section. The radial displacement of the main support is u and that of the auxiliary support is u'. There was an initial clearance between the main support and the auxiliary support. The structure diagram of Smart Spring is shown in Figure 2b.

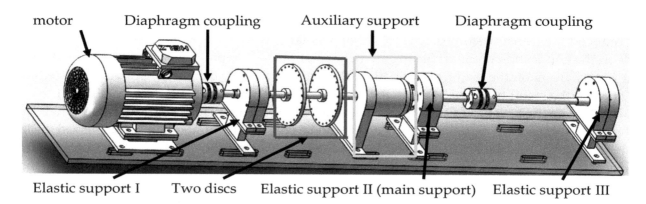

Figure 1. Three-dimensional model of three-support shafting.

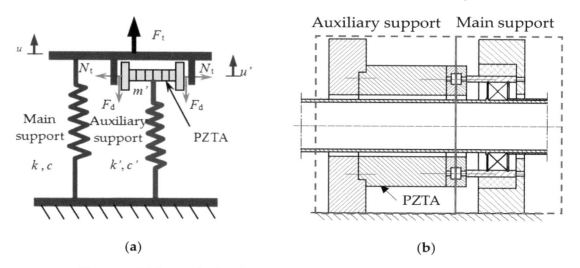

(a)   (b)

Figure 2. (a) Principle sketch and (b) structure diagram of Smart Spring.

The shaft system was a typical variable cross-section continuous rotor system with disk rotor, shaft sleeve, and coupling. Based on the finite element method, the variable cross-section of the three-support shafting was truncated and discretized. As shown in Figure 3, taking no account of the influence of the coupling on the bending vibration of the shaft system in discrete processing, the three-support shafting was discretized into disk, shaft section, and bearing seat, in turn, while the two sections of shaft were regarded as a single optical axis. The dynamic model of three-support shafting with double disks was established. The allowable degrees of freedom in the model were x-axis rotation and the translation of Y and Z. The basic parameters are shown in Table 1.

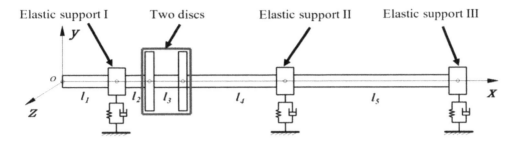

Figure 3. Dynamic model of three-support shafting with double disks.

Table 1. Basic parameters of three-support shafting.

Parameter Name	Value/Unit	Parameter Name	Value/Unit
The density of the shaft ρ	7850/(kg·m^{-3})	Elastic modulus E	2×10^{11}/Pa
Shaft radius r	7.5/mm	Disc radius R	75/mm
Length of shaft l_1	120/mm	Disc width b	8/mm
Length of shaft l_2	70/mm	Support stiffness k_b	1.7×10^5/(N·m^{-1})
Length of shaft l_3	80/mm	Support damping c_b	60/(N·s·m^{-1})
Length of shaft l_4	270/mm	Unbalance magnitude e_0	6.3×10^{-5} kg·m
Length of shaft l_3	420/mm	Auxiliary support stiffness k_a	6×10^5 N·m^{-1}

According to the dynamic model of three-support shafting, the virtual prototype model of shafting was established in ADAMS simulation software, as shown in Figure 4. Combined with the actual working state of the shafting, the constraint conditions of the virtual prototype model of the shafting were set. The eccentricity was added to the flexible shaft to simulate the imbalance of the shaft, and the spring with stiffness and damping was used to simulate the elastic support. A rotation drive was added to the left end face of the flexible shaft. The rotation drive adopted the mode of constant acceleration, and the acceleration was set as 20π rad/s^2. In order to realize the contact and action between the friction disks of the main support and the auxiliary support of the Smart Spring, as shown in Figure 5, two friction disks were established on the virtual prototype model to the equivalent mass of the main and auxiliary supports, and the control force and contact pair were added to the friction disk. Furthermore, the friction force was linearized, which was equal to the product of the friction coefficient and control force after linearization.

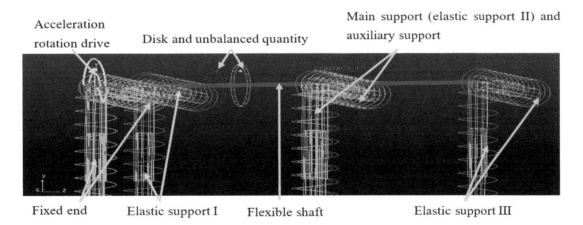

Figure 4. Virtual prototype model of double-disk shafting.

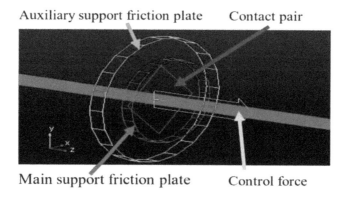

Figure 5. Realization of the Smart Spring function.

ADAMS software focuses on the mechanical dynamics simulation, while MATLAB focuses on the control system simulation. Using the ADAMS and MATLAB joint simulation method, based on the ADAMS/control module, we imported the mechanical system simulation model from ADAMS directly to MATLAB without tedious derivation and the listing of a large number of equations to describe the law of the control system, which greatly simplified the workload of modeling. At the same time, we added the complex control directly to the ADAMS model, which can simulate the whole system at one time, and, even if there are problems, they can be easily solved. The ADAMS virtual prototype model was imported into MATLAB, and the open-loop control joint simulation model of double-disk shafting was obtained, as shown in Figure 6. The joint simulation model consisted of the ADAMS module, input module, and output module. The input module was the control force of the actuator, and the control force was input into ADAMS module. The two output modules were the displacement responses in X direction and Y direction at the middle position of the shafting.

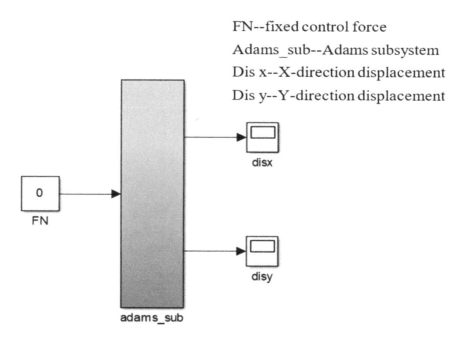

Figure 6. Joint simulation model of open-loop control.

2.2. Analysis of Simulation Results of Shafting Over-Critical

The mid-point of support I and support II was selected as the measuring point, and fixed control forces of 100 N, 200 N, 300 N, 425 N, 800 N, and 1000 N were applied respectively to simulate the shaft system passing through the critical point, and the displacement response at the measuring point was obtained. Due to the isotropy of the rotor, the shafting had similar displacement response curves in X and Y directions. Therefore, the displacement response curves in the Y direction were selected for analysis.

Figure 7 shows the over-critical bending vibration response of a double-disk three-support shafting with Smart Spring support under different fixed control forces. It can be seen from Figure 7 that the application of fixed control force can significantly suppress the critical response peak value and the critical speed shifts to the right. When the fixed control force was less than 200 N, the vibration response decreased in the whole speed range. When t < 6 s, i.e., the speed was lower than 3600 r/min, the vibration could be attenuated by applying 0–1000 N fixed control force. However, when t > 7 s, i.e., the rotating speed was greater than 4200 r/min, and the fixed control force was greater than 200 N, the vibration was increased by applying the fixed control force.

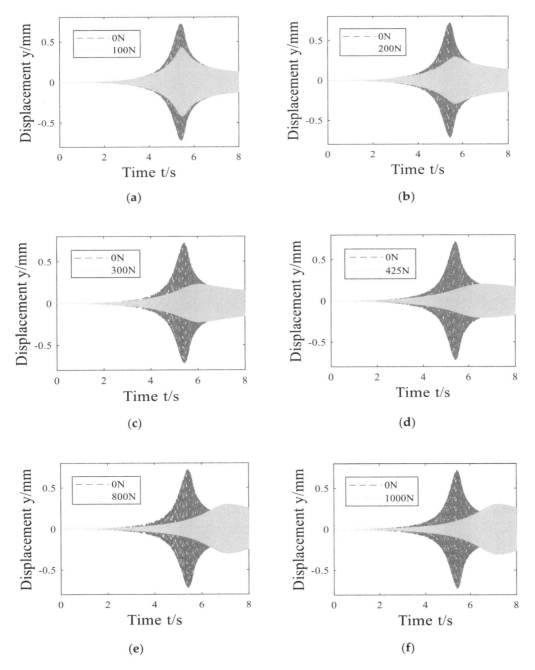

Figure 7. Response of the shaft system under (**a–f**) fixed control forces, respectively 100 N, 200 N, 300 N, 425 N, 800 N and 1000 N of the Smart Spring.

Figure 8 shows the relationship between the critical response peak value and the fixed control force. The results showed that with the increase of the fixed control force, the peak value of the over-critical response first decreased and then increased. When the fixed control force was 0 N, the peak value was 0.722 mm; when the fixed control force was 425 N, the peak value was the minimum, which was 0.205 mm. At this time, the vibration reduction rate was the largest, reaching 71.6%. Therefore, the control force of the Smart Spring was not better when greater, but there was an optimal fixed control force. Figure 9 shows that with the increase of the fixed control force, the first-order critical speed of the double-disk shaft system gradually increased, and tended to be stable when the control force was greater than 700 N, which was close to 4350 r/min, that is, the damping control gradually transited to the stiffness control.

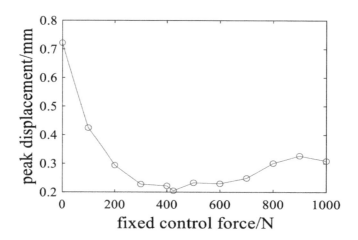

Figure 8. Relationship between peak value of over-critical response of shafting and fixed control force.

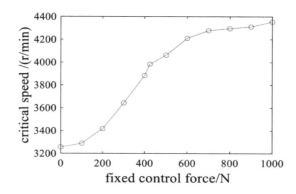

Figure 9. Relationship between first critical speed and fixed control force of shafting.

3. Experimental Verification of Shaft System Over-Critical Vibration Control

3.1. Design of Test System

The test system consisted of the shafting test bed, executive system, data acquisition system, and control system. The actuator system consisted of the piezoelectric ceramic controller and piezoelectric actuator. The data acquisition system included sensor, data collector DH5922N and its supporting software, DHDAS dynamic signal acquisition and analysis system. The control system included controller AD5436 and its supporting software. The layout of the test platform is shown in Figures 10 and 11. In the experiment, the output voltage of the piezoelectric ceramic controller was increased by 15 V, and the corresponding shaft system over-critical response was recorded, and the test results were analyzed.

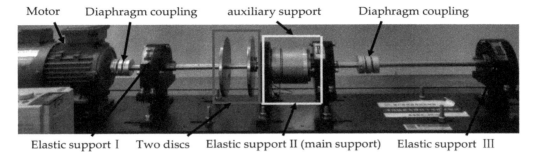

Figure 10. Shafting test bench.

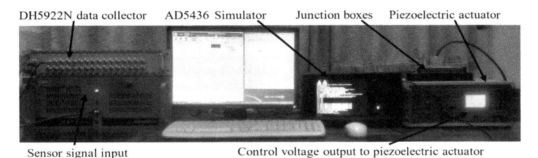

Figure 11. Layout of test data acquisition device.

When the voltage was applied, the PZTA extended to both sides to make the main and auxiliary friction surfaces of the Smart Spring support tight. If the control voltage of the PZTA was increased after jacking, the support of the Smart Spring pair deformed in the axial direction. The axial displacement x_a of the support under different loading voltages (0 V–150 V, step size 15 V) was measured by an eddy current sensor, which was the deformation x_a of the Smart Spring support. Assuming that the axial stiffness k_a of the auxiliary support remained constant with the increase of the force, then:

$$N(t) = k_a \times x_a. \tag{1}$$

In Equation (1) k_a is the axial stiffness of the auxiliary support and x_a is the axial displacement of the auxiliary support.

As shown in Figure 12, the relationship between control force and control voltage was not a precise linear relationship, and there was a certain hysteresis effect. When the control voltage increased from 0 V to 150 V at 15 V intervals and then decreased to 0 V at 15 V intervals, the curves of the control force could not coincide, that is, the piezoelectric ceramics had a hysteresis effect. When the control voltage was 150 V, the maximum control force was 447 N.

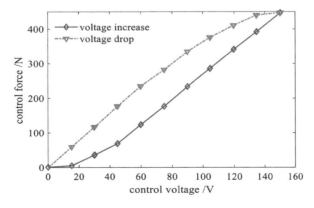

Figure 12. Relationship between control force and control voltage of the Smart Spring support.

3.2. Analysis of Experiment Result

Figure 13 shows the critical time domain response of shafting under various control voltages. It can be seen from the figure that after the control voltage was applied, the peak value of the shaft acceleration over-critical value decreased obviously. With the increase of the control voltage, the vibration peak value decreased more obviously when the shafting was over-critical.

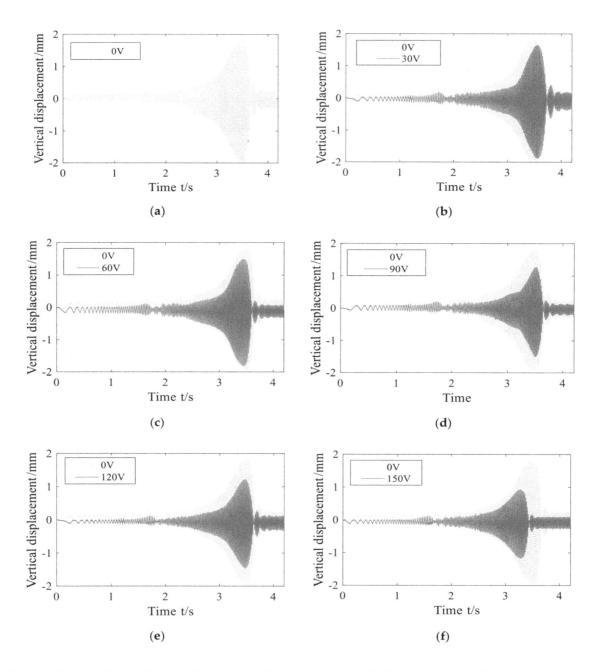

Figure 13. Accelerated over-critical vibration response of shafting under (**a–f**) control voltages, respectively 0 V, 30 V, 60 V, 90 V, 120 V and 150 V.

The relationship between the peak value of shaft acceleration and the control voltage under each control voltage is shown in Figure 14. It can be seen from the figure that the vibration reduction performance of the Smart Spring support was good, and the maximum vibration reduction rate reached 44.2% with the change trend of the vibration peak value with the control voltage.

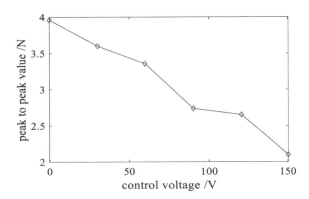

Figure 14. Relationship between peak and peak value of shaft system over-critical vibration response and control voltage.

Comparing Figures 7 and 14, the change trend of the first section of the two figures was consistent, and the displacement response decreased with the increase of the control parameters. However, because of the small axial stiffness of the bearing pedestal and the limited displacement of the piezoelectric actuator, the optimal control parameter in Figure 14 appeared at the maximum value, and the maximum control force was 447 N, so the piezoelectric actuator failed to exert its maximum capacity.

Figure 15 shows the comparison results of the shafting vibration damping ratio under the different control voltages/control forces obtained by test and simulation analyses. It can be seen from the figure that under the trans-critical state of the shafting, the variation trends of the damping rate obtained by the test and simulation were the same, which indicated the accuracy of the joint simulation model. When the fixed control force of the intelligent spring increased from 100 N to 200 N, the vibration reduction rate of the simulation results increased by 21 percentage points, and the vibration reduction rate of the test results increased by 15 percentage points; when the control force of the intelligent spring increased from 200 N to 300 N, the vibration reduction rate of the simulation results and that of the test results increased by three percentage points. The difference between the experimental damping rate and the simulated damping rate was due to the error of the test bench and the limitation of the control force of the piezoelectric actuator. For example, there was a certain gap between the angular acceleration of the motor and the simulated angular acceleration, the axial stiffness of the bearing pedestal was too small, and the displacement of the piezoelectric ceramic actuator was limited.

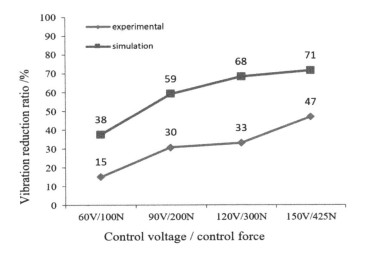

Figure 15. Vibration reduction rate of shafting under different control voltages/forces.

4. Research on Vibration Reduction of Shafting with Function Control Force Exerted by Smart Spring

It can be seen from the previous analysis results that the vibration damping rate of the shafting was different at different speeds when the Smart Spring applied fixed control force. In order to obtain better control effect of shaft system over-critical vibration, a vibration suppression method with function control force was proposed. In other words, the function control force that changes with acceleration and time was applied to the Smart Spring. Therefore, after establishing the joint simulation model of shafting, the relationship between control force F_N and time was studied to determine the optimal control force under constant operating speed, and then the control force function was determined to carry out joint simulation analysis on the critical open-loop control of the Smart Spring applying function control force.

4.1. Establishment of Simulation Model of Three-Support Shafting with Smart Spring

This section takes the three-support shafting test bed shown in Figure 16 as the research object, establishes the virtual prototype model and imports it into MATLAB to obtain the joint simulation model of three-support shafting with Smart Spring.

Auxiliary Secondary

Figure 16. Three-support shafting test bench with Smart Spring pair support.

4.2. Determination of Optimal Fixed Control Force at Constant Speed

The joint simulation model under constant speed as shown in Figure 17 was established. In order to realize the constant speed operation of the shafting, a state variable was added to the input module to input the constant speed. In the speed range of 0–8400 r/min, the simulation of optimal fixed control force at constant speed was carried out with the interval of 600 r/min. The axis orbit of the shafting under different fixed control forces was obtained by simulation, and the control force with the minimum radius of the axis trajectory was selected as the optimal control force F_{op}.

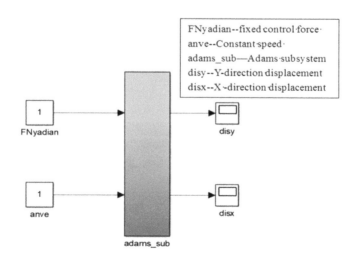

Figure 17. Joint simulation model of applying function control force at constant speed.

Figure 18 shows the axis orbit of the steady-state response under each fixed control force when the shaft speed was 6000 r/min. It can be seen from the figure that the axis trajectories of the steady-state response were all circular. With the increase of the control force, the radius of the steady-state response also changed, which indicated that different fixed control forces had different suppression effects on the steady-state response of constant speed operation.

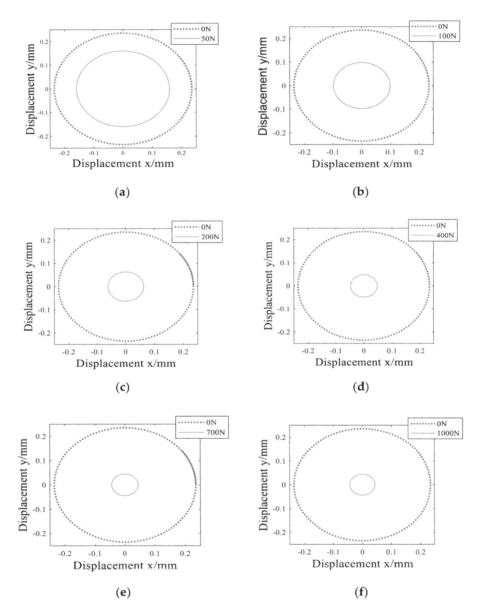

Figure 18. Steady-state response center orbit under (**a**–**f**) control forces at 6000 r/min, respectively 50 N, 100 N, 200 N, 400 N, 700 N and 1000 N.

Since the optimal control forces in the speed range of 0–6000 r/min were all 1000 N, it was no longer necessary to analyze the case where the shaft speed was lower than 6000 r/min. Figure 19 shows the relationship between the fixed control force of the Smart Spring and the steady response of the steady-state response at the speeds of 6000 r/min, 6600 r/min, 7200 r/min, 7800 r/min, and 8400 r/min. It can be seen from the figure that with the increase of speed, the optimal control force F_{op} and the maximum damping rate of constant speed operation became smaller and smaller. When the speed was 6000 r/min, the optimal control force F_{op} was 1000 N, and the maximum vibration reduction rate was about 80%; when the speed was 6600 r/min, the optimal control force F_{op} was 650 N, and the maximum vibration reduction rate was about 58%; when the speed was 7200 r/min, the optimal control force F_{op} was 200 N, and the maximum damping rate was about 17%; when the speed was 8400 r/min, the optimal control force F_{op} was 25 N, and the maximum damping rate was close to zero.

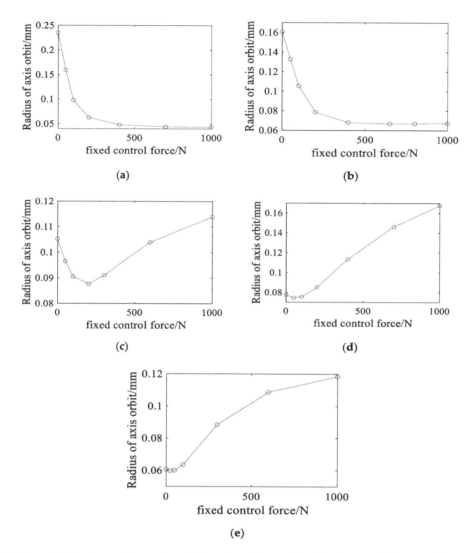

Figure 19. Relationship between the radius of the axis orbit and the fixed control force under (**a**–**e**) speeds, respectively 6000 r/min, 6000 r/min, 6600 r/min, 7200 r/min, 7800 r/min and 8400 r/min.

4.3. Polynomial Fitting of Function Control Force

According to the above analysis, the required control force is related to acceleration and time, so that

$$F_N(t) = a_0 + a_1 t + a_2 t^2 + \cdots + a_m t^m. \tag{2}$$

When the shaft system was accelerated with an angular acceleration of 20π rad/s^2, the times to reach 6000 r/min, 6600 r/min, 7200 r/min, 7800 r/min, and 8400 r/min were 10 s, 11 s, 12 s, 13 s, and 14 s, respectively, and the optimal control forces were 1000 N, 650 N, 200 N, 50 N and 25 N, respectively. Equation (2) is used to fit the functional relationship between control force F_N and acceleration and time. The goal of fitting is to minimize the square sum D of the upper deviation, that is, Equation (3) is the minimum.

$$D = \sum_{i=1}^{N} \delta_i^2 = \sum_{i=1}^{N} \left[F_N(t_i) - F_{OP_i} \right]^2. \tag{3}$$

In Equation (3), the optimal control force is F_{OP_i}, the fixed control force is $F_N(t_i)$ and the variance is δ.

According to the least square theorem, if $N > m + 1$, then there is a unique set of polynomial coefficients $a_0, a_1 \ldots a_m$, which minimize the value of Equation (3), and

$$x_m = \left(A^{\mathrm{T}}A\right)^{-1}A^{\mathrm{T}}b. \tag{4}$$

In Equation (4), $x_m = [a_0\ a_1\ a_2 \cdots a_m]^{\mathrm{T}}$, $b = [F_{\mathrm{OP}_1}\ F_{\mathrm{OP}_2} \cdots F_{\mathrm{OP}_N}]^{\mathrm{T}}$, the value of matrix A is as follows:

$$A = \begin{bmatrix} 1 & t_1 & t_1{}^2 & \cdots & t_1{}^m \\ 1 & t_2 & t_2{}^2 & \cdots & t_2{}^m \\ \vdots & \vdots & \vdots & & \vdots \\ 1 & t_N & t_N{}^2 & \cdots & t_N{}^m \end{bmatrix}.$$

By substituting the time and fixed control force values corresponding to speeds of 6000 r/min, 6600 r/min, 7200 r/min, 7800 r/min, and 8400 r/min into Equation (4), the results show that $a_0 = 143{,}280$, $a_1 = -57{,}510$, $a_2 = 7687.2$, $a_3 = -342.14$, that is, the fitting function of optimal control force and time for $10\ \mathrm{s} \leq t \leq 14\ \mathrm{s}$ is as follows:

$$F_{\mathrm{N}}(t) = 143{,}280 - 57{,}510t + 7687.2t^2 - 342.14t^3. \tag{5}$$

According to Equation (5), the curve shown in Figure 20 is obtained.

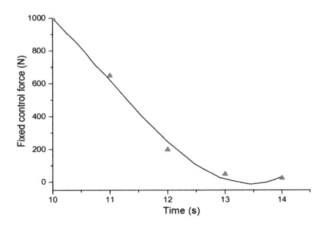

Figure 20. Control force fitting curve for $10\ \mathrm{s} \leq t \leq 14\ \mathrm{s}$.

According to the simulation results of the shaft center trajectory under constant speed, the optimal control force of the shafting was 1000 N in the speed range of 0–6000 r/min, i.e., within 0–10 s, the function control force expression of the optimal control force and time is as follows:

$$F_{\mathrm{N}}(t) = \begin{cases} 1000, & 0 \leq t \leq 10 \\ 143{,}280 - 57{,}510t - 7687.2t^2 - 342.14t^3, & 10 < t \leq 14 \end{cases}. \tag{6}$$

4.4. Analysis of Simulation Results of Supercritical

4.4.1. Analysis of Simulation Results of Over-Critical with Fixed Control Force

The midpoint of the central line between support I and support II was selected as the measuring point, and the displacement response at this point was measured. Figure 21 shows the critical bending vibration response of a three-support shafting with Smart Spring support under different fixed control forces. The fixed control forces applied were 50 N, 100 N, 150 N, 200 N, 250 N, 300 N, 600 N, and 1000 N, respectively. It can be seen from the figure that the application of fixed control force significantly inhibited the peak value of critical response, and the critical speed obviously shifted to the right; when the fixed control force was not greater than 100 N, the vibration response in the whole speed range of 0–8400 r/min (i.e., 0 s < t < 14 s) decreased; when the speed was lower than 7200 r/min (i.e., t < 12 s), the vibration was attenuated by applying 0–1000 N fixed control force, but when the rotating speed was less than 7200 r/min (when the speed was higher than 7800 r/min (i.e., t > 13 s)) and the fixed control force was greater than 150 N, the vibration increased when the fixed control force was applied.

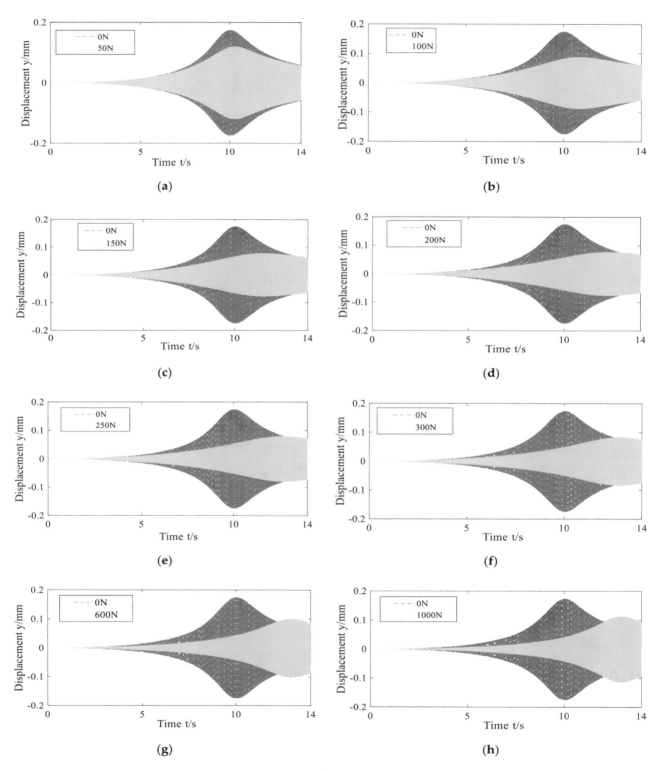

Figure 21. Response of shaft system under (**a–h**) fixed control forces, respectively 50 N, 100 N, 150 N, 200 N, 250 N, 300 N, 600 N and 1000 N.

Figure 22 shows the relationship between the peak displacement response of the measuring point and the fixed control force when the shafting was over-critical. It can be seen from the figure that with the increase in fixed control force, the peak value of critical response first decreased and then increased gradually. When the control force was 0 N, the peak value was 0.1749 mm; when the control force was 200 N, the peak value was the minimum, which was 0.0777 mm, and the maximum vibration reduction rate was 55.6%.

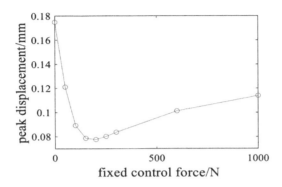

Figure 22. Relationship between peak displacement of measurement point and fixed control force.

As shown in Figure 23, the first critical speed increased gradually and tended to be stable after the control force was greater than 600 N, which was close to 7800 r/min, that is, the damping control gradually transited to the stiffness control.

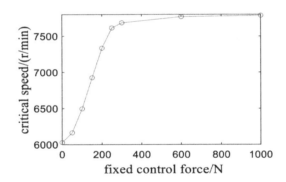

Figure 23. Relationship between first critical speed and fixed control force.

It can be concluded from Figures 22 and 23 that the Smart Spring support has good vibration reduction performance. By applying fixed control force, the bending vibration response of three-support shafting accelerating through critical can be greatly attenuated, but the greater the control force, the better. The optimal fixed control force is 200 N, and the vibration response of shafting is the minimum.

4.4.2. Analysis of Simulation Results of Applying Function Control Force Over-Critical

According to the determined function control force, the joint simulation model established in Figure 16 was modified, and then the function control force was applied to the model. A variable F_N of function control force was established in the MATLAB workspace. F_N contained two columns of data, the first column was time and the second column was the corresponding control force. The two output state variables are the displacement in X direction and Y direction at the midpoint of elastic supports I and II.

By comparing the vibration analysis results of the three-support shafting of the Smart Spring without control force, fixed control force, and function control force, Figures 24 and 25 are obtained. It can be seen from the figure that the critical peak value of the function control force and the best fixed control force was 200 N, but in the whole speed range, the displacement response of the function control force was less than that of the fixed control force of 200 N, which achieved a better vibration reduction effect. Therefore, for the open-loop control of shaft system acceleration over-critical vibration, the vibration reduction effect of Smart Spring with function control force is the best.

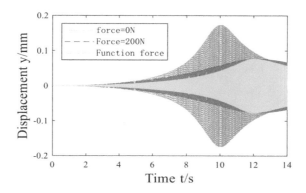

Figure 24. Response curves of the shaft system under different control methods.

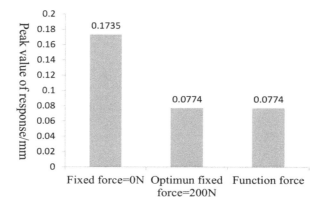

Figure 25. Critical vibration peak value of shafting under different control methods.

5. Conclusions

Previous studies have proved that constant control force exerted by Smart Spring is effective for vibration control of shafting. However, there are few researches on the improvement of control strategy. Based on ADAMS and MATLAB simulation software, this paper analyzes the lateral bending vibration control of the shaft system under critical state, and improves the control strategy of the intelligent spring and proposes a vibration suppression method based on function control force. The main conclusions are as follows:

(1)   In this paper, the authors established a joint simulation model for the double-disk three-support shafting with Smart Spring, and carried out the open-loop control simulation and experimental study of the shafting over-critical vibration. The simulation results showed that the fixed control force exerted by the Smart Spring can greatly reduce the displacement response of the double-disk shaft system accelerating through the critical value. However, the fixed control force is not better when greater, but there is an optimal fixed control force making the vibration control effect the best. The optimal fixed control force is 425 N, and the vibration reduction rate is the maximum, which is 71.6%. The double-disk shafting test system was designed and tested. The test results showed that when the maximum control voltage is 150 V, the vibration reduction rate is the highest, reaching 44.2%, which verified that the Smart Spring support has good control effect on the lateral bending vibration of the three-support shafting when the acceleration is over-critical, and further proved the rationality of the joint simulation method.

(2)   The open-loop control of the spring system was carried out by the joint control function of the spring system. The simulation results showed that with the increase of fixed control force,

the first critical speed gradually increases and tends to be stable, and the peak value decreases first and then increases. The optimal fixed control force is 200 N, of which the maximum vibration reduction rate reaches 55.6%. Compared with the fixed control force, the displacement response of the function control force is significantly less than that of the fixed control force of 200 N, so it has a better vibration reduction effect.

Author Contributions: Conceptualization, M.-M.L. and L.-L.M.; methodology, M.-M.L.; software, M.-M.L.; validation, M.-M.L., L.-L.M. and C.-G.W.; data curation, L.-L.M.; writing—original draft preparation, L.-L.M.; writing—review and editing, M.-M.L.; supervision, C.-G.W.; project administration, R.-P.Z.; funding acquisition, R.-P.Z. All authors have read and agreed to the published version of the manuscript.

References

1. Meng, G. Retrospect and Prospect to the Research on Rotor Dynamics. *Shock Noise* **2002**, *15*, 1–9.
2. Peng, B. Research on Multi-span Shaft Dynamics and Vibration Reduction via Smart Spring Support. Master's Thesis, Nanjing University of Aeronautics and Astronautics, Nanjing, China, 2017.
3. Li, W.Z.; Wang, L.Q. Review on Vibration Control Technology of High-speed Rotor System. *J. Mech. Strength* **2005**, *27*, 44–49.
4. Zhang, T.; Meng, G.; Zhang, Z.X. Active Control of the Rotor System by Electromagnetic Bearing-Extrudod Oil Film Damper. *J. Xi'an Pet. Inst. Nat. Sci. Ed.* **2002**, *6*, 80–83.
5. Xing, J. Research on Active Vibration Control of Rotor System Using Magnetorheological Fluid Damper. Master's Thesis, Beijing University of Chemical Technology, Beijing, China, 2015.
6. Qu, W.Z.; Sun, J.C.; Qiu, Y. Active control of vibration using a fuzzy control method. *J. Sound Vib.* **2004**, *275*, 917–930.
7. Ishimatsu, T.; Shimomachi, T.; Taguchi, N. Active vibration control of flexible rotor using electromagneticdamper. In Proceedings of the IECON'91: 1991 International Conference on Industrial Electronics, Control and Instrumentation, Kobe, Japan, 28 October 1991; IEEE: Piscataway, NJ, USA.
8. Palazzolo, A.B.; Jagannathan, S.; Kascak, A.F.; Montague, G.T.; Kiraly, L.J. Hybrid active vibration control of rotor bearing system using piezoelectric actuators. *J. Vib. Acoust.* **1993**, *115*, 111–119. [CrossRef]
9. De, N. Research on Design Method of Smart Spring Damping System and Its Application in the Drive Shaft System. Ph.D. Thesis, Nanjing University of Aeronautics and Astronautics, Nanjing, China, 2015.
10. Nitzsche, F. The use of smart structures in the realization of effective semi-active Control systems for vibration reduction. *J. Braz. Soc. Mech. Sci. Eng.* **2012**, *XXXIV*, 371–377. [CrossRef]
11. Nitzsche, F.; Feszty, D.; Grappasonni, C.; Coppotelli, G. Whirl-tower Open-loop Experiments and Simulations with an Adaptive Pitch Link Device for Helicopter Rotor Vibration. In Proceedings of the Aiaa/asme/asce/ahs/asc Structures, Structural Dynamics, & Materials Conference, Boston, MA, USA, 8–11 April 2013; pp. 1–13.
12. Afagh, F.F.; Nitzsche, F.; Morozova, N. Dynamic modelling and stability of hingeless helicopter blades with a smart spring. *Aeronaut. J.* **2004**, *108*, 369–377. [CrossRef]
13. Coppotelli, G.; Marzocca, P.; Ulker, F.D.; Campbell, J.; Nitzsche, F. Experimental Investigation on Modal Signature of Smart Spring/Helicopter Blade System. *J. Aircr.* **2008**, *45*, 1373–1380. [CrossRef]
14. Chen, Y. Development of the smart spring for active vibration control of helicopter blades. *J. Intell. Mater. Syst. Sructures* **2004**, *15*, 37–47.
15. Wickramasinghe, V. Experimental evaluation of the smart spring impedance control approach for adaptive vibration suppression. *J. Intell. Mater. Syst. Sructures* **2008**, *19*, 171–179. [CrossRef]
16. Cavalini, A.A., Jr. Vibration attenuation in rotating machines using smart spring mechanism. In *Mathematical Problems in Engineering*; Hindawi Publishing Corporation: London, UK, 2011; pp. 1–14.
17. NI, D.; Zhu, R.-P. Influencing factors of vibration suppression performance for a smart spring device. *J. Vib. Shock* **2012**, *31*, 87–98.
18. Peng, B.; Zhu, R.; Li, M. Bending Vibration Suppression of a Flexible Multispan Shaft Using Smart Spring Support. *Shock Vib.* **2017**, *2017*, 1–12. [CrossRef]

Anti-Disturbance Control for Quadrotor UAV Manipulator Attitude System Based on Fuzzy Adaptive Saturation Super-Twisting Sliding Mode Observer

Ran Jiao [1],*, Wusheng Chou [1,2], Yongfeng Rong [1] and Mingjie Dong [3]

[1]   School of Mechanical Engineering and Automation, Beihang University, Beijing 100191, China; wschou@buaa.edu.cn (W.C.); ryf_2018@buaa.edu.cn (Y.R.)

[2]   The State Key Laboratory of Virtual Reality Technology and Systems, Beihang University, Beijing 100191, China

[3]   School of Mechanical Engineering and Applied Electronics Technology, Beijing University of Technology, Beijing 100124, China; dongmj@bjut.edu.cn

*   Correspondence: jiaoran@buaa.edu.cn

Abstract: Aerial operation with unmanned aerial vehicle (UAV) manipulator is a promising field for future applications. However, the quadrotor UAV manipulator usually suffers from several disturbances, such as external wind and model uncertainties, when conducting aerial tasks, which will seriously influence the stability of the whole system. In this paper, we address the problem of high-precision attitude control for quadrotor manipulator which is equipped with a 2-degree-of-freedom (DOF) robotic arm under disturbances. We propose a new sliding-mode extended state observer (SMESO) to estimate the lumped disturbance and build a backstepping attitude controller to attenuate its influence. First, we use the saturation function to replace discontinuous sign function of traditional SMESO to alleviate the estimation chattering problem. Second, by innovatively introducing super-twisting algorithm and fuzzy logic rules used for adaptively updating the observer switching gains, the fuzzy adaptive saturation super-twisting extended state observer (FASTESO) is constructed. Finally, in order to further reduce the impact of sensor noise, we invite a tracking differentiator (TD) incorporated into FASTESO. The proposed control approach is validated with effectiveness in several simulations and experiments in which we try to fly UAV under varied external disturbances.

Keywords: super-twisting; sliding mode extended state observer; saturation function; fuzzy logic; attenuate disturbance

1. Introduction

Unmanned aerial vehicles (UAVs) have become a popular and active research topic among scholars worldwide [1,2]. It could work at locations where entry is difficult and humans cannot access [3]. Recently, they are not only used in traditional scenes, such as monitoring, aerial photography, surveying, and patrol, but are also employed in new application scenarios requiring physically interacting with external environment. Then, to conduct physical interaction, aerial robots are either equipped with a rigid tool [4] or an n-degree-of-freedom (DoF) robotic arm [5–8], which are called UAV manipulators. As for the UAV equipped with a robotic arm, several disturbances cannot be avoided, such as wind gust and model uncertainties, during task execution. Therefore, to meet the aerial operation task requirements with high reliability, disturbance rejection control problem should be investigated.

Many controllers have been proposed for disturbance rejection. The Robust Model Predictive Control (MPC) method is used for multirotor UAV to provide robust performance under an unknown but bounded disturbance [9]. Meanwhile, the adaptive control structure is built to obtain coefficients of system with uncertainty online. A controller based on Model Reference Adaptive Control is proposed in [10], which performs better on disturbance rejection compared with non-adaptive controllers. However, as what we know, one of the biggest drawbacks of adaptive controllers is that they perform little robust under the bursting phenomenon.

Adopting the disturbance and uncertainty estimation and attenuation (DUEA) strategy, such as Disturbance observer (DOB) [11], unknown input observer (UIO) [12], and extended state observer (ESO) [13], would be a potential solution for disturbance rejection. Meanwhile, there are several types of DOBs applied to UAVs. Work [14] proposes an inner-loop control structure to recover the dynamics of a multirotor combined with additional objects similar to the bare multirotor using DOB-based method. A linear dual DOB is built in [15] to reject modeling error and external disturbance when designing the control system. The authors of [16] propose a DOB-based tracking flight control method for a quadrotor to reject disturbance, which is supposed to be composed of some harmonic elements, and applies it to flight control of the UAV Quanser Qball 2. A disturbance observer with finite time convergence (FTDO), which could conduct online estimation of the unknown uncertainties and disturbances, is incorporated into an hierarchical controller to solve the problem of path tracking of a small coaxial rotor-type UAV [17]. The authors of [18] introduce a robust DOB for an aircraft to compensate the uncertain rotational dynamics into a nominal plant and proposes a nonlinear feedback controller implemented for desired tracking performance.

As for ESO, it was first proposed by Han [19] and has been introduced in many control methods [20]. Traditionally, ESO methods mainly focus on coping with disturbances which slowly change [21]. Nevertheless, it is evident that the disturbance caused by the wind gust sometimes performs drastic change so that it can not be estimated by traditional ESO thoroughly. Thus, an enhanced ESO that could quickly estimate the disturbance is necessary in this field. A high order sliding mode observer is built to estimate unmodeled dynamics and external disturbances for an aerial vehicle to track the trajectory [22]. Moreover, a sliding mode observer is proposed for equivalent-input-disturbance approach to control the under-actuated subsystem of a quadrotor UAV [23]. However, chattering phenomenon is normal in traditional sliding mode observer (SMO) because of the discontinuous sign function. In order to reduce the chattering problem, the super-twisting algorithm is introduced in a ESO to mitigate estimation chattering [24], but there is still a little chattering. In order to solve it, the authors of [25] proposed a new SMO control strategy to reduce the estimated speed chattering of the motor, in which the switching function is replaced by a sigmoid function. The sigmoid function has also been used in SMO for UAV in [26]; however, it is conducted without experiment.

The main contributions of this study are given as follows.

(1) In order to alleviate the estimation chattering problem of traditional SMESO, a new observer named fuzzy adaptive saturation super-twisting extended state observer (FASTESO) is proposed, in which a saturation function is invited to replace the discontinued sign function, meanwhile a super-twisting algorithm is introduced to prevent excessively high observer gain and TD [19] and is incorporated for avoiding directly using acceleration information, which is full of noise.

(2) To stay robustness under disturbances with unknown bounds, fuzzy logic rules are introduced as an adaptive algorithm to adaptively adjust the observer switching gains. Furthermore, it also contributes to the chattering attenuation under high switching gain with low estimation value and observer performance improvement under low switching gain with high estimation value.

(3) The proposed method is verified with effectiveness on a quadrotor UAV manipulator prototype in several simulations and experiments.

The rest of this article is organized as follows. Section 2 introduces some preliminaries of this work. Section 3 describes the kinematic and dynamic models of the quadrotor UAV manipulator. The construction of proposed FASTESO and attitude controller are given in Section 4. Additionally, several simulations are conducted in Section 5. Moreover, Section 6 shows the experiment. Finally, Section 7 presents the conclusion.

2. Preliminaries

In this section, some mathematical preliminaries are provided for understanding the whole paper more easily.

2.1. Notation

The 2-norm of a vector or a matrix is provided by $\|\cdot\|$. $\lambda_{\max}(Z)$ and $\lambda_{\min}(Z)$ represent the maximal and minimum eigenvalue of the matrix Z, respectively. Moreover, the operator $S(\cdot)$ denotes a vector $\kappa = \begin{bmatrix} \kappa_1 & \kappa_2 & \kappa_3 \end{bmatrix}^T$ to a skew symmetric matrix as

$$S(\kappa) = \begin{bmatrix} 0 & -\kappa_3 & \kappa_2 \\ \kappa_3 & 0 & -\kappa_1 \\ -\kappa_2 & \kappa_1 & 0 \end{bmatrix} \tag{1}$$

The sign function is given as

$$sign(\kappa) = \begin{cases} \frac{\kappa}{|\kappa|}, & |\kappa| \neq 0 \\ 0, & |\kappa| = 0 \end{cases} \tag{2}$$

2.2. Quaternion Operations

The unit quaternion $q = \begin{bmatrix} q_0 & q_v \end{bmatrix}^T \in R^4$, $\|q\| = 1$, is used to represent the rotation of the quadrotor in this paper. Several corresponding operations are given as follows.

The quaternion multiplication:

$$q \otimes \sigma = \begin{bmatrix} q_0\sigma_0 - q_v^T\sigma_v \\ q_0\sigma_v + \sigma_0 q_v - S(\sigma_v)q_v \end{bmatrix} \tag{3}$$

The relationship between rotation matrix C_A^B and unit quaternion q is represented by:

$$C_A^B = (q_0^2 - q_v^T q_v)I_3 + 2q_v q_v^T + 2q_0 S(q_v) \tag{4}$$

Take the time derivative of Equation (4), we could obtain

$$\dot{C}_A^B = -S(\omega)C_A^B \tag{5}$$

The derivative of a quaternion and the quaternion error q_e are provided, respectively, as

$$\dot{q} = \begin{bmatrix} \dot{q}_0 \\ \dot{q}_v \end{bmatrix} = \frac{1}{2}q \otimes \begin{bmatrix} 0 \\ \omega \end{bmatrix} = \frac{1}{2}\begin{bmatrix} -q_v^T \\ S(q_v) + q_0 I_3 \end{bmatrix}\omega \tag{6}$$

$$q_e = q_d^* \otimes q \tag{7}$$

where q_d represents the desired quaternion whose conjugate is $q_d^* = \begin{bmatrix} q_{d0} & -q_{dv} \end{bmatrix}^T$. Moreover, ω denotes the angular rate of the system.

$$\omega_e = \omega - C_d^b \omega_d \tag{8}$$

$$\dot{q}_e = \frac{1}{2} q_e \otimes \begin{bmatrix} 0 \\ \omega_e \end{bmatrix} = \frac{1}{2} \begin{bmatrix} -q_{ev}^T \\ S(q_{ev}) + q_{e0} I_3 \end{bmatrix} \omega_e \tag{9}$$

3. Models of UAV

In this section, we present a mathematical description of the quadrotor UAV manipulator system model. The abstract graph is shown in Figure 1, in which the robotic arm is fixed at the geometric center of the UAV.

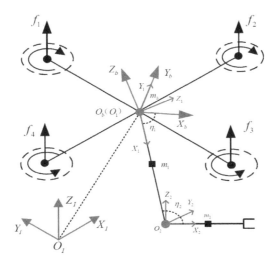

Figure 1. Quadrotor and robotic arm system with coordinate reference frames.

3.1. Kinematic Model

Figure 1 shows several coordinates in UAV that we consider.

O_I: the world-fixed inertial reference frame. O_b: the body-fixed reference frame centered in the quadrotor Center of Gravity (CoG). O_i: the coordinate of robotic arm i, where $(i = 1, 2)$ denotes the link number.

Moreover, several parameters definition are given as follows.

$p^I = [x, y, z]^T$: the absolute position of O_b with respect to O_I. $\Psi - [\varphi, \theta, \psi]^T$: Euler angles used for representing the UAV attitude angle. $\eta = [\eta_1, \eta_2]^T$: the joint angles of the robotic arm relative to the zero position. $\mathbf{H} = [(p^I)^T, \Psi^T, \eta^T]^T \in \mathbf{R}^{8 \times 1}$: a vector used for including all the generalized variables. $\dot{p}^I$: the absolute linear velocity of UAV described in O_I. $\dot{p}^b$: the linear velocity of UAV with respect to O_b. $\omega^I = [\omega_x^I, \omega_y^I, \omega_z^I]^T$: UAV's absolute rotational velocity described in O_I. $\omega^b = [\omega_x^b, \omega_y^b, \omega_z^b]^T$: the rotational velocity of UAV relative to O_b. Then we could get:

$$\begin{cases} \dot{p}^I = R_b \dot{p}^b \\ \omega^I = R_t \dot{\Psi} \\ \omega^b = (R_b)^T \omega^I = (R_b)^T R_t \dot{\Psi} = R_r \dot{\Psi} \end{cases} \tag{10}$$

where $R_t \in \mathbf{R}^{3 \times 3}$ represents the transformation matrix for converting Euler angle rates $\dot{\Psi}$ into ω^I. $R_b \in \mathbf{R}^{3 \times 3}$ denotes the rotation matrix representing the orientation of O_b relative to O_I, and $R_r = (R_b)^T R_t \in \mathbf{R}^{3 \times 3}$ is used to map the derivative of Ψ into UAV angular rate expressed in O_b.

Where

$$\begin{cases} R_b = \begin{bmatrix} c\theta c\psi & s\varphi s\theta c\psi - c\varphi s\psi & c\varphi\, s\theta\, c\psi + s\varphi s\psi \\ c\theta s\psi & s\varphi s\theta s\psi + c\varphi c\psi & c\varphi\, s\theta s\psi - s\varphi c\psi \\ -s\theta & s\varphi c\theta & c\varphi\, c\theta \end{bmatrix} \\ \qquad R_r = \begin{bmatrix} 1 & 0 & -s\theta \\ 0 & c\varphi & s\varphi c\theta \\ 0 & -s\varphi & c\varphi c\theta \end{bmatrix} \end{cases} \tag{11}$$

$c(\cdot)$ and $s(\cdot)$ denote $cos(\cdot)$ and $sin(\cdot)$, respectively.

Moreover, the position and angular rate of the frame O_i, which is fixed to the robotic arm, with respect to O_I are provided by

$$\begin{cases} p_i^I = p^I + R_b p_i^b \\ \omega_i^I = \omega^I + R_b \omega_i^b \end{cases} \tag{12}$$

where $i = (x, y, z)$. The vectors p_i^b, ω_i^b, which are expressed in O_b, represent the position of O_i and the angular rate of the i-th robotic arm frame with respect to O_b, respectively. Additionally, more relationships are provided,

$$\begin{cases} \dot{p}_i^b = J_{pi}\dot{\eta} \\ \omega_i^b = J_{ri}\dot{\eta} \end{cases} \tag{13}$$

where $J_{pi} \in R^{2\times2}$ and $J_{ri} \in R^{2\times2}$ are Jacobian matrices representing the translational and angular velocities of each robotic arm link to the $\dot{\eta}$, respectively. According to Equations (12) and (13), we could get the translational and angular velocity of O_i with respect to O_I as

$$\begin{cases} \dot{p}_i^I = \dot{p} + S(\omega^I)R_b p_i^b + R_b J_{pi}\dot{\eta} \\ \omega_i^I = \omega^I + R_b J_{ri}\dot{\eta} \end{cases} \tag{14}$$

where $S(\cdot)$ represents the skew-symmetric matrix.

3.2. Dynamic Model

In order to obtain the dynamic model of the quadrotor UAV manipulator system, the Euler-Lagrange equation is introduced.

$$\frac{d}{dt}\frac{\partial L}{\partial H_k} - \frac{\partial L}{\partial H_k} = u_k + d_k, L = K - U \tag{15}$$

where $k = (1, ..., 8)$. L: the Lagrangian with kinetic energy K and potential energy U of the integrated system. u_k: the generalized driving force. d_k: external disturbance applied to the system. Additionally, K could be given in detail as

$$\begin{cases} K = K_b + \sum\limits_{i=1}^{2} K_i \\ K_b = \frac{1}{2}m_b(\dot{p}^I)^T \dot{p}^I + \frac{1}{2}(\omega^I)^T R_b I_b R_b^T \omega^I \\ K_i = \frac{1}{2}m_i(\dot{p}_i^I)^T \dot{p}_i^I + \frac{1}{2}(\omega_i^I)^T R_b R_i^b I_i R_i^i R_b^T \omega_i^I \end{cases} \tag{16}$$

where m_b is the quadrotor base mass, m_i is the i-th robotic arm mass ($i = (1, 2)$), I is the inertia matrix, and R_i^b the rotation matrix between the frame fixed to the center of mass of the i-th link and O_b. Then, the total potential is provided as

$$\begin{cases} U = U_b + \sum\limits_{i=1}^{2} U_i \\ U_b = m_b g e_3^T p^I \\ U_i = m_i g e_3^T p_i^I \end{cases} \tag{17}$$

where e_3 denotes unit vector $[0, 0, 1]^T$ and g represents the gravity constant. By substituting Equations (16) and (17) into Equation (15), the dynamic model of the quadrotor UAV manipulator system could be obtained as

$$M(\mathbf{H})\ddot{\mathbf{H}} + C(\mathbf{H}, \dot{\mathbf{H}}) + G(\mathbf{H}) = u + d \tag{18}$$

Moreover, the total kinetic energy can be expressed:

$$K = \frac{1}{2}\dot{\mathbf{H}}^T M(\mathbf{H})\dot{\mathbf{H}} \tag{19}$$

where

$$M(\mathbf{H}) = \begin{bmatrix} M_{11} & M_{12} & M_{13} \\ M_{21} & M_{22} & M_{23} \\ M_{31} & M_{32} & M_{33} \end{bmatrix} \tag{20}$$

Details of the inertia matrix $M(\mathbf{H})$ and Coriolis matrix $C(\mathbf{H}, \dot{\mathbf{H}})$ can be found in [27]. The $G(\mathbf{H})$ could be obtained via partial derivative Equation (21):

$$G(\mathbf{H}) = \frac{\partial U}{\partial \mathbf{H}} \tag{21}$$

Express Equation (18) in detail as

$$\begin{bmatrix} M_{11} & M_{12} & M_{13} \\ M_{21} & M_{22} & M_{23} \\ M_{31} & M_{32} & M_{33} \end{bmatrix} \begin{bmatrix} \ddot{p}^I \\ \ddot{\Psi} \\ \ddot{\eta} \end{bmatrix} + \begin{bmatrix} G_1 \\ G_2 \\ G_3 \end{bmatrix} + \begin{bmatrix} C_{11} & C_{12} & C_{13} \\ C_{21} & C_{22} & C_{23} \\ C_{31} & C_{32} & C_{33} \end{bmatrix} \begin{bmatrix} \dot{p}^I \\ \dot{\Psi} \\ \dot{\eta} \end{bmatrix} = \begin{bmatrix} u_t \\ u_r \\ u_l \end{bmatrix} + \begin{bmatrix} d_t \\ d_r \\ d_l \end{bmatrix} \tag{22}$$

where u_t, u_r, u_l represent the generalized control inputs corresponding to p^I, Ψ, η. u_l is not used, as we do not consider a dynamic robotic arm in this work. Vector $d = [d_t, d_l, d_r]^T$ are the lumped external disturbance. As for the quadrotor attitude loop subsystem, we would transform u_r to the quadrotor inputs $f_v = [\omega_1^2, \omega_2^2, \omega_3^2, \omega_4^2]^T$, which is a positive correlation vector with forces produced from the quadrotor propellers. Moreover, ω_i represents rotor rate of quadrotor propeller ($i = 1, 2, 3, 4$). Then, convert the generalized control inputs to the propeller rate:

$$u_r = R_r{}^T \Xi f_v \tag{23}$$

$$\Xi = \begin{bmatrix} \frac{\sqrt{2}l}{2}\Lambda_T & \frac{\sqrt{2}l}{2}\Lambda_T & -\frac{\sqrt{2}l}{2}\Lambda_T & -\frac{\sqrt{2}l}{2}\Lambda_T \\ \frac{\sqrt{2}l}{2}\Lambda_T & -\frac{\sqrt{2}l}{2}\Lambda_T & -\frac{\sqrt{2}l}{2}\Lambda_T & \frac{\sqrt{2}l}{2}\Lambda_T \\ -l\Lambda_C & l\Lambda_C & -l\Lambda_C & l\Lambda_C \end{bmatrix} \tag{24}$$

where Λ_T and Λ_C are the thrust and drag coefficients, respectively. Additionally, l denotes the distance from each motor to the quadrotor CoG.

According to Equation (22), we could obtain in detail the dynamics of the attitude loop subsystem which is the base of next section:

$$M_{21}\ddot{p}^I + M_{22}\ddot{\Psi} + M_{23}\ddot{\eta} + C_{21}\dot{p}^I + C_{22}\dot{\Psi} + C_{23}\dot{\eta} + G_2 = u_r + d_r \tag{25}$$

4. Method

Based on the UAV model, the proposed control strategy is described in detail in this section. As shown in Figure 2, quadrotor attitude control problem could be briefly divided into two components: FASTESO, used for estimating and compensating the disturbance, plays a role of the feedforward loop, and Backstepping controller, built for regulating the orientation to track the desired attitude timely, plays a role of the feedback loop. Meanwhile, TD, saturation function, and fuzzy logic methods are incorporated into the whole control method.

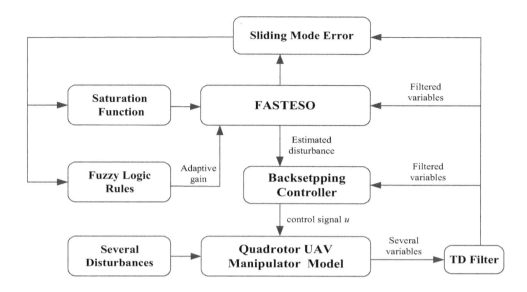

Figure 2. Whole control structure of the proposed method.

4.1. FASTESO

UAVs with robotic arms have more model uncertainty than traditional UAVs. Additionally, UAVs usually face some other disturbances such as external wind when flying outside. Moreover, the control performance of closed loop system will be largely determined by the observation performance. Therefore, in this part, in order to enhance the performance of feedback controller, a new SMO named FASTESO is built to estimate the lumped disturbance exerted on quadrotor UAV manipulator in finite time.

As for the traditional SMO, the sign function is generally adopted as the control function, and owing to switch time and space lag the SMO performs serious chattering problem [28,29]. In order to alleviate the chattering phenomenon, the sign function is replaced by a saturation function in this work. Moreover, the chattering phenomenon would also be invited by large switching gain because when the control signal crosses the sliding surface, larger switching gain would produce faster and bigger switching control parts and perform terrible chattering phenomenon. Additionally, the invited saturation function has a slower performance than the sign function on arithmetical speed, so there would be a time delay relatively. To further settle the problems of chattering and time delay mentioned before, fuzzy logic rules are employed in FASTESO to adjust the switching gains according to the states of the sliding surface.

4.1.1. Construction of Traditional Super-Twisting Extended State Observer (STESO)

In this part, according to our previous work [30], a traditional SMESO named STESO for quadrotor UAV manipulator is constructed. In this work, the situation, in which the robotic arm keeps constant while the UAV faces external disturbances during the flight, is mainly considered: ($\dot{\eta} = \ddot{\eta} = 0$). That is a general scenario that the quadrotor UAV manipulator often encounters when conducting aerial tasks. Nevertheless, usually there is unwanted vibration of the equipped robotic arm caused by the high-speed motors and propellers. To consider all mentioned disturbances, the model (25) could be reconstructed as

$$(M_{21} + \Delta M_{21})\ddot{p}^I + (M_{22} + \Delta M_{22})\ddot{\Psi} + C_{21}\dot{p}^I + (M_{23} + \Delta M_{23})\Delta\ddot{\eta} + C_{22}\dot{\Psi} + C_{23}\Delta\dot{\eta} + G_2 = u_r + d_r \quad (26)$$

where ΔM_{21}, ΔM_{22}, ΔM_{23} represent model uncertainties. $\Delta\dot{\eta}$ and $\Delta\ddot{\eta}$, considered as a portion of model uncertainty, denote the residuals from the temporary variation of the robotic arm parameters $\dot{\eta}, \ddot{\eta}$.

In order to build the observer, reconstruct the attitude dynamic Equation (26) as

$$M_{22}\ddot{\Psi} = u_r + d_r - M_{21}\ddot{p}^I - \Delta M_{21}\ddot{p}^I - \Delta M_{22}\ddot{\Psi} - (M_{23} + \Delta M_{23})\Delta\ddot{\eta} - C_{23}\Delta\dot{\eta} - C_{21}\dot{p}^I - C_{22}\dot{\Psi} - G_2 \quad (27)$$

Introduce the lumped disturbance as

$$d_r{}^* = d_r - \Delta M_{21}\ddot{p}^I - \Delta M_{22}\ddot{\Psi} - (M_{23} + \Delta M_{23})\Delta\ddot{\eta} - C_{23}\Delta\dot{\eta} \tag{28}$$

Combining Equations (27) and (28), we get

$$M_{22}\ddot{\Psi} = u_r + d_r{}^* - M_{21}\ddot{p}^I - C_{21}\dot{p}^I - C_{22}\dot{\Psi} - G_2 \tag{29}$$

where the variable $\ddot{p}^I$ is measured directly from the sensors or velocity differential is too noisy to use, so TD is introduced here to estimate the system acceleration for noise alleviation. Additionally, $\dot{p}^I$ and $\dot{\Psi}$ measured from the sensors are also a little noisy, in that case, TD would be used to reduce noise, too. Moreover, the items M_{21}, C_{21} and G_2 could be obtained in advance according to the presented UAV model. As for dynamics model (29), considering the feedback linearization method, we could reformulate the control input as

$$u_r = u_r^* + M_{21}\ddot{p}^I + C_{21}\dot{p}^I + C_{22}\dot{\Psi} + G_2 \tag{30}$$

Combining Equations (29) and (30), we get

$$M_{22}\ddot{\Psi} = u_r^* + d_r{}^* \tag{31}$$

When building the STESO, it is supposed that every channel is independent from each other. Therefore, only one channel will be presented here and the other two are completely identical. As for model Equation (31), the one-dimensional model for STESO design is provided as

$$J_i\ddot{\Psi}_i = u_{ri}^* + d_{ri}{}^* \tag{32}$$

Introduce a new extended state vector $\zeta = [\zeta_1 \quad \zeta_2]^T$, where $\zeta_1 = J_i\dot{\Psi}_i$ and $\zeta_2 = d_{ri}{}^*$, $(i = \varphi, \theta, \psi)$. Reconstruct the model (32) as

$$\begin{cases} \dot{\zeta}_1 = u_{ri}^* + \zeta_2 \\ \dot{\zeta}_2 = \chi \end{cases} \tag{33}$$

where χ denotes the derivative of $d_{ri}{}^*$, supposed by $|\chi| < f^+$. It means that the differentiation of lumped disturbance is bounded.

Then, according to the work in [30], build STESO for the observable system (33) as

$$\begin{cases} \dot{z}_1 = z_2 + u_{ri}{}^* + \alpha_1|e_1|^{\frac{1}{2}}sign(e_1) \\ \dot{z}_2 = \alpha_2 sign(e_1) \end{cases} \tag{34}$$

where z_1 and z_2 are estimations of ζ_1 and ζ_2, respectively. $e_1 = \zeta_1 - \hat{\zeta}_1$ and $e_2 = \zeta_2 - \hat{\zeta}_2$ denote estimation errors. We could obtain that the system estimation errors e_1 and e_2 would converge to zero within finite time under suitable observer gains α_1, α_2. Moreover, the details of the convergence analysis and parameter selection rules could be found in [30].

4.1.2. Saturation Function

To mitigate the chattering problem, we replace the sign function, usually expressing the discontinuous control in the observer, with a saturation function. The traditional STESO Equation (34) could be rewritten as

$$\begin{cases} \dot{z}_1 = z_2 + u_{ri}{}^* + \alpha_1|e_1|^{\frac{1}{2}}sign(e_1) \\ \dot{z}_2 = k_f * \alpha_2 sat(e_1) \end{cases} \tag{35}$$

where k_f is the adaptive proportional factor of switching gain generated from fuzzy logic rules which will be introduced in next part. $sat()$ represents the saturation function whose curve is shown in Figure 3, where particularly e is the difference between the real system and the estimated one. k_e is the output of the saturation function. We could adjust the sliding mode effect by regulating values of e and k_e. Obviously the chattering would be reduced via increasing e and decreasing k_e. If e is big enough and k_e is small enough, the high-frequency chattering could be avoided. However, meanwhile the robustness of the sliding mode method would also be reduced. Therefore the tuning of these parameters is a tradeoff between the chattering alleviation performance and the system robustness. Therefore, e and k_e should be considered overall according to real applications.

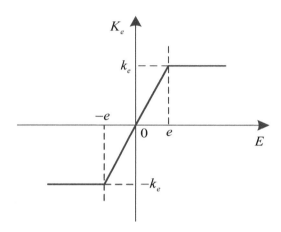

Figure 3. Sketch of the saturation function.

4.1.3. Adaptive Switching Gains with Fuzzy Logic Rules

The fuzzy logic controller was proposed many years ago [31]. In this part, the fuzzy rules are designed according to the fuzzy control theory for effectively optimizing the switching gain based on the states of the sliding-mode surface.

Normally, for purpose of enhancing the ability of observer on anti-disturbance and undertaking the generation of sliding mode, the sliding mode switching gain is often chosen to be too large, but this would further enhance the chattering noise in the disturbance estimation. Therefore, it is feasible to introduce the intelligent method to effectively estimate the switching gain according to the sliding mode arrival condition to alleviate the system chattering problem. For example, as the system state is far from the sliding surface, it means that the absolute value of difference $|e|$ is relatively large, the proportional factor k_f should be enlarged to drive the system state back. Similarly, as the system state is close to the sliding surface, proportional factor k_f should be smaller.

The proposed fuzzy logic system in this work consists of one input variable and one output variable which would be quantized first. The input and output variables are divided into four fuzzy subsets. Let $|e| = \left\{ \begin{array}{cccc} ZR & PS & PM & PB \end{array} \right\}$ denote the input variable and $k_f = \left\{ \begin{array}{cccc} ZR & PS & PM & PB \end{array} \right\}$ represent the output variable, respectively. Each rank is depicted by a membership function of trigonometric function, as shown in Figures 4 and 5, where the fuzzy language is defined as ZR(zero), PS(positive small), PM(positive middle), and PB(positive big).

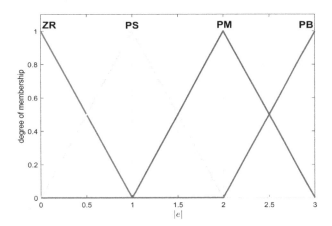

Figure 4. Fuzzy input membership functions.

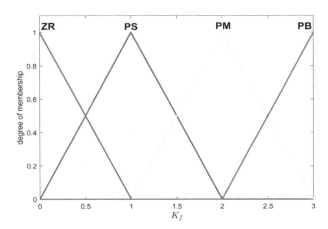

Figure 5. Fuzzy output membership functions.

Then define the fuzzy rules as

Rule 1: If $|e|$ is PB, then k_f is PB.

Rule 2: If $|e|$ is PM, then k_f is PM.

Rule 3: If $|e|$ is PS, then k_f is PS.

Rule 4: If $|e|$ is ZR, then k_f is ZR.

Finally, as for the fuzzy outputs, according to the work in [32], the center of gravity algorithm is adopted as the defuzzification method to convert the fuzzy subset duty cycle changes to real numbers, which is given as

$$k_f = \frac{\int_0^3 k_f \cdot \mu_{out} \cdot dk_f}{\int_0^3 \mu_{out} \cdot dk_f} \tag{36}$$

where μ_{out} represents the resulting output membership function. It takes every rule into account performing the union of the resulting output membership function $\mu_{out,i}$ of each *Rule* i ($i = 1, ..., 4$), which means the maximum operation between them.

4.2. Attitude Controller

In order to achieve high-precision attitude stabilization in the presence of external wind and model uncertainties, a backstepping controller is built here combined with FASTESO for UAV attitude system. The main objective of this part is to guarantee that the state of attitude q and angular rate w converge to the reference values q_d and w_d in real time. As this part is identical to our previous work [30] except the UAV model, which would not impact the controller design. Therefore we are not going to describe it in detail.

According to the authors of [30], the control signal vector u_r is designed as

$$u_r = -M_{22}(S(\omega_e)C_d^b\omega_d - C_d^b\dot{\omega}_d) + M_{21}\ddot{p}^I + M_{23}\ddot{\eta} + C_{21}\dot{p}^I + C_{22}\dot{\Psi}$$
$$+ C_{23}\dot{\eta} + G_2 - M_{22}K_{b1}\dot{q}_{ev} - q_{ev} - K_{b2}\tilde{\omega}_e - \hat{d}_r \tag{37}$$

where $\hat{d}_r$ is the estimated result of disturbance from FASTESO. Additionally, like what we did at FASTESO, TD will also be adopted here to process several variables in Equation (37) to reduce noise. Moreover, K_{b1}, K_{b2} are controller gains to be designed, and ω_e, $\tilde{\omega}_e$, and q_{ev} are defined in Equation (38), details of which can be found in [30].

$$\begin{cases} \omega_e = \omega - C_d^b\omega_d \\ \tilde{\omega}_e = \omega_e + K_{b1}q_{ev} \\ q_e = \begin{bmatrix} q_0 & q_{ev} \end{bmatrix}^T \\ q_e = q_d^* \otimes q \end{cases} \tag{38}$$

5. Simulation

In this section, the performance of proposed FASTESO-based control method is validated on the PX4/Gazebo platform [33], which provides simulations of physics close to the real world. A manually designed scalable disturbance, which includes a sudden change in some special points, is defined in Equation (39) and adopted in the simulations. Moreover, its amplitude could be adjusted by a. The main nominal coefficients of the quadrotor base are generated by the online toolbox of Quan and Dai [34] given in Table 1.

$$d_e = a * (0.25\sqrt{(|0.02\sin(0.1\pi t)|)} + 0.015\sin(0.3\pi t - 5) - 0.02) \tag{39}$$

Table 1. Quadrotor base coefficients used in simulation.

Coefficients	Description	Value
m_b	Mass of the quadrotor base	1.8 kg
J_{bx}	Pitch inertia	1.319×10^{-2} kg · m^2
J_{by}	Roll inertia	1.319×10^{-2} kg · m^2
J_{bz}	Yaw inertia	2.283×10^{-2} kg · m^2
l	Quadrotor base size	0.205 m

5.1. CASE 1 (Observers Performance Comparison)

In this section, several comparison simulations on observer performance of disturbance estimation are conducted.

5.1.1. Domestic Observers Comparison

To verify the effectiveness of the proposed fuzzy logic rules for adaptively adjusting the observer gain to cope with disturbances with different amplitudes, comparative simulations are conducted for a quadrotor UAV manipulator performing a hovering flight with the manually designed scalable disturbance (39) between FASTESO and ASTESO combined with a backstepping controller (BAC). The ASTESO is defined as a simplified version of FASTESO, which is not equipped with adaptive gain under fuzzy logic rules. Their coefficients are given as FASTESO: $\alpha_1 = diag(0.4, 0.4, 0.5)$; $\alpha_2 = diag(0.7, 0.7, 0.9)$; $e = 0.0001$; $k_e = 0.05$. ASTESO: $\alpha_1 = diag(0.4, 0.4, 0.5)$; $\alpha_2 = diag(0.7, 0.7, 0.9)$. BAC: $K_{b1} = diag(5, 5, 8)$; $K_{b2} = diag(3, 3, 4)$. The simulations are divided into two groups classified by different scalars of disturbances, $a = 10, 20$. The corresponding results could be found in Figures 6–9.

As shown in Figure 6, both FASTESO and ASTESO have good estimation performance under scalar $a = 10$. The resultant observer gain of FASTESO is described in Figure 7. Nevertheless, the ASTESO starts to break down when the disturbance is larger, which could be found in Figure 8. Under $a = 20$, the ASTESO fails to follow the larger disturbance, but the proposed FASTESO still performs well. This is because the fixed parameter of ASTESO is not big enough to estimate the disturbance with the amplitude in Figure 8. As shown in Figure 9, the adaptive gain of FASTESO also shows why it is effective under disturbances with different amplitudes. To summarize, this part shows that the proposed FASTESO could estimate a wider range of disturbance thanks to the fuzzy logic rules for adaptive observer gain.

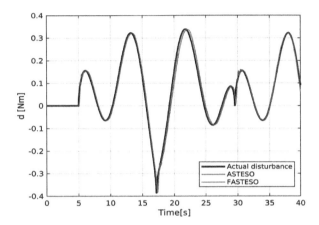

Figure 6. Estimation results from ASTESO and fuzzy adaptive saturation super-twisting extended state observer (FASTESO) with scalar $a = 10$.

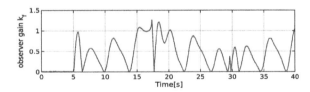

Figure 7. Observer gain of FASTESO during adaptation process with scalar $a = 10$.

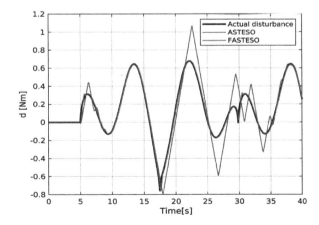

Figure 8. Estimation results from ASTESO and FASTESO with scalar $a = 20$.

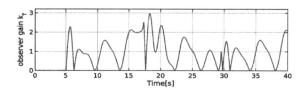

Figure 9. Observer gain of FASTESO during adaptation process with scalar $a = 20$.

5.1.2. Various Observers Comparison

To demonstrate the effectiveness of the introduction of the saturation function into SMO, a simulation similar with last part is conducted under FASTESO with an applied disturbance under $a = 5$ shown in Figure 10. Meanwhile the 2nd-order linear ESO (ESO2) and STESO are also invited for comparison. The parameters of the mentioned observers are listed. FASTESO:$\alpha_1 = diag(0.4, 0.4, 0.5)$; $\alpha_2 = diag(0.7, 0.7, 0.9)$; $e = 0.0001$; $k_e = 0.05$. ESO2:$k_1 = diag(4.5, 4.5, 6.5)$; $k_2 = diag(22, 22, 28)$; $k_3 = diag(5, 5, 8)$; STESO:$\alpha_1 = diag(0.2, 0.2, 0.3)$; $\alpha_2 = diag(0.3, 0.3, 0.4)$; BAC: $K_{b1} = diag(5, 5, 8)$; $K_{b2} = diag(3, 3, 4)$.

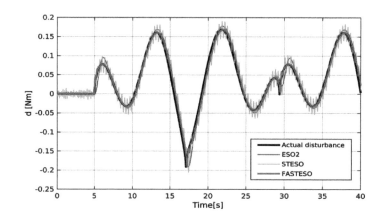

Figure 10. Estimation results from ESO2, STESO, and FASTESO with scalar $a = 5$.

From the response results in Figure 10, we could find that FASTESO has a perfect performance on estimation even at the special point with a sudden change. Although the curve under ESO2 could usually follow the actual value, it performs an overestimation near the peak. As for STESO, it could follow the fast variation of disturbance; however, it has a terrible chattering problem which is also the main reason for the introduction of saturation function in this work. To summarize, the FASTESO performs well at all range of disturbance ($a = 5, 10, 20$). Meanwhile, it follows better than other traditional observers under varied disturbance.

5.2. CASE 2 (Composite Comparison)

To further illustrate the effectiveness of the whole proposed control strategy, we would show all-sided simulation results from Section 5.1.2. Moreover, a single backstepping controller without any observer is also conducted on quadrotor UAV manipulator system for comparison. Three-axis components of disturbance torque $d_r = \begin{bmatrix} d_{r\varphi} & d_{r\theta} & d_{r\psi} \end{bmatrix}^T$ are equal and shown in Figure 10, which are applied on the UAV simultaneously. The result curves of both attitude and angular rate are shown in Figures 11–16.

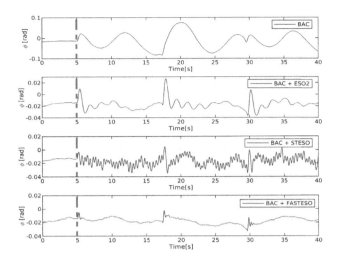

Figure 11. Results of roll angle with unknown external torque disturbance under backstepping controller combined with nothing, ESO2, STESO and FASTESO.

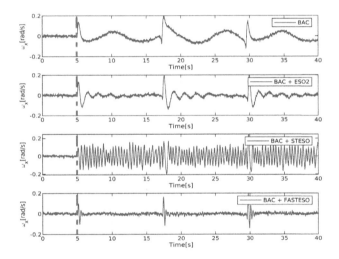

Figure 12. Results of angular rate ω_x with unknown external torque disturbance under backstepping controller combined with nothing, ESO2, STESO and FASTESO.

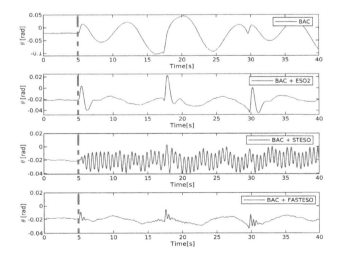

Figure 13. Results of pitch angle with unknown external torque disturbance under backstepping controller combined with nothing, ESO2, STESO and FASTESO.

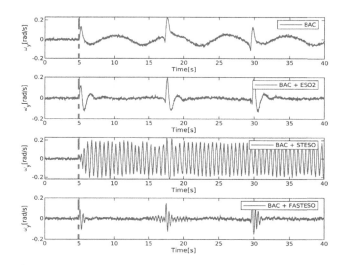

Figure 14. Results of angular rate ω_y with unknown external torque disturbance under backstepping controller combined with nothing, ESO2, STESO and FASTESO.

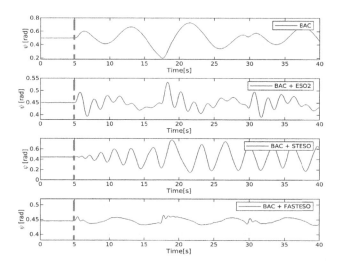

Figure 15. Results of yaw angle with unknown external torque disturbance under backstepping controller combined with nothing, ESO2, STESO and FASTESO.

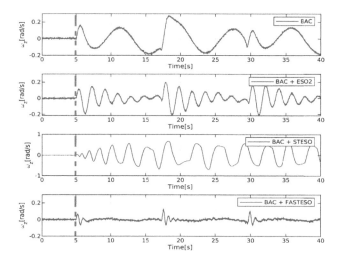

Figure 16. Results of angular rate ω_z with unknown external torque disturbance under backstepping controller combined with nothing, ESO2, STESO and FASTESO.

We could easily find that, without an observer, the single BAC has biggest attitude offset and angular rate fluctuation, which proves its bad robustness under disturbances. Specifically, the attitude curves even follows the applied torque disturbance well and angular rate has a dramatic change at special points. As for BAC+ESO2, although the estimation value follows the disturbance well to some extent, it has a performance of drastic change on both attitude and angular rate at peak point owing to the over estimation near the peak. As for BAC+STESO, in spite of the good performance of STESO on following the disturbance, it has a terrible chattering phenomenon and leads to a mess on UAV system. Compared to other situations, the UAV under BAC+FASTESO has the best performance on both attitude offset and angular rate fluctuation with varied torque disturbance.

6. Experiment

In this section, to verify the effectiveness of the whole proposed method, the quadrotor UAV manipulator is assigned to conduct a hovering task beside an electrical fan, which could generate several level gear wind as torque disturbance acting on the UAV system. BAC+FASTESO is adopted for UAV disturbance rejection test, meanwhile a single BAC without any observer is also conducted for comparison. The experimental scene is shown in Figure 17. The whole proposed control scheme is built in Pixhawk [35]. The parameters are chosen as follows. For BAC+FASTESO: $\alpha_1 = diag(0.2, 0.2, 0.3)$; $\alpha_2 = diag(0.3,0.3,0.4)$ $e = 0.001$; $k_e = 0.1$; $K_{b1} = diag(3,3,5)$; $K_{b2} = diag(2,2,2.5)$. For BAC: $K_{b1} = diag(6,6,9)$; $K_{b2} = diag(3.5,3.5,5)$.

The fan is placed to mainly apply the external torque in roll channel. The results of UAV hovering under first and second gear wind are shown in Figures 18–23. We found that the attitude of hovering UAV changes following the variation of wind gear. Moreover, the attitude vibration offset is reduced by help of FASTESO compared to the situation with single BAC. Meanwhile, the angular rate fluctuation is also reduced. They intuitively verify the effectiveness of FASTESO. Additionally, Figures 22 and 23 show the control output generated only from BAC in the control situations single BAC and BAC+FASTESO, respectively. We could find that the control signal is around zero with FASTESO; however, the control signal is impacted so much owing to external disturbance in situation without any observer. Moreover, the controller gain in BAC+FASTESO could keep smaller to make the whole system stable thanks to the disturbance estimation and compensation from FASTESO, which also contributes to the better performance of BAC+FASTESO. Meanwhile, the estimation of the lumped disturbance including external wind and model uncertainties is shown in Figure 20. Although we do not know the ground truth of the torque disturbance generated from the fan and model uncertainties to verify if the estimation result in Figure 20 is right or not, according to Figures 20 and 22, we could see that their results are almost the same, which also demonstrates the effectiveness of FASTESO to an extent.

Figure 17. Experimental scene.

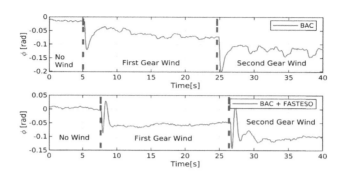

Figure 18. Experimental results of attitude with BAC and BAC+FASTESO under external wind.

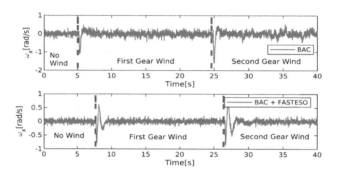

Figure 19. Experimental results of angular rate with BAC and BAC+FASTESO under external wind.

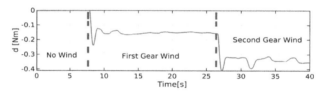

Figure 20. Disturbance estimation result from FASTESO.

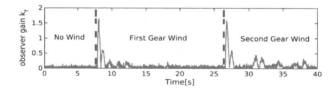

Figure 21. Observer gain of FASTESO during adaptation process.

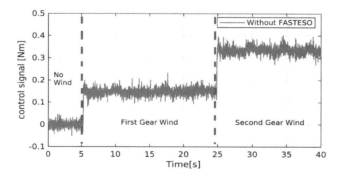

Figure 22. Control output signal only from BAC without FASTESO.

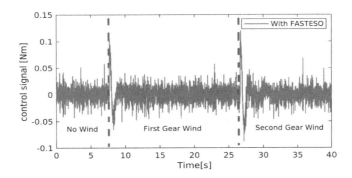

Figure 23. Control output signal only from BAC with FASTESO.

7. Conclusions

In this study, a new observer named FASTESO is proposed, which is incorporated with the saturation function, TD and fuzzy logic rules. Combined with backstepping controller, the whole control scheme for the quadrotor UAV manipulator system is finally built for rejection of lumped disturbance including external wind and model uncertainties. The effectiveness of the whole control structure is veritied in both several simulations and experiments.

In the future, we would pay more attention to the estimation separation of the model uncertainties from the external disturbance.

Author Contributions: Conceptualization, R.J.; writing–original draft preparation, R.J.; software, R.J. and Y.R.; validation, R.J. and Y.R.; methodology, R.J. and M.D.; writing—review and editing, R.J., W.C. and M.D.; supervision, W.C. All authors have read and agreed to the published version of the manuscript.

Abbreviations

Additional abbreviation illustration:

FASTESO   Fuzzy adaptive saturation super-twisting extended state observer

References

1. Premachandra, C.; Ueda, D.; Kato, K. Speed-Up Automatic Quadcopter Position Detection by Sensing Propeller Rotation. *IEEE Sens. J.* **2018**, *19*, 2758–2766. [CrossRef]
2. Nakajima, K.; Premachandra, C.; Kato, K. 3D environment mapping and self-position estimation by a small flying robot mounted with a movable ultrasonic range sensor. *J. Electr. Syst. Inf. Technol.* **2017**, *4*, 289–298. [CrossRef]
3. Premachandra, C.; Otsuka, M.; Gohara, R.; Ninomiya, T.; Kato, K. A study on development of a hybrid aerial/terrestrial robot system for avoiding ground obstacles by flight. *IEEE/CAA J. Autom. Sin.* **2018**, *6*, 327–336. [CrossRef]
4. Ryll, M.; Muscio, G.; Pierri, F.; Cataldi, E.; Antonelli, G.; Caccavale, F.; Bicego, D.; Franchi, A. 6D interaction control with aerial robots: The flying end-effector paradigm. *Int. J. Robot. Res.* **2019**, *38*, 1045–1062. [CrossRef]
5. Kim, S.; Seo, H.; Shin, J.; Kim, H.J. Cooperative aerial manipulation using multirotors with multi-dof robotic arms. *IEEE/ASME Trans. Mechatronics* **2018**, *23*, 702–713. [CrossRef]
6. Jiao, R.; Dong, M.; Chou, W.; Yu, H.; Yu, H. Autonomous Aerial Manipulation Using a Hexacopter Equipped with a Robotic Arm. In Proceedings of the 2018 IEEE International Conference on Robotics and Biomimetics (ROBIO), Kuala Lumpur, Malaysia, 12–15 December 2018; pp. 1502–1507.
7. Ruggiero, F.; Trujillo, M.A.; Cano, R.; Ascorbe, H.; Viguria, A.; Peréz, C.; Lippiello, V.; Ollero, A.; Siciliano, B. A multilayer control for multirotor UAVs equipped with a servo robot arm. In Proceedings of the 2015

IEEE international conference on robotics and automation (ICRA), Seattle, WA, USA, 26–30 May 2015; pp. 4014–4020.

8. Jiao, R.; Chou, W.; Ding, R.; Dong, M. Adaptive robust control of quadrotor with a 2-degree-of-freedom robotic arm. *Adv. Mech. Eng.* **2018**, *10*, 1687814018778639. [CrossRef]

9. Alexis, K.; Papachristos, C.; Siegwart, R.; Tzes, A. Robust model predictive flight control of unmanned rotorcrafts. *J. Intell. Robot. Syst.* **2016**, *81*, 443–469. [CrossRef]

10. Achtelik, M.; Bierling, T.; Wang, J.; Höcht, L.; Holzapfel, F. Adaptive control of a quadcopter in the presence of large/complete parameter uncertainties. In Proceedings of the Infotech@ Aerospace 2011, St. Louis, MO, USA, 29–31 March 2011; p. 1485.

11. Chen, W.H.; Yang, J.; Guo, L.; Li, S. Disturbance-observer-based control and related methods—An overview. *IEEE Trans. Ind. Electron.* **2015**, *63*, 1083–1095. [CrossRef]

12. Cristofaro, A.; Johansen, T.A. Fault tolerant control allocation using unknown input observers. *Automatica* **2014**, *50*, 1891–1897. [CrossRef]

13. Pu, Z.; Yuan, R.; Yi, J.; Tan, X. A class of adaptive extended state observers for nonlinear disturbed systems. *IEEE Trans. Ind. Electron.* **2015**, *62*, 5858–5869. [CrossRef]

14. Kim, S.; Choi, S.; Kim, H.; Shin, J.; Shim, H.; Kim, H.J. Robust control of an equipment-added multirotor using disturbance observer. *IEEE Trans. Control Syst. Technol.* **2017**, *26*, 1524–1531. [CrossRef]

15. Wang, C.; Song, B.; Huang, P.; Tang, C. Trajectory tracking control for quadrotor robot subject to payload variation and wind gust disturbance. *J. Intell. Robot. Syst.* **2016**, *83*, 315–333. [CrossRef]

16. Chen, M.; Xiong, S.; Wu, Q. Tracking flight control of quadrotor based on disturbance observer. *IEEE Trans. Syst. Man, Cybern. Syst.* **2019**. [CrossRef]

17. Mokhtari, M.R.; Cherki, B.; Braham, A.C. Disturbance observer based hierarchical control of coaxial-rotor UAV. *ISA Trans.* **2017**, *67*, 466–475. [CrossRef]

18. Wang, L.; Su, J. Robust disturbance rejection control for attitude tracking of an aircraft. *IEEE Trans. Control Syst. Technol.* **2015**, *23*, 2361–2368. [CrossRef]

19. Han, J. From PID to active disturbance rejection control. *IEEE Trans. Ind. Electron.* **2009**, *56*, 900–906. [CrossRef]

20. Liu, J.; Vazquez, S.; Wu, L.; Marquez, A.; Gao, H.; Franquelo, L.G. Extended state observer-based sliding-mode control for three-phase power converters. *IEEE Trans. Ind. Electron.* **2016**, *64*, 22–31. [CrossRef]

21. Madoński, R.; Herman, P. Survey on methods of increasing the efficiency of extended state disturbance observers. *ISA Trans.* **2015**, *56*, 18–27. [CrossRef]

22. Muñoz, F.; Bonilla, M.; Espinoza, E.S.; González, I.; Salazar, S.; Lozano, R. Robust trajectory tracking for unmanned aircraft systems using high order sliding mode controllers-observers. In Proceedings of the 2017 International Conference on Unmanned Aircraft Systems (ICUAS), Miami, FL, USA, 13–16 June 2017; pp. 346–352.

23. Cai, W.; She, J.; Wu, M.; Ohyama, Y. Disturbance suppression for quadrotor UAV using sliding-mode-observer-based equivalent-input-disturbance approach. *ISA Trans.* **2019**, *92*, 286–297. [CrossRef]

24. Shi, D.; Wu, Z.; Chou, W. Super-twisting extended state observer and sliding mode controller for quadrotor UAV attitude system in presence of wind gust and actuator faults. *Electronics* **2018**, *7*, 128. [CrossRef]

25. Kim, H.; Son, J.; Lee, J. A high-speed sliding-mode observer for the sensorless speed control of a PMSM. *IEEE Trans. Ind. Electron.* **2010**, *58*, 4069–4077.

26. Rong, Y.; Jiao, R.; Kang, S.; Chou, W. Sigmoid Super-Twisting Extended State Observer and Sliding Mode Controller for Quadrotor UAV Attitude System With Unknown Disturbance. In Proceedings of the 2019 IEEE International Conference on Robotics and Biomimetics (ROBIO), Dali, China, 6–8 December 2019; pp. 2647–2653.

27. Lippiello, V.; Ruggiero, F. Cartesian impedance control of a UAV with a robotic arm. *IFAC Proc. Vol.* **2012**, *45*, 704–709. [CrossRef]

28. Zhang, Z.; Xu, H.; Xu, L.; Heilman, L.E. Sensorless direct field-oriented control of three-phase induction motors based on "Sliding Mode" for washing-machine drive applications. *IEEE Trans. Ind. Appl.* **2006**, *42*, 694–701. [CrossRef]

29. Kang, K.L.; Kim, J.M.; Hwang, K.B.; Kim, K.H. Sensorless control of PMSM in high speed range with iterative sliding mode observer. In Proceedings of the Nineteenth Annual IEEE Applied Power Electronics Conference and Exposition, 2004 (APEC'04), Anaheim, CA, USA, 22–26 February 2004; Volume 2, pp. 1111–1116.

30. Jiao, R.; Wang, Z.; Chu, R.; Dong, M.; Rong, Y.; Chou, W. An Intuitional End-to-End Human-UAV Interaction System for Field Exploration. *Front. Neurorobot.* **2019**, *13*, 117. [CrossRef] [PubMed]

31. Lee, C.C. Fuzzy logic in control systems: Fuzzy logic controller. I. *IEEE Trans. Syst. Man, Cybern.* **1990**, *20*, 404–418. [CrossRef]

32. Palm, R. Robust control by fuzzy sliding mode. *Automatica* **1994**, *30*, 1429–1437. [CrossRef]

33. Meier, L.; Honegger, D.; Pollefeys, M. PX4: A node-based multithreaded open source robotics framework for deeply embedded platforms. In Proceedings of the 2015 IEEE international conference on robotics and automation (ICRA), Seattle, WA, USA, 26–30 May 2015; pp. 6235–6240.

34. Quan, Q. Flight Performance Evaluation of UAVs. 2018. Available online: http://flyeval.com/ (accessed on 15 April 2020).

35. Meier, L.; Tanskanen, P.; Heng, L.; Lee, G.H.; Fraundorfer, F.; Pollefeys, M. PIXHAWK: A micro aerial vehicle design for autonomous flight using onboard computer vision. *Auton. Robot.* **2012**, *33*, 21–39. [CrossRef]

Permissions

All chapters in this book were first published in MDPI; hereby published with permission under the Creative Commons Attribution License or equivalent. Every chapter published in this book has been scrutinized by our experts. Their significance has been extensively debated. The topics covered herein carry significant findings which will fuel the growth of the discipline. They may even be implemented as practical applications or may be referred to as a beginning point for another development.

The contributors of this book come from diverse backgrounds, making this book a truly international effort. This book will bring forth new frontiers with its revolutionizing research information and detailed analysis of the nascent developments around the world.

We would like to thank all the contributing authors for lending their expertise to make the book truly unique. They have played a crucial role in the development of this book. Without their invaluable contributions this book wouldn't have been possible. They have made vital efforts to compile up to date information on the varied aspects of this subject to make this book a valuable addition to the collection of many professionals and students.

This book was conceptualized with the vision of imparting up-to-date information and advanced data in this field. To ensure the same, a matchless editorial board was set up. Every individual on the board went through rigorous rounds of assessment to prove their worth. After which they invested a large part of their time researching and compiling the most relevant data for our readers.

The editorial board has been involved in producing this book since its inception. They have spent rigorous hours researching and exploring the diverse topics which have resulted in the successful publishing of this book. They have passed on their knowledge of decades through this book. To expedite this challenging task, the publisher supported the team at every step. A small team of assistant editors was also appointed to further simplify the editing procedure and attain best results for the readers.

Apart from the editorial board, the designing team has also invested a significant amount of their time in understanding the subject and creating the most relevant covers. They scrutinized every image to scout for the most suitable representation of the subject and create an appropriate cover for the book.

The publishing team has been an ardent support to the editorial, designing and production team. Their endless efforts to recruit the best for this project, has resulted in the accomplishment of this book. They are a veteran in the field of academics and their pool of knowledge is as vast as their experience in printing. Their expertise and guidance has proved useful at every step. Their uncompromising quality standards have made this book an exceptional effort. Their encouragement from time to time has been an inspiration for everyone.

The publisher and the editorial board hope that this book will prove to be a valuable piece of knowledge for researchers, students, practitioners and scholars across the globe.

List of Contributors

Marco Ceccarelli
LARM2: Laboratory of Robot Mechatronics, University of Roma Tor Vergata, Via del Politecnico 1, 00133 Rome, Italy
Beijing Advanced Innovation Center for Intelligent Robots and Systems, Beijing Institute of Technology, Beijing 100081, China

Matteo Russo
The Rolls-Royce UTC in Manufacturing and On-Wing Technology, University of Nottingham, Nottingham NG8 1BB, UK

Cuauhtemoc Morales-Cruz
LARM2: Laboratory of Robot Mechatronics, University of Roma Tor Vergata, Via del Politecnico 1, 00133 Rome, Italy
Instituto Politécnico Nacional, GIIM: Group of Research and Innovation in Mechatronics, Av. Juan de Dios Bátiz, 07700 Mexico City, Mexico

Stefano Rodinò, Elio Matteo Curcio, Antonio di Bella, Mattia Persampieri and Michele Funaro
Department of Mechanical, Energy and Management Engineering (DIMEG), University of Calabria, 87036 Rende, Italy

Ming Liu, Mantian Li, Pengfei Wang, Wei Guo and Lining Sun
State Key Laboratory of Robotics and System, Harbin Institute of Technology, Harbin 150001, China

Fusheng Zha
State Key Laboratory of Robotics and System, Harbin Institute of Technology, Harbin 150001, China
Shenzhen Academy of Aerospace Technology, Shenzhen 518057, China

Rabab Benotsmane and László Dudás
Department of Information Technology, University of Miskolc, 3515 Miskolc, Hungary

György Kovács
Institute of Logistics, University of Miskolc, 3515 Miskolc, Hungary

Younsse Ayoubi, Med Amine Laribi, Marc Arsicault and Saïd Zeghloul
Dept. GMSC, Pprime Institute, CNRS, University of Poitiers, ENSMA, UPR 3346 Poitiers, France

Quang Vinh Doan and Tien Dung Le
The University of Danang—University of Science and Technology, 54 Nguyen Luong Bang street, Danang 550000, Vietnam

Anh Tuan Vo and Ngoc Hoai An Nguyen
Electrical and Electronic Engineering Department, The University of Danang—University of Technology and Education, Danang 550000, Vietnam

Hee-Jun Kang
School of Electrical Engineering, University of Ulsan, Ulsan 44610, Korea

Xiaopeng Yan and Baijin Chen
State Key Laboratory of Materials Processing and Die & Mould Technology, School of Materials Science and Engineering, Huazhong University of Science and Technology, Wuhan 430074, China

Han Han, Yanhui Wei and Xiufen Ye
College of Automation, Harbin Engineering University, Harbin 150001, China

Wenzhi Liu
College of Information and Communication Engineering, Harbin Engineering University, Harbin 150001, China

Konrad Johan Jensen, Morten Kjeld Ebbesen and Michael Rygaard Hansen
Department of Engineering Sciences, University of Agder, 4879 Grimstad, Norway

Giuseppe Carbone
Department of Mechanical, Energy and Management Engineering (DIMEG), University of Calabria, 87036 Rende, Italy
CESTER, Technical University of Cluj-Napoca, 400114 Cluj-Napoca, Romania

Eike Christian Gerding and Burkard Corves
IGMR, RWTH Aachen University, 52062 Aachen, Germany

Daniele Cafolla
IRCCS Istituto Neurologico Mediterraneo Neuromed, 86077 Pozzilli, Italy

Matteo Russo
Faculty of Engineering, University of Nottingham, Nottingham NG7 2RD, UK

Giovanni Boschetti
Department of Management and Engineering, University of Padova, 36100 Vicenza, Italy

Miao-Miao Li, Liang-Liang Ma, Chuan-Guo Wu and Ru-Peng Zhu
National Key Laboratory of Science and Technology on Helicopter Transmission, Nanjing University of Aeronautics and Astronautics, Nanjing 210016, China

Ran Jiao and Yongfeng Rong
School of Mechanical Engineering and Automation, Beihang University, Beijing 100191, China

Wusheng Chou
School of Mechanical Engineering and Automation, Beihang University, Beijing 100191, China
The State Key Laboratory of Virtual Reality Technology and Systems, Beihang University, Beijing 100191, China

Mingjie Dong
School of Mechanical Engineering and Applied Electronics Technology, Beijing University of Technology, Beijing 100124, China

Index

Printed in the USA
CPSIA information can be obtained
at www.ICGtesting.com
LVHW082347150324
774517LV00005B/769